Advances in Bioinformatics

Vijai Singh • Ajay Kumar
Editors

Advances in Bioinformatics

Second Edition

Springer

Editors
Vijai Singh
Department of Biosciences
Indrashil University
Mehsana, Gujarat, India

Ajay Kumar
Biotechnology
Rama University
Kanpur, Uttar Pradesh, India

ISBN 978-981-99-8400-8 ISBN 978-981-99-8401-5 (eBook)
https://doi.org/10.1007/978-981-99-8401-5

This Springer imprint is published by the registered company Springer Nature Singapore Pte Ltd.
The registered company address is: 152 Beach Road, #21-01/04 Gateway East, Singapore 189721, Singapore

Printed on acid-free paper

Foreword

I am thrilled to have the opportunity to write the introduction for *Advances in Bioinformatics* volume II, a timely addition to this rapidly evolving field.

The inception of bioinformatics was driven by the volume of biological data that reached a point where it necessitated management in terms of storage, analysis, output, and communication. Initially, bioinformatics practitioners predominantly considered protein structures. However, as reductionist methods became more sophisticated, an exponential increase in genomics, transcriptomics, and pathway data was observed, from single cells to multicellular systems.

This book encompasses both fundamental and advanced aspects of bioinformatics, covering key developments such as gene discovery, genome analysis, genomics, transcriptomics, proteomics, metabolomics, structural bioinformatics, metabolic flux analysis, drug discovery, drug repurposing, and much more. It also delves into contemporary bioinformatics, including the analysis of non-coding RNA, next-generation sequencing, genome-scale modelling, genome editing, high-throughput drug screening, precision medicine, preventive medicine, automation, artificial intelligence, and machine learning.

I am delighted to acknowledge the commendable efforts of Dr. Vijai Singh and Dr. Ajay Kumar, who, with the support of Springer Nature, have painstakingly crafted this outstanding volume.

This book serves as an invaluable resource not only for newcomers to the field of bioinformatics but also for students, researchers, scientists, clinicians, practitioners, policymakers, and stakeholders who seek to harness the potential of bioinformatics, spanning from foundational science to practical applications.

School of Biotechnology, Jawaharlal Nehru University
New Delhi, India

Pawan K. Dhar

Preface

Recent development in bioinformatics is being used in almost all domains of biological sciences. Bioinformatics uses computation for extracting knowledge from biological data and uses retrieval, collection, storage, manipulation and modelling for analysis, prediction, imaging, visualization by using computation power, algorithm, and software. Bioinformatics is currently used for gene discovery, genome analysis, genomics, proteomics, metabolic flux analysis, drug discovery, drug repurposing, and many more.

This book presents the latest developments in bioinformatics, highlighting the importance of bioinformatics in gene discovery, genome analysis, genomics, transcriptomics, proteomics, metabolomics, structural bioinformatics, metabolic flux analysis, drug discovery, drug repurposing, and many more. This book offers several recent topics are currently used in bioinformatics including analysis of non-coding RNA, next-generation sequencing, gene synthesis, genome-scale modelling, genome editing, high-throughput drug screening, precision medicine, preventive medicine, automation and artificial intelligence, and machine learning.

This book offers an excellent and informative text on bioinformatics, benefitted by simple to understand and easy-to-read format. This book uses a rich literary text of excellent depth, clarity, and coverage. It highlights a number of aspects of bioinformatics in a way that can help future investigators, researchers, students, and stakeholders to perform their research with greater ease. This book provides a primer for basic knowledge from which scientific knowledge can grow, widen, and accelerate bioinformatics research in many areas.

Mehsana, Gujarat, India — Vijai Singh
Kanpur, Uttar Pradesh, India — Ajay Kumar

Acknowledgement

Vijai Singh I am delighted to express my sincere gratitude and deep appreciation to Dr. J.S. Yadav, Director (Research), Indrashil University, India, for extending his outstanding support and motivation to complete this book. I would like to give many thanks to co-editor Dr. Ajay Kumar of this book who gave me outstanding personal and professional support as well as inspiration to finish this book.

I am delighted to thank all the authors for their excellent contributions to this book. I would like to thank Mrs. Swati Sharma (Associate Editor—Biomedicine) and Mrs. Aishwarya Thyagarajan (Production Editor) from Springer for their excellent management of this project.

I would like to thank Prof. Rakesh Rawal, Prof. Bharat Maitreya. Prof. Pawan K. Dhar, Dr. Poonam Bhargava, Dr. Madhvi Joshi, Dr. Bhabatosh Das, Dr. Pablo Carbonell, Dr. Rupesh Maurya, Dr. Satya Prakash, Dr. Vimal C. Pandey, Dr. Suresh Ramakrishna, Dr. Dinh-Toi Chu, Dr. Mukesh Kumar Awasthi, and those whose names do not feature here but have directly or indirectly contributed in shaping this project.

I greatly appreciate the support of my students Dr. Nisarg Gohil. Mr. Khushal Khambhati, and Dr. Gargi Bhattacharjee, whose discussion and comments helped to shape this book.

I wish to express my gratitude to my beloved wife Pritee Singh for her endless support, patience, and inspiration. Lots of affection for my kids Aaradhya and Ayush who missed me during this project. I would like to warmly thank the faculty and staff of Indrashil University for providing a great working environment.

I am aware that even despite our best efforts, the first version always comes with some error that may have crept in the compilation. I would be delighted to receive feedback from readers to further improve the future book. Last but not least, my sincere thanks to GOD for his supreme POWER for endowing me to live with joy and victory in the shape of this book.

Ajay Kumar This book is an outcome of utter studies and literature survey. It is an honour and pleasure to express my profound gratitude to my family, friends, and colleagues who always have good words for me inculcating a source of strength and inspiration for taking this venture. I would like to give many thanks to co-editor

Dr. Vijai Singh of this book who gave me outstanding personal and professional support as well as inspiration for accomplishment of this book.

I am especially grateful to Dr. Prateek Singh (Director, Rama University Kanpur, India), Dr. Pranav Singh (Director, Rama University), Dr. C. S. Raghuvanshi (Dean, Faculty of Engineering and Technology, Rama University), and Mr. R.K Yadav (Controller of Examination, Rama University) who never stopped me to take this challenge for developing ideas and supported me to develop a strong foundation.

I also take this opportunity to thank my worthy and learned colleagues in the Department of Biotechnology, Faculty of Engineering and Technology, Rama University, whose knowledge and experience have eased and developed the confidence to take up my work and finally taking to finish the note.

I would also like to mention sincere cooperation from Mr. Indrajeet Singh and Er. Fariya Khan who always extended their helping hands for discussions and cooperation. And last but not least, I feel deeply and highly obliged to my beloved wife Shraddha and my children Nishit and Anshika who not only motivated me to author this book but showed patience when deprived them of my much-needed attention during this course of time so that my book could see the light of the day. It would not be out of place to seek the blessings of my elder brothers Er. Prem Chandra and Dr. Ram Kumar, who always had been a guiding figure in my passion.

I am also thankful to Almighty God who has given me the wisdom to edit this book. Finally, I dedicate this book to my parents for their countless blessings to accomplish the target.

Contents

Editors and Contributors

About the Editors

Vijai Singh is an associate professor and head of the Department of Biosciences, School of Science at Indrashil University, Rajpur, Mehsana, Gujarat, India. He was an assistant professor in the Department of Biological Sciences and Biotechnology at the Institute of Advanced Research, Gandhinagar, India, and an assistant professor in the Department of Biotechnology at the Invertis University, Bareilly, India. Prior to that, he was a Postdoctoral Fellow in the Synthetic Biology Group at the Institute of Systems and Synthetic Biology, Paris, France, and the School of Energy and Chemical Engineering at the Ulsan National Institute of Science and Technology, Ulsan, South Korea. He received his Ph.D. in Biotechnology (2009) from the National Bureau of Fish Genetic Resources, Uttar Pradesh Technical University, Lucknow, India, with a research focus on the development of molecular and immunoassays for diagnosis of *Aeromonas hydrophila*. His research interests are focused on building novel biosynthetic pathways for the production of medically and industrially important biomolecules. Additionally, his laboratory is working on CRISPR-Cas9 tools for genome editing. He has more than 10 years of research and teaching experience in synthetic biology, metabolic engineering, bioinformatics, microbiology, and industrial microbiology.

Ajay Kumar as a professor and head of the Department of Biotechnology at the Faculty of Engineering and Technology at Rama University Uttar Pradesh, Kanpur. Dr. Kumar has successfully served more than 20 years in research and teaching. He has experience in genomics and proteomics, bioprocess engineering, bioinformatics, microbiology, industrial microbiology, genetic engineering, fermentation technology, and food biotechnology. He has held several key positions in well-renowned Universities and Engineering Institutes. Dr. Kumar received his M. Tech. (Biotechnology) from the Institute of Engineering and Technology, Lucknow, India, and Ph. D. from the ICAR-Central Institute for Research on Goats, Mathura, India. His expertise lies in computational vaccine and drug development, cancer biology genomics and proteomics, and fermentation. He is also a member of the Board of

Study and Academic Council of Rama University, Kanpur, and Regional Food Research Analysis Center, Department of horticulture and food processing, Lucknow, Uttar Pradesh. Being a member of a professional body such as the International Association of Engineers (IAENG) and INSA, he has rendered consultancy services in vaccine research.

Contributors

Ghloamareza Abdi Department of Biotechnology, Persian Gulf Research Institute, Persian Gulf University, Bushehr, Iran

Ghlomareza Abdi Department of Biotechnology, Persian Gulf Research Institute, Persian Gulf University, Bushehr, Iran

Gholamreza Abdi Department of Biotechnology, Persian Gulf Research Institute, Persian Gulf University, Bushehr, Iran

Shabroz Alam Department of Biotechnology, Era University, Lucknow, Uttar Pradesh, India

Asad Amir Department of Biotechnology and Microbiology, Meerut Institute of Engineering and technology, Meerut, Uttar Pradesh, India

Mustafeez Mujtaba Babar Department of Basic Medical Sciences, Shifa College of Pharmaceutical Sciences, Shifa Tameer-e-Millat University, Islamabad, Pakistan
Advanced Drug Delivery and Regenerative Biomaterials, Stanford University School of Medicine, Stanford University, Palo Alto, CA, USA

Mukul Barwant Department of Botany, Sanjivani Arts, Commerce and Science College, Ahmednagar, Maharashtra, India

Nhat Le Bui Faculty of Applied Sciences, International School, Vietnam National University, Hanoi, Vietnam
Center for Biomedicine and Community Health, International School, Vietnam National University, Hanoi, Vietnam

Dinh-Toi Chu Faculty of Applied Sciences, International School, Vietnam National University, Hanoi, Vietnam
Center for Biomedicine and Community Health, International School, Vietnam National University, Hanoi, Vietnam

Ngo Anh Dao Center for Biomedicine and Community Health, International School, Vietnam National University, Hanoi, Vietnam

Van-Quy Do Center for Biomedicine and Community Health, International School, Vietnam National University, Hanoi, Vietnam

Krishna Bihari Dubey Department of Computer Science and Engineering, ABES Institute of Technology, Ghaziabad, Uttar Pradesh, India

Sushma Dubey Department of Biotechnology, Kalinga University, Naya Raipur, Chhattisgarh, India

Juveriya Israr Faculty of Biosciences, Institute of Biosciences and Technology, Shri Ramswaroop Memorial University, Barabanki, Uttar Pradesh, India
Department of Biotechnology, Era University, Lucknow, Uttar Pradesh, India

Mukul Jain Cell and Developmental Biology Laboratory, Centre of Research for Development, Parul University, Vadodara, Gujarat, India
Department of Life Sciences, Parul Institute of Applied Sciences, Parul University, Vadodara, Gujarat, India

Amit Joshi Department of Biochemistry, Kalinga University, Naya Raipur, Chhattisgarh, India

Vikas Kaushik Science Habitat, Markham, ON, Canada

Ajay Kumar Department of Biotechnology, Rama University, Kanpur, Uttar Pradesh, India

Prashant Kumar Department of Bioinformatics, Kalinga University, Naya Raipur, Chhattisgarh, India

Indra Mani Department of Microbiology, Gargi College, University of Delhi, New Delhi, India

Anshu Mathuria Department of Microbiology, Gargi College, University of Delhi, New Delhi, India

Mehak Department of Microbiology, Gargi College, University of Delhi, New Delhi, India

Hara Prasad Mishra Department of Pharmacology, University College of Medical Sciences, University of Delhi, Delhi, India

Santosh Kumar Mishra Department of Life Sciences, School of Basic Sciences and Research, Sharda University, Greater Noida, Uttar Pradesh, India

Sankalp Misra Faculty of Biosciences, Institute of Biosciences and Technology, Shri Ramswaroop Memorial University, Barabanki, Uttar Pradesh, India

Duy Ha Nguyen Vietnam Military Medical University, Hanoi, Vietnam

Mai-Anh Nguyen Center for Biomedicine and Community Health, International School, Vietnam National University, Hanoi, Vietnam

Nil Patil Cell and Developmental Biology Lab, Centre of Research for Development, Parul University, Vadodara, Gujarat, India
Department of Life Sciences, Parul Institute of Applied Sciences, Parul University, Vadodara, Gujarat, India

Prabir Kumar Paul Department of Biotechnology, Manav Rachna International Institute of Research and Studies, Faridabad, Haryana, India

Jayakumar Rajadas Advanced Drug Delivery and Regenerative Biomaterials Laboratory, Cardiovascular Institute and Pulmonary and Critical Care Medicine, Stanford University School of Medicine, Stanford University, Palo Alto, CA, USA

Sahabjada Siddiqui Department of Biotechnology, Era University, Lucknow, Uttar Pradesh, India

Avinash Singh Department of Biotechnology, Meerut Institute of Engineering & Technology, Meerut, Uttar Pradesh, India

Indrajeet Singh Department of Biotechnology, Rama University, Kanpur, Uttar Pradesh, India

Vijai Singh Department of Biosciences, School of Science, Indrashil University, Mehsana, Gujarat, India

Maryam Abbasi Tarighat Cell and Developmental Biology Laboratory, Centre of Research for Development, Parul University, Vadodara, Gujarat, India

Mohd Tariq Department of Life Sciences, Parul Institute of Applied Sciences, Parul University, Vadodara, Gujarat, India

Mugdha Tendulkar K. J. Somaiya Medical College and Research Centre, Mumbai, Maharashtra, India

Reshma Tendulkar Vivekanand Education Society's, College of Pharmacy, Mumbai, Maharashtra, India

Hue Vu Thi Center for Biomedicine and Community Health, International School, Vietnam National University, Hanoi, Vietnam
Faculty of Applied Sciences, International School, Vietnam National University, Hanoi, Vietnam

Yen Vy Nguyen Thi Center for Biomedicine and Community Health, International School, Vietnam National University, Hanoi, Vietnam

Linh Thao Tran Center for Biomedicine and Community Health, International School, Vietnam National University, Hanoi, Vietnam

Thuy-Duong Vu Center for Biomedicine and Community Health, International School, Vietnam National University, Hanoi, Vietnam

Tanya Waseem Department of Pharmaceutical Chemistry, Shifa College of Pharmaceutical Sciences, Shifa Tameer-e-Millat University, Islamabad, Pakistan

Chia-Ching Wu Department of Cell Biology and Anatomy, College of Medicine, National Cheng Kung University, Tainan, Taiwan
International Center for Wound Repair and Regeneration, National Cheng Kung University, Tainan, Taiwan
Department of Biomedical Engineering, National Cheng Kung University, Tainan, Taiwan

Revolutionizing Genomics: Exploring the Potential of Next-Generation Sequencing

1

Ghloamareza Abdi, Maryam Abbasi Tarighat, Mukul Jain, Reshma Tendulkar, Mugdha Tendulkar, and Mukul Barwant

Abstract

Next-generation sequencing (NGS) technologies have revolutionized the field of genomics by enabling high-throughput, cost-effective, and rapid DNA sequencing on an unprecedented scale. This introduction offers a synopsis of NGS and its profound implications across diverse areas of biological research and medical diagnostics. The fundamental principles underlying NGS, including library preparation, sequencing-by-synthesis, and data generation, are outlined. The different NGS platforms, such as Illumina, Ion Torrent, and Oxford Nanopore, as well as their

The original version of this chapter was revised. A correction to this chapter can be found at https://doi.org/10.1007/978-981-99-8401-5_18

G. Abdi
Department of Biotechnology, Persian Gulf Research Institute, Persian Gulf University, Bushehr, Iran

M. A. Tarighat (✉)
Faculty of Nano and Bio Science and Technology, Persian Gulf University, Bushehr, Iran
e-mail: matarighat@pgu.ac.ir

M. Jain
Faculty of Nano and Bio Science and Technology, Persian Gulf University, Bushehr, Iran

Department of Life Sciences, Parul Institute of Applied Sciences, Parul University, Vadodara, Gujarat, India

R. Tendulkar
Vivekanand Education Society's, College of Pharmacy, Mumbai, Maharashtra, India

M. Tendulkar
K. J. Somaiya Medical College and Research Centre, Mumbai, Maharashtra, India

M. Barwant
Department of Botany, Sanjivani Arts, Commerce and Science College, Ahmednagar, Maharashtra, India

V. Singh, A. Kumar (eds.), *Advances in Bioinformatics*,
https://doi.org/10.1007/978-981-99-8401-5_1

respective strengths and limitations, are discussed. Recent advancements in sequencing technologies, such as single-cell sequencing, long-read sequencing, and spatial transcriptomics, are explored, expanding the capabilities of NGS and facilitating comprehensive genomic investigations. Subsequently, the applications of NGS in genomics, transcriptomics, epigenomics, metagenomics, and personalized medicine are examined. The accelerated discovery of genetic variants, gene expression patterns, DNA methylation profiles, and microbial communities through NGS is emphasized. Moreover, the role of NGS in uncovering disease mechanisms, identifying therapeutic targets, and enabling precision medicine approaches is discussed. Furthermore, the computational challenges associated with NGS data analysis, including read alignment, variant calling, and data interpretation, are addressed. The pivotal role of bioinformatics and data analysis pipelines in transforming raw sequencing data into biologically meaningful insights is highlighted. Additionally, the integration of NGS data with other omics datasets and the emerging field of multi-omics integration, providing a holistic view of biological systems, are briefly touched upon.

Additionally, the impact of NGS on clinical diagnostics, encompassing the detection of genetic disorders, cancer genomics, infectious disease surveillance, and pharmacogenomics, is elucidated. The potential of NGS-based liquid biopsies and non-invasive prenatal testing in revolutionizing clinical practice is underscored. Lastly, the challenges and considerations associated with NGS, such as data storage, privacy concerns, ethical considerations, and the importance of standardization and quality control measures, are addressed. The significance of interdisciplinary collaborations among scientists, clinicians, and bioinformaticians in harnessing the full potential of NGS and driving innovation in genomic research and healthcare is emphasized. In conclusion, this comprehensive introduction to next-generation sequencing provides an overview of the technology, its applications, and its impact across various fields. By empowering researchers and clinicians with unprecedented genomic information, NGS has the potential to revolutionize our understanding of biological systems, unravel disease complexities, and facilitate personalized approaches to healthcare.

Keywords

Next-generation sequencing · High-throughput sequencing · DNA sequencing · Genomics · Bioinformatics · Personalized medicine

1.1 Introduction

Next-generation sequencing (NGS) technologies allow for high-throughput, rapid, and precise determination of the nucleotide order within DNA/RNA molecules. Since the development of contemporary sequencing technologies, the identification of nucleic acid sequences has become a frequent and essential tool across all domains of biological investigation. The best use of genomic resources can be made of bioinformatics platforms, diverse computational tools, and databases to identify promising vaccine targets for future experimental validation. Researchers are working to create a potent vaccine against the SARS-CoV-2 infection to combat this epidemic, which has claimed a significant number of lives around the globe, with the completion of genome sequencing efforts for multiple coronavirus strains, interest in a thorough analysis of the roles played by the proteins and their 3D structures has grown (Chatterjee et al. 2021). Bioinformatics employs computational, mathematical, and statistical techniques to gather, organize, and analyze huge and complicated genetic sequencing data as well as related biological data, particularly in the context of genomics and molecular pathology. A bioinformatics pipeline is a group of bioinformatics algorithms that are used to handle NGS data in a pre-set order. Next-generation sequencing (NGS)-based molecular tests have transformed the field of medicine by enabling personalized diagnosis, risk assessment, and treatment for patients with both cancer and non-neoplastic disorders. These advanced sequencing technologies generate substantial amounts of quantitative and intricate sequencing data, requiring clinical laboratories to employ resource-intensive data processing pipelines. These pipelines are crucial for analyzing the data and detecting clinically significant genetic alterations (Roy et al. 2018). Accordingly, the core competencies of the field of bioinformatics are the interpretation and application of these biological data. Utilizing a variety of programming languages and mathematical and statistical methods, bioinformatics tools organize, analyze, and interpret biological data at the molecular, cellular, and genomic levels. NGS and bioinformatics' combined capacity is crucial for epidemiological study, medical diagnosis, and therapy. The "next-generation sequencing" (NGS) method of massively parallel sequencing offers incredibly high throughput, scalability, and speed. This method can be used to determine the nucleotide sequence of entire genomes or individual DNA or RNA segments. According to NGS, which enables laboratories to carry out a variety of tasks and explore biological systems at a level previously unimaginable, the biological sciences have undergone a revolution. The depth of data required to respond to today's complicated genomics inquiries is greater than what is currently possible with traditional DNA sequencing technologies. NGS has met that demand and developed into a common tool for responding to these questions. Currently, next-generation sequencing is actively used in genetic and genomic studies. For experimental biologists, developing the bioinformatics skills necessary to assess and comprehend the enormous volumes of sequencing data produced by next-generation sequencers is becoming increasingly important. A high-throughput method for determining the precise arrangement of nucleotides in a DNA molecule is next-generation sequencing. By offering high throughput sequencing at significantly lower costs, recent technology advancements in next-generation sequencing have moved the field closer to the objective of recreating all genomes within a community. Although the amount of raw sequence data available has increased significantly thanks to these next-generation sequencing methods, there are still a number of new informatics issues that need to be resolved in order to advance metagenomics' condition and realize its potential (Scholz et al. 2012).

Several high-throughput sequencing (HTS) or Next-Generation Sequencing (NGS) platforms that are based on different cyclic-array sequencing implementations have emerged over the last several years. Cyclic-array sequencing is the process of iteratively sequencing a dense array of DNA characteristics by enzymatic manipulation and imaging-based data collecting (Magi et al. 2010; Mitra and Church 1999). The landscape of genetic medicine has been significantly changed by next-generation sequencing (NGS) technology, which employs a massively parallel sequencing paradigm. NGS systems' high throughput capabilities have led to an exponential accumulation of sequence data that has beyond our existing technological capacity to handle and analyze genomic data completely. There is an increasing need for the integration of discrete NGS data into clinical settings due to the fast falling cost of sequencing per base and the development of affordable benchtop laboratory sequencers in the realm of personalized medicine. Through employing reverse-transcriptase polymerase chain reaction to transform RNA molecules into complementary DNA (cDNA) molecules, NGS also sequences RNA molecules. Due to the large level of sequence redundancy at a locus, high-throughput NGS gives quantitative information (depth of coverage) in addition to the sequence itself, unlike Sanger sequencing. Because of this characteristic of NGS data, laboratories can use various bioinformatics algorithms to identify a wide variety of genetic changes from a single NGS run on a sample (Roy et al. 2016). The combination of bioinformatics and next-generation sequencing (NGS) technologies has emerged as a powerful tool for detecting, characterizing, and analyzing human diseases. NGS-generated sequences offer several advantages over conventional methods, such as improved accuracy in pathogen detection, characterization of resistance mutations or genes, identification of vaccine escape variants, assessment of recombination or reassortment, and analysis of virulence and pathogenicity factors. Consequently, NGS and bioinformatics have become essential components of research and public health laboratories worldwide (Schmidt et al. 2016). Furthermore, the decreasing costs and computational requirements of NGS, along with improvements in sequencing error rates and simplified laboratory approaches, have made NGS and bioinformatics more accessible and in-demand (Toledo-Rueda et al. 2018).

However, harnessing the potential of NGS data requires significant expertise and competence due to its complex and nuanced nature. Moreover, the application of NGS and bioinformatics methodologies as routine surveillance and tracking tools in public health laboratories necessitates specialized information technology (IT) infrastructure and quality management systems (Salje et al. 2017). Thus, when establishing NGS and bioinformatics laboratories, careful selection of bioinformatics tools and analyses becomes crucial. Additionally, it is essential to have personnel with expertise in analysis pipelines, wet lab techniques, sequencing platforms, and familiarity with the pathogens of interest. Adequate computational and IT infrastructure, including networks and storage systems, is also necessary (Faria et al. 2016). All these factors play a critical role in developing NGS and bioinformatics capabilities in research or public health settings. Therefore, minor variations in nucleic acid extraction and sequencing methods, as well as the diverse capacities of sequencing platforms, become significant considerations in capacity building (Gire et al. 2014). Understanding the advantages and limitations of different sequencing and wet lab methods is essential (Grard et al. 2012). Moreover, the abundance of bioinformatics tools currently available and the rapid expansion of the field pose challenges in standardizing analyses across laboratories and teams (Maljkovic et al. 2020). Genetic diagnosis has become critically important in medical practice as it can definitively diagnose a wide range of clinically diverse disorders. It enables more precise disease prognosis and guides the selection of optimal treatment options for affected individuals. The ability to examine the human genome at various levels, from chromosomal to single-base changes, greatly enhances its current potential. Next-generation sequencing (NGS) or massively parallel sequencing (MPS) are commonly used terms to describe this technology, encompassing a broad range of methodologies (Shendure and Ji 2008). NGS allows for the rapid and cost-effective generation of vast amounts of data in each instrument run, enabling parallel analysis of multiple samples. As a result, several brands have emerged in the NGS market, including Illumina, Ion Torrent (Thermo Fisher Scientific), BGI Genomics, PacBio, and Oxford Nanopore Technologies, each offering unique solutions to address the challenge of handling massive sequencing data (Ameur et al. 2019). Although the classification of second-generation sequencing as based on large parallel and clonal amplification of molecules (polymerase chain reaction (PCR)) is not universally agreed upon in the literature (Pereira et al. 2020).

Many NGS-based methods that investigate genetic variation and its association with specific phenotypes adopt case-control study designs with unrelated individuals. However, these study designs are susceptible to population stratification bias (PSB) since patients and controls may have different genetic ancestries (Freedman et al. 2004; Kanzi et al. 2020). The integration of NGS-based genetic analysis strategies into clinical diagnostics and genetic medicine has been greatly facilitated by the quality of data provided by NGS, coupled with reasonable costs, improved data handling capabilities (Posey 2019), increased computational power, and effective bioinformatics analysis tools (Koboldt et al. 2013; Kanzi et al. 2020). A transformative technology known as next-generation sequencing (NGS) is reshaping the field of human molecular genetic testing (Moorthie et al. 2013). It enables sequencing reactions to be parallelized in unprecedented ways, enabling highly multiplexed testing paradigms with relatively quick turnaround times and lower costs (Mardis 2013). Bioinformatics analytics based on NGS aim to convert signals into data, data into understandable information, and information into actionable knowledge (Metzker 2010; Oliver et al. 2015).

We want focuses on various NGS approaches, platform awareness, Metagenomics, the use of NGS in agriculture, and the diagnostic capabilities of NGS.

1.1.1 The Evolution of DNA Sequencing

The field of DNA sequencing witnessed a significant breakthrough with the introduction of Sanger sequencing, also known as dideoxy sequencing, by Frederick Sanger and his colleagues in 1977 (Fig. 1.1). This pioneering method involved the

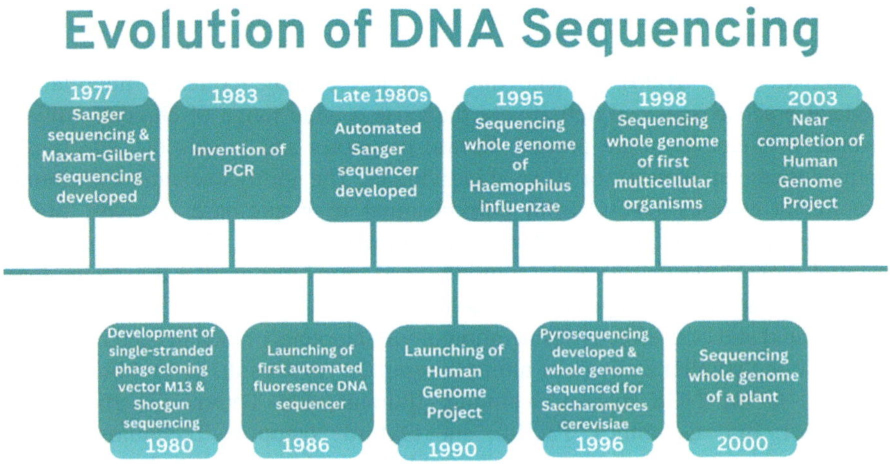

Fig. 1.1 Evolution of DNA sequencing technologies over time

use of chain-terminating nucleotides and gel electrophoresis to determine DNA sequences. Sanger sequencing quickly became the gold standard for DNA sequencing and played a pivotal role in key scientific discoveries, including the Human Genome Project (Sanger et al. 1977). However, due to its time-consuming nature and high costs, Sanger sequencing had limitations when it came to large-scale sequencing projects. The next major milestone in DNA sequencing arrived with the emergence of next-generation sequencing (NGS) technologies. NGS, also referred to as high-throughput sequencing, revolutionized the field by enabling parallel sequencing of millions of DNA fragments, significantly enhancing sequencing speed and throughput (Metzker 2010). Over the years, various NGS platforms with distinct sequencing chemistries and approaches were developed. Among them, Illumina's sequencing-by-synthesis technology, introduced in 2006, gained widespread adoption due to its accuracy, scalability, and cost-effectiveness (Shendure and Ji 2008). Other notable NGS platforms include Roche's 454 pyrosequencing, Ion Torrent's semiconductor sequencing, and Pacific Biosciences' single-molecule real-time (SMRT) sequencing. The latest advancement in DNA sequencing is the emergence of third-generation sequencing technologies, often referred to as long-read sequencing. These innovative technologies, such as Oxford Nanopore Technologies' nanopore sequencing and PacBio's SMRT sequencing, produce significantly longer DNA reads spanning thousands to tens of thousands of nucleotides. Long-read sequencing has overcome the limitations of short-read sequencing in resolving complex genomic regions, repetitive elements, and structural variations (Jain et al. 2015).

Moreover, DNA sequencing advancements have led to substantial cost reductions. The cost of sequencing a human genome has plummeted from millions of dollars with the first human genome sequence to a few thousand dollars today, making it more accessible for research and clinical applications. This continuous evolution of DNA sequencing technologies has revolutionized various fields, including genomics, personalized medicine, and evolutionary biology. It has facilitated the identification of disease-causing genetic mutations, the study of microbial communities, the exploration of ancient DNA, and the investigation of complex traits. The vast amount of DNA sequence data generated has propelled the development of computational tools and algorithms for data analysis and interpretation, further expanding the impact of DNA sequencing in biology and medicine.

1.2 Next-Generation Sequencing Platforms

Since the introduction of next-generation sequencing (NGS) technology in 2005, the number of high-throughput sequencing systems with varying prices, chemistries, capacities, and applications has significantly increased. Among the available platforms, Illumina is the sole provider offering a wide range of platforms (Table 1.1) suitable for diverse settings, ranging from small labs, schools, and clinical labs to enormous high-throughput sequencing facilities. Alongside their adaptable MiSeq platform, Illumina has introduced other platforms such as GAIIx and MiSeqDx, the first in vitro diagnostic testing platform approved by the Food and Drug Administration. To cater to different cost and capacity requirements, Illumina has developed NextSeq, NovaSeq, MiniSeq, and iSeq platforms. Despite higher error rates compared to Illumina systems, the Ion Torrent/Ion S5 platform, acquired by Life Technologies, is still popular due to its affordability and user-friendly operation. Pioneering the single-molecule sequencing industry, Pacific Biosciences (PacBio) leads with its PacBioRS/RSII and the latest Sequel platform, capable of producing average read lengths of 10 kb (Maljkovic et al. 2020). In 2014, Oxford Nanopore unveiled the MinION, a compact single-molecule sequencer the size of a flash drive. The manufacturer's approach of involving the scientific community in determining the necessary hardware and software for the device attracted a significant user base. Subsequent software improvements have focused on addressing the platform's increased error rates, which range from 13 to 20% (Maljkovic et al. 2020).

The choice of sequencing platform depends significantly on the goals of a laboratory's research. Smaller targeted platforms like MiSeq, NextSeq, or Ion Torrent have proven successful in whole-genome sequencing of bacteria or viruses (Salje et al. 2017; Stewart-Ibarra et al. 2018). The various approaches used by current NGS platforms impact the number, quality, and choice of sequencing applications. The standard NGS workflow involves the extraction of genomic DNA from test samples, followed by library preparation involving DNA fragmentation, adaptor ligation, adaptor sequencing, sample enrichment, and ultimately sequencing (Buermans and Den Dunnen 2014; Kanzi et al. 2020). Certain genome sequencing applications, such as those involving bacteria with highly repetitive genome structures or modular plasmid structures, require more reliable platforms capable of delivering longer sequence reads (PacBio) or reads of moderate length and greater depth (HiSeq, NovaSeq) (LaBreck et al. 2018). The choice of platform also considers the expertise and skill levels of personnel, in addition to the study

Table 1.1 List of NGS sequencing platforms and their expected throughputs, error types and error rates. Each platform has distinct advantages owing to cost, error rate, and read length (Scholz et al. 2012)

NGS sequence	Platform	Run time (h)	Read length (bp)	Throughput per run (Mb)	Error type	Error rate (%)
Roche	454 FLX+	18–20	700	900	Indel	1
	454 FLX Titanium	10	400	500	Indel	1
	454 GS	10	400	50	Indel	1
	Illumina GAIIx	14	2 × 150	96,000	Substitution	>0.1
	HiSeq 2000	8	2 × 100	400,000	Substitution	>0.1
	HiSeq 2000 V3	10	2 × 150	<600,000	Substitution	>0.1
	MiSeq	1	2 × 150	1000	Substitution	>0.1
Life technologies	SOLiD 4	12	50 × 35	71,000	A-T Bias	>0.06
	SOLiD 5500xl	8	75 × 35 PE 60 × 60 MP	155,000	A-T Bias	>0.01
Ion torrent	PGM 314 Chip	3	100	10	Indel	1
	PGM 316 Chip	3	100+	100	Indel	1
	PGM 318 Chip	3	200	1000	Indel	1
Pacific biosciences	RS	14/ 8 Smart Cells	1500	45/SC	Insertions	15

objectives. In the lab, Ion Torrent is known for its user-friendly and straightforward operation, while the challenges of data analytics require employees with the necessary bioinformatics training. On the other hand, MiSeq provides data storage and platform bioinformatics support through a user-friendly graphical interface, albeit requiring more training. The connectivity of the sequencing platform and the availability of training are crucial factors, especially in regions such as Africa, South America, Central America, and Asia. Labs in these areas must also consider the accessibility of chemicals, the simplicity of setup, operation, and maintenance of sequencing platforms, as well as the availability of trained laboratory and bioinformatics experts (Maljkovic et al. 2020).

1.3 Library Preparation and Sequencing Workflow

Over the past few decades, we have witnessed the revolutionization in genomic research owing to the advent of NGS. Parallel to the evolution of sequencing technologies, we can also observe a massive breakthrough in the techniques for assembling nucleic acids for establishing NGS libraries (Quail et al. 2008). The core aspect of establishing NGS library is assembling the nucleic acid target such as RNA or DNA as per the compatibility of the sequencing technique employed (Head et al. 2014). The fundamental steps in preparing the target include:

1. Fragmenting the sequences of the concerned target to a required length
2. Transformation of the target into a dsDNA
3. Linking of oligonucleotide adapters to the ends of the desired target fragments.
4. Quantitation of product for library sequencing

The critical aspect of constructing NGS libraries heavily depends upon the overall size of the target DNA fragments in the final sequenced library. The nucleic acid fragmentation is achieved with the help of three approaches – physical, chemical, and enzymatic (Marine et al. 2011). The library size can be altered based on the size of the insert (refers to the library region among adapter sequences), as the adapter sequence length is constant throughout. The range of insert size for greater efficiency can be ascertained by the constraints imposed by the sequencing implementation and NGS instrumentation adopted. In case of the adoption of Illumina technology, cluster generation influences the optimal insert size, which involves libraries undergoing denaturation, dilution, and distribution upon the surface of the flow cell in two-dimensions further subjecting it to amplification. It is known that shorter products tend to amplify more proficiently than longer products (Knierim et al. 2011). On the other hand, longer inserts give rise to highly diffused clusters than short inserts. Till today, the technology has managed to construct sequenced libraries employing Illumina instrumentation comprising as much as 1500 bases in length.

Post-library construction steps involves refining the library size and elimination of adapter dimers and other related library preparation molecules. The adapter dimers are known to generate as a consequence of their self-ligation in the absence of an appropriate library insert. They are known to form clusters efficiently, thereby occupying valuable space on the flow cell without aiding retrieving valuable knowledge (Sakharkar et al. 2004). Contemporarily, refinement of the libraries is carried out with the help of either magnetic-bead-based clean up or on agarose gels. The preliminary precaution to avoid the generation of adapter dimers is to guarantee adequate starting material. The statistics suggest that the probability of the development of adapter dimers is significantly high when starting material is limiting the process.

We meticulously analyze the steps involved in the construction as well as the challenges imposed while heightening our efficiency during the preparation of NGS libraries.

1.3.1 Sample Collection and DNA Extraction

There is plethora of factors to consider before constructing libraries from DNA like-mass of the starting material.

Whether the application is for de novo sequencing or resequencing, many more.

The construction of libraries is heavily susceptible to bias that originates from genomic material comprising exceptionally high or low Guanine-Cytosine (GC) content. These biases can be eliminated by meticulously selecting polymerases of PCR amplification, buffers, etc. Preparing libraries from DNA for whole genome sequencing or from specified target fragments within genomes, PCR amplicons, and ChIP-seq experiments tend to follow the general procedure as mentioned (Seguin-Orlando et al. 2013).

Once the DNA has been fragmented and extracted, there is a need to blunt the ends of the concerned fragments. Following blunting, the ends are 5′ phosphorylated with a mixture of 3 enzymes namely—Klenow Large Fragment, T4 polynucleotide kinase, and T4 DNA polymerase (Dabney and Meyer 2012). The 3′ ends of the fragment undergo tailing with adenylate residues with the help of Klenow Fragment (exo-) or Taq polymerase. Regarding adenylate tailing, thermostable Taq polymerase exhibits higher efficiency than Klenow Fragment (exo-). Klenow Fragment (exo-) is used where heating is not required as in the case of mate-pair library construction (Oyola et al. 2012).

1.3.2 Library Preparation Methods

It is vital to decide upon the core objective of the library preparation before ascertaining the appropriate library protocol. The successful library preparation necessitates the completion of the following steps

1. Fragmentation
2. End repair
3. Phosphorylation of 5′ ends
4. Adenine-tailing of 3′ ends for ligation with sequence adapters
5. Adapter ligation
6. PCR Amplification

After DNA fragmentation, it becomes necessary to prepare the genomic material for sequencing to ensure the continuation of the whole procedure. It is an essential step to blunt the extracted fragment ends. Upon blunting, phosphorylation, and A-tailing further enhance the efficiency of the sequencing process. During the ligation of sequence adapters, the optimal adapter:fragment ratio is ~10:1 (Adey et al. 2010). This ratio is influenced by the copy number of the genomic material and the molarity of the starting material. Another factor to be considered is the number of sequence adapters to add (Wang et al. 2012). A higher quantity of adapters disrupts the whole procedure by favoring the formation of adapter dimers, further complicating their separation and refinement during the PCR amplification. Generally, column or bead-based clean-ups are carried out after end repair and tailing. After ligation, adopting bead-based clean-ups for eliminating excess adapter dimers is done.

There has been immense research on sequencing genomes from single cells. The contemporary approach involves amplifying the whole genome with Multiple Displacement Amplification (MDA). The MDA involves the employment of random primers comprising a highly processive strand displacing polymerase—phi29 (Dean et al. 2001). Though this technique reproduces enough material to construct the whole library, it gravely suffers from considerable bias due to non-linear amplification. Recent research suggested the addition of a quasilinear preamplification step which reduced the bias considerably, hence rocketing the efficiency of MDA (Zong et al. 2012). To aid library construction from 1 to 96 single cells per run is achieved by Fluidigm (California, United States) by using a technology platform adhering to microfluidics and small compartmentalization.

1.3.3 Quality Control and Library Quantification

Library Quantification is one of the most critical steps for ensuring the achievement of optimal cluster density and uniform sample pooling during sequencing. The cluster density influences the run performance, especially the total data output and quality in non-patterned flow cells (Head et al. 2014). For patterned flow cells, flow cell occupancy affects the coordinates of the data passing filter. Underloading results in the production of high data quality while compromising the data output. On the other hand, overloading leads to poor run performance disrupting the whole cascading process. It leads to lower Q30 scores, sequence artifacts, and low data output. The preliminary cause behind under and overloading is inaccurate library quantification practices. The following practices ensure greater efficiency of the sequencing procedure—qPCR, fluorometric quantification and bioanalyzer/fragment analyzer, and equivalent instruments.

1.3.3.1 qPCR

This technique involves selective quantification of the full-length library fragments. It is achieved by using primers annealing to the p5 and p7 sequences. The sequences bearing both p5 and p7 sequences can only attach to the flow cell in order to develop clusters. The Illumina qPCR guide suggests employing KAPA qPCR kits comprising 6 DNA standards which help to generate the standard curve (Hung et al. 2018). Numerous qPCR kits are accessible for varied qPCR chemistries like DNA binding dyes (SYBR Green), hydrolysis probes (TaqMan probes), etc. qPCR primers are designed to work with all Illumina adapters.

1.3.3.2 Fluorometric Quantification

Fluorometric systems like PicoGreen and Qubit are known to utilize fluorescence-based dyes that specifically attach to RNA, single-stranded DNA (ssDNA), and dsDNA. This technique is favorable when libraries comprising broad fragment size distribution like Nextera XT, has to be sequenced. This technique risks the overestimation of the concentration of the library as it measures all the dsDNA in the sample pool (Bentley et al. 2008). It comprises partially constructed fragments and residual primer dimers obtained from PCR.

1.3.3.3 Bioanalyzer/Fragment Analyzer

Automated electrophoresis systems based on microfluidics are employed on a large scale in the library construction workflows. To name a few, Fragment Analyzer, Bioanalyzer, TapeStation are primarily employed for quality control. It includes inspecting the distribution of library size and so on. For quantification of Illumina libraries comprising narrow size distributions, TruSeq-targeted RNA expression libraries, AmpliSeq and TruSeq, and Ñ¢ small RNA libraries are employed. These instruments are unsuitable for quantifying other library types as they tend to decrease the accuracy with increasing library fragment size distribution.

1.3.4 Sequencing Workflow Overview

The next-generation sequencing workflow comprises three fundamental steps:

Step 1: Library preparation
Step 2: Sequencing
Step 3: Data analysis

It is essential to segregate and decontaminate the concerned nucleic acid after DNA extraction as the extraction techniques may lead to the development of inhibitors in the sample, thereby hindering the whole process. It is recommended to strictly follow an extraction protocol assigned to the concerned sample taken. Following extraction, many of the NGS workflows demand a quality control (QC) step. UV spectrophotometry should be employed for the assessment of the sample purity and fluorometric techniques should be adopted for quantification of the sample.

Step 1: Library preparation

The fundamental step for the success of the construction of NGS library is library construction. This step encompasses preparing the nucleic acid samples (here, DNA) to be compatible with the concerned sequencer (McCarthy 2010). The construction of sequenced libraries generally involves fragmenting DNA samples followed by the addition of a specific adapter to its ends. In the Illumina sequencing workflow, the fragments bear complementary sequences aiding the DNA fragments to link to the flow cell. These fragments further undergo amplification and purification. With a view to conserve resources, multiple libraries are pooled altogether and undergo sequencing in the same run. This procedure is recognized as multiplexing. During the ligation of adapter sequencers, different 'barcodes' which are unique index sequences assigned to every library. These unique sequences are utilized to individualize libraries throughout data analysis.

Step 2: Sequencing

The libraries are loaded with prepared samples and placed on the concerned sequencers. These fragments then undergo cluster generation, which involves the amplification of DNA clusters leading to the yield of millions of copies of the concerned fragments. In Illumina sequencers, clustering occurs automatically.

Chemically altered nucleotides are known to attach DNA template strands based on the principle of complementarity in a procedure known as Sequencing-By-Synthesis (SBS). These chemically transformed nucleotides bear a fluorescent tag along with a reversible terminator. The role of the terminator is to obstruct the integration of the following base (Merker et al. 2018). The fluorescent signal hints at the nucleotide that has been bound to aid the cleavage of the terminator to incorporate the following base (Jain et al. 2015). This same process repeats itself on the reverse strand of the DNA after the forward strand has been read. This procedure is known as pair-end sequencing.

Step 3: Data analysis

Base calling is a procedure involving the instrument software ascertaining the nucleotides and predicts the overall accuracy of the base calls (Xu and Seki 2020). The data analysis involves importing the sequencing data into a standard analysis tool.

1.3.5 Challenges and Best Practices in Library Preparation

Library preparation is itself a crucial step, along with being an integral part of the NGS workflow. The standardized implementation of quality control practices is exceptionally significant in ensuring greater efficiency in the workflow (Hess et al. 2020). The challenges are posed at each step mentioned above and must be tackled with utmost care to ensure high-quality sequencing results.

Extraction of an adequate quantity of DNA from the starting material while refraining from disturbing the inhibitors can indeed prove to be intimidating. It is also determined by the complexity of the sample material (Van Dijk et al. 2014). There is an alarming possibility of error due to carry-over contamination in the sequencing runs. Handling of enormous amount of data also presents challenges for us (Hoople et al. 2017). NGS data storage requires considerable IT resources and extensive data retrieval systems.

Library construction faces three significant challenges – contamination, protocol complexity, and cost. For instance, the workflow of Illumina TruSeq Nano includes a ten-step process in order to attach barcodes and adapters to nucleotides. This might give rise to errors owing to the complexity of the protocol. The bead-based purification techniques are also erogenous and may lead to the failure of the whole library preparation (Hrdlickova et al. 2017).

Another inherent problem faced is of contamination of the sample. Their prime sources of contamination are the preamplification steps followed when the concentration of the starting material is low. The overall expenses incurred due to laboratory equipment, employment of the trained personnel and the reagents used are pretty high (Hoeijmakers et al. 2013). The cost of library preparation for different sequencing procedures varies. Recent research suggests tagmentation to be an effective approach to minimize the costs by overlapping the fragmentation and ligation procedures in the workflow.

Solutions to these obstacles can be achieved by adopting automated workflow, which will carry out complex protocols ensuring high reproducibility and lower human intervention hence drastically reducing the probability of errors. Reduction in human involvement also nullifies the chances of contamination. Miniaturization in the usage of reagents should be adopted to achieve further education in the costs.

1.4 Applications of Next-Generation Sequencing

During this era of technological revolution, the advent of NGS has significantly molded our approach toward sequencing. This technology has made the sequence of millions of DNA fragments possible with an accelerated pace and incomparable accuracy. It has immensely benefited the extensive genomic studies carried out for numerous purposes. NGS has a broad array of applications in different fields hence highlighting its importance (Table 1.2). The most prominent use of NGS is the detection of undiscovered pathogenic organisms in the genomes of living organisms. Statistically, less than 2% of the total human genomic content includes most of the pathogenic material known as the exome (Behjati and Tarpey 2013).

The sequencing of this genomic region leads to the cost-effectiveness of NGS. Moreover, NGS has incomparable advantages over conventional sequencing methods (Fig. 1.2). To name a few, cost-effectiveness, greater accuracy, and high throughput are some of the most remarkable benefits. Here, we discuss the applications of NGS and its wide-ranging aspects in different fields.

1.4.1 Genomic Sequencing

Sequencing of the whole human genome has been a tedious task as well as an important milestone in mankind's quest for finding solutions to many questions in genetics. In the past years, Sanger sequencing technology has been utilized for over a decade to decipher the whole human genome and to formulate the final draft. As against this, this very same task is completed within a day using the NGS technology. NGS is utilized to ascertain the DNA or RNA sequence for the detailed study of genetic transformation that occurs upon the entry of a pathogenic organism into the body (Harris et al., 2015). It enables the sequencing of millions of genes at a single time in multiple samples under consideration. It also analyzes the varied kinds of genetic attributes in a single sequencing run. It is an enormously parallel sequencing technology offering high-throughput, accuracy, and time conservation (Delaneau et al. 2019). The mere advent of such a technology that satisfies all the conditions of

Table 1.2 Advantage and disadvantages of different types of sequencing platforms (Maljkovic et al. 2020)

S. no.	Sequencing platform	Year released	Applications	Advantages	Disadvantages
1	Sanger ABI 3730xl	2002	Amplicon sequencing	Long readings, high quality, and cheap Top qualong reads, low quality, high quality	Low throughput, high cost, substitution mistakes, and the need for clean sequenced material to generate high-quality sequence data
2	PacBio RSII	2010	Viral genome, microbial genome, eukaryotic genome, human/exome genomics, RNAseq/ transcriptomics, complex population sequencing, epigenetics	Used while researching methylomes	Indels, a big lab footprint, and high cost
3	Ion Torrent/ PGM318	2010	Amplicon sequencing, viral genome, microbial genome, eukaryotic genome, human/ exome genomics, diagnostics, pathogen surveillance	Simple mechanism, upgradeable, and inexpensive instrument	Homopolymer difficulties lead to a greater error rate, longer hands-on time, fewer overall reads, higher cost per MB, and indel issues
4	ABI SOLiD 5500xl/ Wildfire	2010	Amplicon sequencing, viral genome, microbial genome, eukaryotic genome, human/ exome genomics, RNAseq/ transcriptomics, methylation studies, single-nucleotide polymorphism/ variation studies, pathogen surveillance	High precision, independent flow cell lanes, and the capacity to recover from unsuccessful sequencing cycles	Platform longevity, quick reads, more gaps in assemblies, uneven data dispersion, and high capital costs
5	Illumina MiSeq	2011	Amplicon sequencing, viral genome, microbial	Low cost/MB, quick runtime, moderate cost/	Substitution errors cause a higher error

(continued)

Table 1.2 (continued)

S. no.	Sequencing platform	Year released	Applications	Advantages	Disadvantages
			genome eukaryotic genome, human/ exome genomics, RNAseq/ transcriptomics, single nucleotide polymorphism/ variation studies, diagnostics, pathogen surveillance	instrument and runs, and versatile	rate as the sequencing reaction progresses
6	Oxford Nanopore MinION	2014	Amplicon sequencing, viral genome, microbial genome, eukaryotic genome human/ exome genomics, RNAseq/ transcriptomics, single nucleotide polymorphism/ variation studies, metagenomics, epigenetics, pathogen surveillance	Longest individual reads, a user community that is reachable, and a portable USB size	Deletions, a lower throughput than competing devices, and low single-read pass accuracy
7	Illumina NextSeq 500	2015	Amplicon sequencing, viral genome, microbial genome, eukaryotic genome, human/ exome genomics, RNAseq/ transcriptomics, methylation studies, metagenomics, single nucleotide polymorphism/ variation studies, ChIP-seq, metatranscriptomics, diagnostics, pathogen surveillance	High potential sequence yield, user-friendly, and extensible	Expensive, requires strong indexing abilities, and has problems with replacement errors The mistake rate rises as the sequencing reaction progresses
8	Illumina NovaSeq 6000	2017	Viral genome, microbial genome eukaryotic genome, human/exome genomics, RNAseq/	High sequence yield potential with no usage limitations	High cost, high DNA concentration, high indexing skill required,

(continued)

Table 1.2 (continued)

S. no.	Sequencing platform	Year released	Applications	Advantages	Disadvantages
			transcriptomics, methylation studies, metagenomics, single nucleotide polymorphism/ variation studies, ChIP-seq, metatranscriptomics		substitution error concerns As the sequencing procedure continues, the error rate rises and the frequency of duplicate reads increases
9	PacBio Sequel	2016	Viral genome, microbial genome eukaryotic genome, human/exome genomics, RNAseq/ transcriptomics, complex population sequencing, epigenetics	Rapid, desktop-sized, and long reads	Costly, moderate throughput
10	Oxford Nanopore PromethION	2018	Amplicon sequencing, viral genome, microbial genome eukaryotic genome, human/ exome genomics, RNAseq/ transcriptomics, single nucleotide polymorphism/ variation studies, metagenomics, epigenetics, pathogen surveillance	More output than MinION, longer scans per individual, a reachable user base, and scalable	Low single-read pass accuracy and delete concerns

an ideal procedure minimizing losses makes NGS a unique and pivotal process in the era of technological advancements.

1.4.2 Transcriptome Analysis

In the preceding decades, after the advent of microarray technology and whole-genome sequencing, many methods were used to carry out transcriptome analysis. Previously, mRNA was analyzed using microarray or RT-PCR techniques. The problem with both of these techniques was that microarray technology failed to

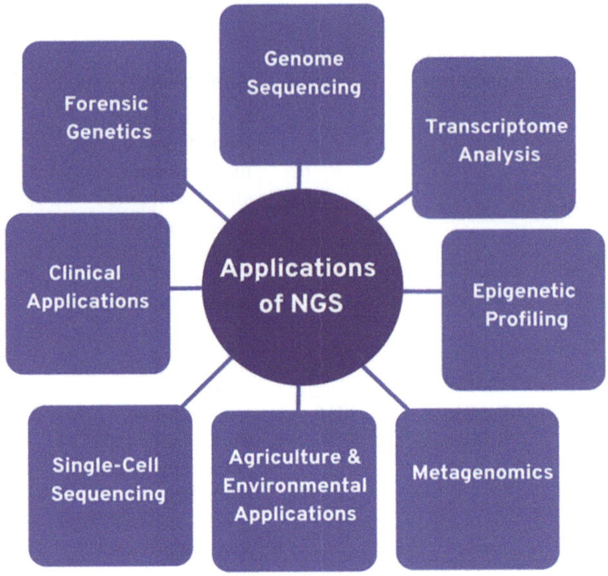

Fig. 1.2 Common applications of next-generation sequencing

possess an exquisite sensitivity. In contrast, RT-PCR was quite an expensive process for the whole human genome. Alternatively, NGS not only presents high throughput, accuracy, and swiftness but also offers genome annotation, and ascertaining of non-coding RNA (Voelkerding et al. 2009). Human genome sequencing took place for more than 15 years at the expense of millions of dollars. In the NGS era, this supposedly daunting task was completed within eight days at the cost of one million dollars (Lohr 2011). Another practical application of NGS includes pyrosequencing relied on the technology of sequencing-by-synthesis (SBS) (Denoeud et al. 2008). In pyrosequencing, the transcriptomic variant is called as RNA-seq or short-read massively parallel sequencing (Wang et al. 2008). RNA-seq is a novel, rapid transcriptomic profiling technology in this era (Wang et al. 2010).

1.4.3 Epigenetic Profiling

The title of one of the earliest adopters of NGS can be designated to the epigenetic profiling community. Research-based on NGS has provided extensive information regarding the epigenetic profiling of different cell types. In the preceding decades, the process of DNA methylation gained much attention due to the discovery of 5-hydroxymethyl-cytosine owing to its exemplary role in pluripotency and epigenetic reprogramming (Meaburn and Schulz 2012). These increasing advancements in this field give rise to various methodological procedures ensuring high coverage, accuracy, and single-base resolution profiling in a small number of mammalian cells. Epigenetic profiling aims at disease stratification and observing the genomic transformations in the genetic material of living organisms.

1.4.4 Metagenomics

Metagenomics primarily deals with the investigation of the genetic material obtained because of NGS. The samples can be obtained from the environment or of clinical origin. There are two types of significant approaches in metagenomics—target sequencing and metagenomic shotgun sequencing. Metagenomic NGS (mNGS) basically studies the sample and appropriately assigns the microorganisms to their standard genomes and which also characterizes their amount in the sample (Gu et al. 2019). The ability to identify and classify the nuclei acid belonging to different taxa makes the process unique and informative.

1.4.5 Single-Cell Sequencing

Almost every area of genomic research has been intervened by single-cell sequencing. The discovery of genomic amplification of RNA and DNA dates to the early 1990s. The process was cumulatively tedious, time-consuming, and extremely expensive rendering the whole procedure inefficient (Anaparthy et al. 2019). The advent of massively-parallel short-read sequencing changed the whole scenario of genetics. Conventionally, NGS analyzes the whole genome of a particular group of cells, whereas single-cell sequencing keeps a track of the genome from individual cells. Presently, single-cell sequencing is adopted to study the genome of DNA-methylome, scDNA-seq, and transcriptome (scRNA-seq) obtained from individual cells (Wang et al. 2023). They recognize novel mutations in cancer cells and examine the epigenomic modifications occurring during embryonic development. They are also used to monitor how specific genes are expressed in a homogenous cell population.

1.4.6 Clinical Applications

Clinical advancements aim at improving the quality of life for living organisms. The clinical uses associated with NGS provide us with a wide array of benefits. NGS operates at different levels, each level possessing its unique importance (Santos et al. 2017). Whole genome sequencing has far reached applications in the research background than clinical settings. The clinical setting is more concerned with constitutional genetic diseases rather than somatic cancer mutations. It plays a pivotal role in the prognosis of rare genetic diseases (Gorgannezhad et al. 2018).

NGS assay can be employed for exome sequencing. Apart from diagnosing diseases, NGS identifies mutation targets for therapy in various types of hereditary cancers (Zhang et al. 2017a, b). They are also responsible for the tests for tumor mutation burden and instability of microsatellites. They are also used for testing variations or mutations from cell-free circulating DNA, most commonly known as liquid biopsy (Cohen et al. 2017). Different liquid biopsy studies have been carried out for varied types of tumors. NGS-based liquid biopsy is utilized in non-invasive prenatal testing. Hence, NGS has far-reaching positive effects on humankind (Giannopoulou et al. 2018).

1.4.7 Forensic Genetics

Contemporarily, forensic DNA profiles comprise of size measurements which are interpreted in the number of repeats known as short-term tandem repeats (STR) markers. Due to the highly decreased costs of sequencing technologies as a consequence of NGS, it has led to the formulation of NGS assays by researchers for forensic DNA applications. These assays sequence STR markers which rocket up the efficiency to differentiate individuals in a highly complex sample mixture (Børsting and Morling 2015). Moreover, alternate markers like single nucleotide polymorphisms (SNPs) can seamlessly integrate into the casework laboratories. These unlock potential in discovering ancestral prediction in numerous unsolved cases. NGS has made possible immense progress in the field of forensic sciences.

1.5 Data Analysis and Bioinformatics

Sequencing genomes leads to the creation of an enormous amounts of data. The accurate storage of this data for better retrieval in the future proves to be a daunting task. In this technological era, the analytical tools present today carry out the important work of data analysis, thus aiding in retrieving information for future perusal (Table 1.3). NGS Data Analysis comprises of three basic stages – primary, secondary, and tertiary data analysis (Fig. 1.3).

1. Primary data analysis
   Real-time analysis (RTA) software works along the cycles of imaging and sequencing chemistry. It also provides base calls and linked quality scores exhibiting the primary structure of DNA or RNA strands. RTA software accomplishes the primary data analysis automatically.
2. Secondary data analysis
   This step of analysis involves the alignment of DNA or RNA fragments into an entire sequence hence enabling us to recognize genetic variants among them.
3. Tertiary data analysis
   This data analysis involves interpreting the genetic variations recognized through the knowledge of basic sciences in order to diagnose a particular disease or prevent them.

1.5.1 Data Processing and Quality Control

Recently we have witnessed an enormous surge in the volume of datasets generated owing to the colossal progress in technology, especially NGS. This creates a need to

Table 1.3 Common bioinformatics tools for NGS data analysis

NGS data analysis		
S. no.	Steps involved in data analysis	Bioinformatics tools
1.	Cleaning of NGS data	FastQC
		ClinQC
		Filtlong
		Nanofilt
		MiniBar
		Porechop
2.	Read mapping	FastQC
		htSeqTools
		Trimmomatic
		SAMStat
		CLC Genomics Workbench v11 (Qiagen), GEM3
		Novoalign
3.	Alignment	STAR
		TopHat2
		HISAT2
4.	Exploration of NGS data	R Software
		Illumina
		ABI/SOLiD
		Roche
		SanGeniX
		Galaxy
		Bowtie
		BWA
5.	Variant calling	CLC Genomics Workbench v11 (Qiagen)
		Genome Analysis Tool Kit
		FreeBayes
		HaplotypeCaller (GATK-HC)
		LoFreq
		SAMtools
		Platypus
		VarScan
6.	Deeper analyses	
	For chromosome building	Chromosomer
	For mapping tools	HISAT
	For de novo assembly	Trinity
		Soap de novo
	For gene expression profiling	Deseq2

create a data storage pipeline that is able to store, analyze, and retrieve large volumes of information (Gogol-Döring and Chen 2012). According to the statistical analysis, Illumina X-Ten System has the potential to generate datasets of volume two

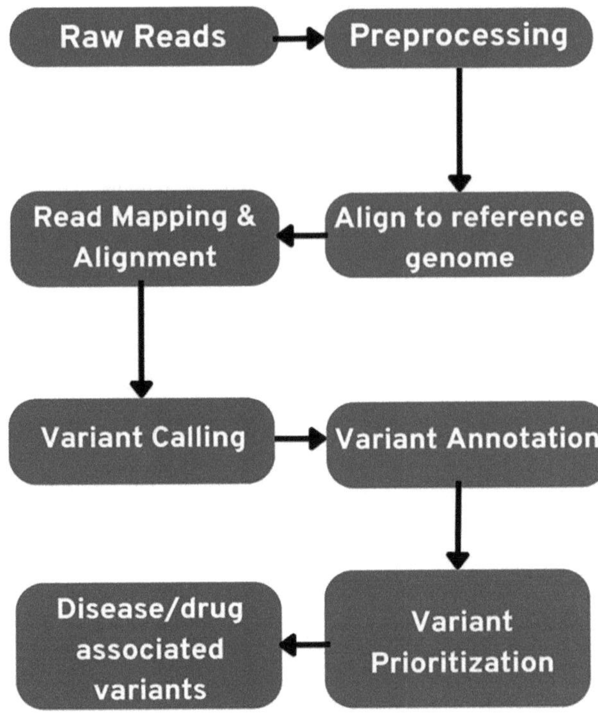

Fig. 1.3 Bioinformatics Workflow for NGS Data Analysis

petabases per year, which is extremely arduous to handle. Moreover, the lower costs for genome sequencing further enhance the progression of larger projects to completion, hence generating more data. These conditions hint us towards using the optimized algorithms for the usual tasks carried out during NGS. Big data algorithms are one of the most highly comprehensive algorithmic techniques which are employed to increase the efficiency of the project undertaken. It also highlights the fact of transforming the usual data-storing paradigm considering the volume and the required handling of the data generated. Generally, high-performance computational techniques like graphics processing units (GPUs), CPU clusters, field programmable gate arrays (FPGAs), and clouds are used (Bahassi and Stambrook 2014).

It is highly critical to verify the quality of the mapping process. The overall percentage of mapped reads is indicative of the sequencing precision and the existence of contaminated genetic material (here, DNA).

1.5.2 Read Mapping and Alignment

Once the data is obtained from the mentioned pre-processing steps, there is a need to undergo read mapping or alignment. There are two primary pathways to be followed depending on the availability of the genomic sequence.

When we need to compare one organism's genome with a standard genome, it is possible to express the transcripts by mapping the reference genomic sequence, popularly known as genome mapping (McCombie et al. 2019). When it is mapped with a transcriptome, it is called transcriptome mapping. This process of mapping requires no pre-requisite information about the transcribed regions or the sequence in which the splicing of exons has taken place. This novel approach gives rise to unannotated transcripts.

When researching on genome without the reference genome, there is a need to assemble the transcripts into longer contigs, known as de novo assembly (Kumar et al. 2019). These contigs can be considered transcriptomes which are expressed after remapping the reads for quantification. Numerous bioinformatics tools like RNA-seq read alignment program (HISAT2), TopHat, etc. are employed.

1.5.3 Variant Calling and Structural Variation Analysis

Variant Calling is primarily the procedure of recognizing single nucleotide polymorphisms (SNPs), insertions, and deletions from the obtained high-quality data as a consequence of NGS. In order to evaluate the precision of the variant calls, it becomes inevitable to have accessible standard datasets comprising already known true variants. Numerous such benchmarking datasets have been made available to the public in the past few years. The most widely used datasets are Platinum genome datasets for NA12878, Genome in a Bottle (GIAB), etc. (Hu et al. 2021). The GIAB dataset has considerably improved in the last few decades by incorporating data from multiple short-read and linked-read sequencing. It has also expanded the reference from one to seven samples.

1.5.4 Transcriptome Analysis Tools

Transcriptome analysis allows the researchers to characterize the transcriptional activity (coding and non-coding). It also enables us to focus on the subsets of some specific target genes and transcripts and profiles millions of genes concurrently to generate a picture of the functions of the cell. These gene expression analyses provide insights into the actively expressed genes and transcripts.

The various tools for transcriptome analysis are—gene expression microarrays, RNA-seq, and qRT-PCR. RNA-seq technology includes software like Solexa, SOLiD, PyroMark ID, 454, etc. (Midha et al. 2019). These tools provide better gene expression, single nucleotide variation, alternative sequencing, detection of fusion genes and absolute quantification.

1.5.5 Epigenetic Data Analysis

The three standard NGS-based techniques that are available for epigenetic analysis include ChIP-seq, ATAC-seq, and methyl-seq. Illumina is an array-based epigenetic analysis tool that is extremely accurate. These tools are robust, easy to operate & cost-effective. Epigenetic studies unravel variations observed in expressed phenotypes, transcription errors, and inactivation of X-chromosomes. Methyl-seq primarily investigates the methylation status of the genomic sequence with single nucleotide resolution (Athanasopoulou et al. 2022). This method involves bisulfite treatment converting cytosine residues to uracil. In this technique, the methylated residues are left untransformed.

ChIP-seq involves the union of chromatin immunoprecipitation (ChIP) along with NGS to recognize binding sites of DNA-linked proteins present in the genome. It is usually used to map transcription factors and histone modifications (Weirather et al. 2016). ATAC-seq is an assay for transposase-accessible sequencing of chromatin that ascertained the regions of accessibility of chromatin and also maps DNA binding proteins to recognize active enhancers, promoters, and other cis-regulatory molecules.

1.5.6 Metagenomics Data Analysis

The fundamental stage in metagenomics is to execute quality control, as it is a necessity to eliminate technical errors arising from the analysis. The core aim of this can be designated to annihilate unwanted adapter sequences, excessively short or low-quality reads, etc. Numerous programs like FastQC and MultiQC for short-read analysis are employed (Ewels et al. 2016). LongQC and MinionQC are some of the software used for long-reds analysis (Fukasawa et al. 2020). This annihilation of undesirable data sequences significantly reduces computational time and cost.

In metagenomics, the primary step aims at the elimination of unwanted sequences (Lanfear et al. 2019). This important task is done through a potent bioinformatics tool called Trimmomatic. This tool is specifically designed to eliminate undesirable low-quality adapters and reads. Cutadapt is another tool that recognizes and eliminates adapters and other sequence types.

1.5.7 Integrative Analysis and Interpretation

In the current scenario, NGS studies are known to integrate a biological approach and couple sequencing data obtained with other types of information (Park et al. 2021). For example, protein-protein interaction (PPI) networks and protein family pathways, etc. in an integrative analysis (Hutter and Zenklusen 2018). The knowledge that is experimentally validated enhances analysis models and thereby enriches the integrative analysis approach. These analyses help us to extract important information from the extracted high-quality NGS data (Chuang et al. 2007). With a perspective of dealing with the enormous amount of data and its complexity, deep learning methodologies, and machine learning have become the essence of data analysis in NGS workflow (Zhang et al. 2017a, b). Technological advancements suggest that the use of Deep Neural Networks (DNN) has skyrocketed the overall efficiency of these data analysis processes (Luo et al. 2019; Jiao et al. 2020). DNN-based approach for large-scale volume data has proved to be a better alternative for predictions of mutations in the genomes (Table 1.3). Such technological marvels will be invented incessantly in the future, which will further revolutionize the existing paradigm and the perspective of genome sequencing.

1.6 Challenges and Opportunity in Future of NGS

NGS technologies have rapidly transitioned from being solely a research tool to a diverse clinical platform (Cradic et al. 2014). This achievement has been made possible by acknowledging the limitations of the technology and addressing them through the implementation of clinical assays. Challenges such as short read durations, high error rates, time-consuming or expensive protocols, and bioinformatics shortcomings have been resolved to varying degrees, enabling successful utilization of the technology in clinical settings. However, despite these advancements, several obstacles still hinder the realization of increased and improved levels of clinical value. To enhance confidence in the clinical application of NGS technologies, it is crucial to develop improved gold standards that allow for better performance characterization of tools specifically designed for structural and copy number variation studies. Long-term solutions to these problems will also include longer reads. Another challenge in clinical sequencing is haplotype phasing. Genotype information is often unphased, which means that details regarding the chromosome from which a variant originated are not recorded. It can be useful to have this information in order to identify compound heterozygous events among other things. Due to their lack of resolution, laboriousness, or price, traditional phasing techniques have limited practical utility. NGS may be able to overcome phasing with several algorithmic techniques that are being developed (Desai and Jere 2012). A current unknown in genomic profiling is the characterization of bigger genetic abnormalities. For insertions or deletions that are less than 50 bases long, the word "indel" is frequently employed, while structural or copy number variants are used to describe longer occurrences. This seemingly random distinction signifies a region of doubt in variant calling, where performance measurements are less definitely defined, in part because there aren't enough gold standard datasets (Oliver et al. 2015). Although there are many software programs that can identify these variances, they are not thought to be as developed as programs that can identify minor variants. Tools commonly exhibit discrepancies, and no single set of algorithms is considered adequate for fully profiling an individual's structural and copy number variations (Eslami et al. 2017). An important challenge arises from the recent update of the human reference genome GRCh38 by the Genome Reference Consortium. This release, the first in over 4 years, introduces several changes compared to the previous version. Notable updates include a more comprehensive representation of pericentromeric regions, alternate sequence representation for variable regions, and the correction of numerous bases previously identified as errors or minor alleles. These significant differences have implications for the various resources that annotate the genomic sequence, which are vital components of genomics-based workflows. Clinical laboratories may face the task of reannotating legacy results to ensure compatibility with the new genome release, potentially leading to an increased workload depending on the specific application. Additionally, it may be necessary to reanalyze individuals with previously unidentified illnesses using the new genome to evaluate whether any sequence or annotation modifications affect read mapping and variant calls (Oliver et al. 2015). The ability to more accurately assess the functional significance of identified variations presents perhaps the biggest obstacle to clinical sequencing efforts. Particularly in exome- or genome-wide familial research, the abundance of genomic sequencing data creates rapidly rising numbers of variations of uncertain significance (VUS). Due to their potential as disease-causing agents or therapeutic targets, these variations have significant clinical value (Ritchie et al. 2014). Additionally, because they may be inherited by family members or future offspring, their impact goes beyond only the patient who is afflicted. In the past, factors like cosegregation, population frequency, and functional analysis have been used to better characterize this variation; however, these data are frequently sparse, and the large numbers of VUS produced make low-throughput methods of functional characterization difficult to use. Because most bioinformatics techniques up to this point have focused on the coding region of the genome, thus disregarding 99% of variation, noncoding variants are particularly difficult to analyze. In this domain, more recent aggregative approaches are starting to go beyond coding sequences and include critical data from significant ongoing programs like ENCODE (Kircher et al. 2014).

Phenotypic information is being used more and more to improve variant prioritization based on expected functional relevance. Such methods estimate the likelihood that a gene is involved in causing an observable trait by comparing an individual's phenotype to information found in illness and phenotypic ontologies (Robinson et al. 2014). The ability to prioritize causative variations has significantly improved as a result of early efforts in this field. Beyond this, early-stage clinical annotation projects like ClinVar will support the phenotypic characterization of such variation based on evidence and the distribution of the ensuing knowledge (Singleton et al. 2014; Oliver et al. 2015).

1.6.1 Metagenomics

Metagenomics, fueled by the development of NGS, has undergone a significant transformation in sequencing and analyzing metagenomic data. Technological advancements, improved throughput, and reduced sequencing costs have reshaped the metagenomics landscape. The primary objective of metagenome sequencing projects is to comprehensively characterize a community by answering questions regarding its composition and activities. Efforts are made to understand the community structure, including the taxonomic breakdown, relative abundance of species, genic contribution of each member, and intra-species heterogeneity. As new NGS technologies continue to emerge, the field of metagenomics has adapted to accommodate the unique sequencing data they generate. However, the volume of NGS data and the relatively short reads pose challenges in analyzing and maximizing their scientific value. Illumina, along with SOLiD, has been the primary platform for deep-coverage sequencing and analysis of shotgun metagenomes, despite the application of 454 pyrosequencing in some metagenomic investigations. However, the short length of Illumina reads (currently between 36 and 150 bp) presents challenges and limitations for many read-based analyses (Scholz et al. 2012).

Worldwide, novel algorithms are being developed to address sequence alignment, assembly, and read annotation in metagenomics, allowing the processing of millions to billions of very short reads. However, the selection of industry-accepted best practices for analysis has been delayed, even though it would provide valuable resources. Therefore, a core team of bioinformaticians with expertise in installing, updating, and utilizing the latest tools is essential for any NGS investigation in metagenomics (Loy et al. 2002). Additionally, sufficient data storage, typically several hundred gigabytes per sample, must be budgeted when using NGS platforms. For in-depth investigations into functional gene metagenomics, researchers are interested in studying gene families with specific enzymatic functions. Microarray and 16S community profiling methods have been used for such studies, but direct sequencing of the community is now more efficient and faster (He et al. 2010; Iwai et al. 2010). While the ultimate goal is to reconstruct every genome within a given environment, the computing complexity involved makes it impractical. Alternatively, reads can be assembled into contigs to perform taxonomic classification and functional assignments, or read-based reconstruction can be carried out to determine the functional and taxonomic components of the metagenome, each with its inherent limitations. Metagenome assembly poses unique challenges, requiring significant memory resources to reassemble the genomes found within a metagenome. The assembly of low-abundance genomes may also prove challenging, compounded by the wide range of genome abundances within a sample (Scholz et al. 2012).

1.6.2 Shotgun Sequencing of Metagenomes

NGS technologies have revolutionized the exploration of complex microbial communities at lower costs and higher throughput compared to Sanger-based sequencing. This has enabled the characterization of diverse and complex microbial communities, including soil and pelagic microbial communities and animal-associated microflora in the human intestinal tract, human saliva, and cow rumen (Deng et al. 2008; Turnbaugh et al. 2007; Willner et al. 2011; Zhang et al. 2012; Brulc et al. 2009; Scholz et al. 2012).

1.6.3 Future Prospect of Metagenomics

Looking ahead, NGS allows us to delve into the genetic makeup of complex communities in unprecedented ways. However, the complexity of metagenomic materials presents challenges and analytical bottlenecks, despite the development of specialized tools for high-throughput sequencers. Both read-based analysis and assembly-based strategies have their limitations, but both approaches should be explored to gain a comprehensive understanding of metagenome projects. Computational resources for assembly, annotation, and analysis are currently the greatest obstacles to metagenomics initiatives. It is expected that sequencing centers will increasingly focus on providing bioinformatics resources and expertise to the community. Nevertheless, the vast amount of sequencing data will exceed available resources unless new algorithms for assembly and analysis are developed (Scholz et al. 2012).

1.7 Conclusion

In conclusion, next-generation sequencing (NGS) has revolutionized the field of genomics, enabling researchers to obtain unprecedented insights into the intricacies of genetic information. This transformative technology has surpassed traditional sequencing methods in terms of speed, throughput, and cost-effectiveness, making it widely accessible and applicable in diverse areas of research and medicine. NGS has played a pivotal role in advancing our understanding of genomics by generating massive amounts of sequencing data in a relatively short time. This wealth of information has allowed for comprehensive studies of genomes, transcriptomes, epigenomes, and metagenomes, leading to breakthrough discoveries and insights into the complexities of life. One of the key strengths of NGS lies in its broad range of applications. It has significantly impacted fields such as medical genetics, where it has facilitated the identification of disease-causing variants, the elucidation of complex genetic disorders, and the development of personalized medicine approaches. NGS has also proven invaluable in cancer genomics, enabling the characterization of tumor genomes, identification of therapeutic targets, and monitoring of treatment response. Moreover, NGS has greatly contributed to microbial genomics, providing a deeper understanding of microbial communities, pathogenicity, and antibiotic resistance mechanisms. It has revolutionized evolutionary biology by enabling the study of population genetics, phylogenetics, and the dynamics of genetic variation over time. The analysis of NGS data presents its own set of challenges, including data storage, quality control, bioinformatics analysis, and data interpretation. However, ongoing advancements in computational tools and bioinformatics algorithms continue to address these challenges, improving data processing, analysis, and interpretation. In summary, next-generation sequencing has ushered in a new era of genomics, empowering researchers and clinicians with the ability to explore the intricacies of the genome and unlock its secrets. With its versatility, scalability, and ever-increasing affordability, NGS will undoubtedly continue to drive ground breaking discoveries, transform medical diagnostics, and pave the way for precision medicine and personalized healthcare in the years to come.

References

Adey A, Morrison HG, Asan XX, Kitzman JO, Turner EH, Shendure J et al (2010) Rapid, low-input, low-bias construction of shotgun fragment libraries by high-density in vitro transposition. Genome Biol 11:1–17

Ameur A, Kloosterman WP, Hestand MS (2019) Single-molecule sequencing: towards clinical applications. Trends Biotechnol 37(1):72–85

Anaparthy N, Ho YJ, Martelotto L, Hammell M, Hicks J (2019) Single-cell applications of next-generation sequencing. Cold Spring Harb Perspect Med 9(10):a026898

Athanasopoulou K, Boti MA, Adamopoulos PG, Skourou PC, Scorilas A (2022) Third-generation sequencing: the spearhead towards the radical transformation of modern genomics. Life (Basel) 12(1):30

Bahassi EM, Stambrook PJ (2014) Next-generation sequencing technologies: breaking the sound barrier of human genetics. Mutagenesis 29(5):303–310

Behjati S, Tarpey PS (2013) What is next generation sequencing? Arch Dis Child Educ Pract Ed 98(6):236–238

Bentley DR, Balasubramanian S, Swerdlow HP, Smith GP, Milton J, Brown CG, Roe PM et al (2008) Accurate whole human genome sequencing using reversible terminator chemistry. Nature 456(7218):53–59

Børsting C, Morling N (2015) Next generation sequencing and its applications in forensic genetics. Forensic Sci Int Genet 18:78–89

Brulc JM, Antonopoulos DA, Berg Miller ME, Wilson MK, Yannarell AC, Dinsdale EA, White BA et al (2009) Gene-centric metagenomics of the fiber-adherent bovine rumen microbiome reveals forage specific glycoside hydrolases. Proc Natl Acad Sci U S A 106(6):1948–1953

Buermans HPJ, Den Dunnen JT (2014) Next generation sequencing technology: advances and applications. Biochim Biophys Acta 1842(10):1932–1941

Chatterjee R, Ghosh M, Sahoo S, Padhi S, Misra N, Raina V, Suar M, Son YO (2021) Next-generation bioinformatics approaches and resources for coronavirus vaccine discovery and development—a perspective review. Vaccine 9(8):812

Chuang HY, Lee E, Liu YT, Lee D, Ideker T (2007) Network-based classification of breast cancer metastasis. Mol Syst Biol 3(1):140

Cohen JD, Javed AA, Thoburn C, Wong F, Tie J, Gibbs P, Lennon AM et al (2017) Combined circulating tumor DNA and protein biomarker-based liquid biopsy for the earlier detection of pancreatic cancers. Proc Natl Acad Sci U S A 114(38):10202–10207

Cradic KW, Murphy SJ, Drucker TM, Sikkink RA, Eberhardt NL, Neuhauser C, Grebe SK et al (2014) A simple method for gene phasing using mate pair sequencing. BMC Med Genet 15(1): 1–8

Dabney J, Meyer M (2012) Length and GC-biases during sequencing library amplification: a comparison of various polymerase-buffer systems with ancient and modern DNA sequencing libraries. Biotechniques 52(2):87–94

Dean FB, Nelson JR, Giesler TL, Lasken RS (2001) Rapid amplification of plasmid and phage DNA using phi29 DNA polymerase and multiply-primed rolling circle amplification. Genome Res 11(6):1095–1099

Delaneau O, Zagury JF, Robinson MR, Marchini JL, Dermitzakis ET (2019) Accurate, scalable and integrative haplotype estimation. Nat Commun 10(1):5436

Deng W, Xi D, Mao H, Wanapat M (2008) The use of molecular techniques based on ribosomal RNA and DNA for rumen microbial ecosystem studies: a review. Mol Biol Rep 35:265–274

Denoeud F, Aury JM, Da Silva C, Noel B, Rogier O, Delledonne M, Artiguenave F (2008) Annotating genomes with massive-scale RNA sequencing. Genome Biol 9(12):1–12
Desai AN, Jere A (2012) Next-generation sequencing: ready for the clinics? Clin Genet 81(6): 503–510
Eslami EM, Chiatante G, Miroballo M, Tang J, Ventura M, Amemiya CT, Alkan C et al (2017) Discovery of large genomic inversions using long range information. BMC Genomics 18(1): 1–12
Ewels P, Magnusson M, Lundin S, Käller M (2016) MultiQC: summarize analysis results for multiple tools and samples in a single report. Bioinformatics 32(19):3047–3048
Faria NR, Azevedo RD, Kraemer MU, Souza R, Cunha MS, Hill SC, Vasconcelos PF et al (2016) Zika virus in the Americas: early epidemiological and genetic findings. Science 352(6283): 345–349
Freedman ML, Reich D, Penney KL, McDonald GJ, Mignault AA, Patterson N, Gabriel SB, Topol EJ, Smoller JW, Pato CN, Pato MT, Petryshen TL, Kolonel LN, Lander ES, Sklar P, Henderson B, Hirschhorn JN, Altshuler D (2004) Assessing the impact of population stratification on genetic association studies. Nat Genet 36:388–393
Fukasawa Y, Ermini L, Wang H, Carty K, Cheung MS (2020) LongQC: a quality control tool for third generation sequencing long read data. G3 (Bethesda) 10(4):1193–1196
Giannopoulou L, Kasimir-Bauer S, Lianidou ES (2018) Liquid biopsy in ovarian cancer: recent advances on circulating tumor cells and circulating tumor DNA. Clin Chem Lab Med 56(2): 186–197
Gire SK, Goba A, Andersen KG, Sealfon RS, Park DJ, Kanneh L, Sabeti PC et al (2014) Genomic surveillance elucidates Ebola virus origin and transmission during the 2014 outbreak. Science 345(6202):1369–1372
Gogol-Döring A, Chen W (2012) An overview of the analysis of next generation sequencing data. In: Next generation microarray bioinformatics: methods and protocols, pp 249–257
Gorgannezhad L, Umer M, Islam MN, Nguyen NT, ShiddikY MJ (2018) Circulating tumor DNA and liquid biopsy: opportunities, challenges, and recent advances in detection technologies. Lab Chip 18(8):1174–1196
Grard G, Fair JN, Lee D, Slikas E, Steffen I, Muyembe JJ, Leroy EM et al (2012) A novel rhabdovirus associated with acute hemorrhagic fever in central Africa. PLoS Pathog 8(9): e1002924
Gu W, Miller S, Chiu CY (2019) Clinical metagenomic next-generation sequencing for pathogen detection. Annual Review of Pathology: Mechanisms of Disease 14:319–338
Harris SA, Harris EA (2015) Herpes simplex virus type 1 and other pathogens are key causative factors in sporadic Alzheimer's disease. J Alzheimers Dis 48(2):319–353
He Z, Deng Y, Van Nostrand JD, Tu Q, Xu M, Hemme CL, Zhou J et al (2010) GeoChip 3.0 as a high-throughput tool for analyzing microbial community composition, structure and functional activity. ISME J 4(9):1167–1179
Head SR, Komori HK, LaMere SA, Whisenant T, Van Nieuwerburgh F, Salomon DR, Ordoukhanian P (2014) Library construction for next-generation sequencing: overviews and challenges. Biotechniques 56(2):61–77
Hess JF, Kohl TA, Kotrová M, Rönsch K, Paprotka T, Mohr V, Paust N et al (2020) Library preparation for next generation sequencing: a review of automation strategies. Biotechnol Adv 41:107537
Hoeijmakers WA, Bártfai R, Stunnenberg HG (2013) Transcriptome analysis using RNA-Seq. In: Malaria: methods and protocols, pp 221–239
Hoople GD, Richards A, Wu Y, Kaneko K, Luo X, Feng GS, Pisano AP (2017) Gel-seq: whole-genome and transcriptome sequencing by simultaneous low-input DNA and RNA library preparation using semi-permeable hydrogel barriers. Lab Chip 17(15):2619–2630
Hrdlickova R, Toloue M, Tian B (2017) RNA-Seq methods for transcriptome analysis. Wiley Interdiscip Rev RNA 8(1):e1364

Hu T, Chitnis N, Monos D, Dinh A (2021) Next-generation sequencing technologies: an overview. Hum Immunol 82(11):801–811

Hung SS, Meissner B, Chavez EA, Ben-Neriah S, Ennishi D, Jones MR, Steidl C et al (2018) Assessment of capture and amplicon-based approaches for the development of a targeted next-generation sequencing pipeline to personalize lymphoma management. J Mol Diagn 20(2): 203–214

Hutter C, Zenklusen JC (2018) The cancer genome atlas: creating lasting value beyond its data. Cell 173(2):283–285

Iwai S, Chai B, Sul WJ, Cole JR, Hashsham SA, Tiedje JM (2010) Gene-targeted-metagenomics reveals extensive diversity of aromatic dioxygenase genes in the environment. ISME J 4(2): 279–285

Jain M, Fiddes IT, Miga KH, Olsen HE, Paten B, Akeson M (2015) Improved data analysis for the MinION nanopore sequencer. Nat Methods 12(4):351–356

Jiao W, Atwal G, Polak P, Karlic R, Cuppen E, Danyi A, Stein LD et al (2020) A deep learning system accurately classifies primary and metastatic cancers using passenger mutation patterns. Nat Commun 11(1):728

Kanzi AM, San JE, Chimukangara B, Wilkinson E, Fish M, Ramsuran V, De Oliveira T (2020) Next generation sequencing and bioinformatics analysis of family genetic inheritance. Front Genet 11:544162

Kircher M, Witten DM, Jain P, O'roak BJ, Cooper GM, Shendure J (2014) A general framework for estimating the relative pathogenicity of human genetic variants. Nat Genet 46(3):310–315

Knierim E, Lucke B, Schwarz JM, Schuelke M, Seelow D (2011) Systematic comparison of three methods for fragmentation of long-range PCR products for next generation sequencing. PLoS One 6(11):e28240

Koboldt DC, Steinberg KM, Larson DE, Wilson RK, Mardis ER (2013) The next-generation sequencing revolution and its impact on genomics. Cell 155(1):27–38

Kumar KR, Cowley MJ, Davis RL (2019) Next-generation sequencing and emerging technologies. Semin Thromb Hemost 45(7):661–673

LaBreck PT, Rice GK, Paskey AC, Elassal EM, Cer RZ, Law NN, Merrell DS et al (2018) Conjugative transfer of a novel staphylococcal plasmid encoding the biocide resistance gene, qacA. Front Microbiol 9:2664

Lanfear R, Schalamun M, Kainer D, Wang W, Schwessinger B (2019) MinIONQC: fast and simple quality control for MinION sequencing data. Bioinformatics 35(3):523–525

Lohr S (2011) Jobs tried exotic treatments to combat cancer, book says. New York Times

Loy A, Lehner A, Lee N, Adamczyk J, Meier H, Ernst J, Wagner M et al (2002) Oligonucleotide microarray for 16S rRNA gene-based detection of all recognized lineages of sulfate-reducing prokaryotes in the environment. Appl Environ Microbiol 68(10):5064–5081

Luo P, Ding Y, Lei X, Wu FX (2019) deepDriver: predicting cancer driver genes based on somatic mutations using deep convolutional neural networks. Front Genet 10:13

Magi A, Benelli M, Gozzini A, Girolami F, Torricelli F, Brandi ML (2010) Bioinformatics for next generation sequencing data. Genes 1(2):294–307

Maljkovic BI, Melendrez MC, Bishop-Lilly KA, Rutvisuttinunt W, Pollett S, Talundzic E, Jarman RG et al (2020) Next generation sequencing and bioinformatics methodologies for infectious disease research and public health: approaches, applications, and considerations for development of laboratory capacity. J Infect Dis 221(Suppl_3):S292–S307

Mardis ER (2013) Next-generation sequencing platforms. Annu Rev Anal Chem 6:287–303

Marine R, Polson SW, Ravel J, Hatfull G, Russell D, Sullivan M, Wommack KE et al (2011) Evaluation of a transposase protocol for rapid generation of shotgun high-throughput sequencing libraries from nanogram quantities of DNA. Appl Environ Microbiol 77(22):8071–8079

McCarthy A (2010) Third generation DNA sequencing: pacific biosciences' single molecule real time technology. Chem Biol 17(7):675–676

McCombie WR, McPherson JD, Mardis ER (2019) Next-generation sequencing technologies. Cold Spring Harb Perspect Med 9(11):a036798

Meaburn E, Schulz R (2012) Next generation sequencing in epigenetics: insights and challenges. Semin Cell Dev Biol 23(2):192–199
Merker JD, Wenger AM, Sneddon T, Grove M, Zappala Z, Fresard L, Ashley EA et al (2018) Long-read genome sequencing identifies causal structural variation in a Mendelian disease. Genet Med 20(1):159–163
Metzker ML (2010) Sequencing technologies—the next generation. Nat Rev Genet 11(1):31–46
Midha MK, Mengchu W, Chiu K-P (2019) Long-read sequencing in deciphering human genetics to a greater depth. Hum Genet 138(11–12):1201–1215
Mitra RD, Church GM (1999) In situ localized amplification and contact replication of many individual DNA molecules. Nucleic Acids Res 27(24):e34–e39
Moorthie S, Hall A, Wright CF (2013) Informatics and clinical genome sequencing: opening the black box. Genet Med 15(3):165–171
Oliver GR, Hart SN, Klee EW (2015) Bioinformatics for clinical next generation sequencing. Clin Chem 61(1):124–135
Oyola SO, Otto TD, Gu Y, Maslen G, Manske M, Campino S, Quail MA et al (2012) Optimizing Illumina next-generation sequencing library preparation for extremely AT-biased genomes. BMC Genomics 13:1–12
Park Y, Heider D, Hauschild AC (2021) Integrative analysis of next-generation sequencing for next-generation cancer research toward artificial intelligence. Cancers (Basel) 13(13):3148
Pereira R, Oliveira J, Sousa M (2020) Bioinformatics and computational tools for next-generation sequencing analysis in clinical genetics. J Clin Med 9(1):132
Posey JE (2019) Genome sequencing and implications for rare disorders. Orphanet J Rare Dis 14(1):153
Quail MA, Kozarewa I, Smith F, Scally A, Stephens PJ, Durbin R, Turner DJ et al (2008) A large genome center's improvements to the Illumina sequencing system. Nat Methods 5(12): 1005–1010
Ritchie GR, Dunham I, Zeggini E, Flicek P (2014) Functional annotation of noncoding sequence variants. Nat Methods 11(3):294–296
Robinson PN, Köhler S, Oellrich A, Wang K, Mungall CJ, Lewis SE et al (2014) Improved exome prioritization of disease genes through cross-species phenotype comparison. Genome Res 24(2): 340–348
Roy S, LaFramboise WA, Nikiforov YE, Nikiforova MN, Routbort MJ, Pfeifer J, Pantanowitz L et al (2016) Next-generation sequencing informatics: challenges and strategies for implementation in a clinical environment. Arch Pathol Lab Med 140(9):958–975
Roy S, Coldren C, Karunamurthy A, Kip NS, Klee EW, Lincoln SE, Carter AB et al (2018) Standards and guidelines for validating next-generation sequencing bioinformatics pipelines: a joint recommendation of the Association for Molecular Pathology and the College of American Pathologists. J Mol Diagn 20(1):4–27
Sakharkar MK, Chow VT, Kangueane P (2004) Distributions of exons and introns in the human genome. In Silico Biol 4(4):387–393
Salje H, Lessler J, Maljkovic Berry I, Melendrez MC, Endy T, Kalayanarooj S, Cummings DA et al (2017) Dengue diversity across spatial and temporal scales: local structure and the effect of host population size. Science 355(6331):1302–1306
Sanger F, Air GM, Barrell BG, Brown NL, Coulson AR, Fiddes CA, Hutchison CA, Slocombe PM, Smith M (1977) Nucleotide sequence of bacteriophage phi X174 DNA. Nature 265:687–695
Santos R, Ursu O, Gaulton A, Bento AP, Donadi RS, Bologa CG, Overington JP et al (2017) A comprehensive map of molecular drug targets. Nat Rev Drug Discov 16(1):19–34
Schmidt K, Mwaigwisya S, Crossman LC, Doumith M, Munroe D, Pires C, Livermore DM et al (2016) Identification of bacterial pathogens and antimicrobial resistance directly from clinical urines by nanopore-based metagenomic sequencing. J Antimicrob Chemother 72(1):104–114
Scholz MB, Lo CC, Chain PS (2012) Next generation sequencing and bioinformatic bottlenecks: the current state of metagenomic data analysis. Curr Opin Biotechnol 23(1):9–15

Seguin-Orlando A, Schubert M, Clary J, Stagegaard J, Alberdi MT, Prado JL, Orlando L et al (2013) Ligation bias in illumina next-generation DNA libraries: implications for sequencing ancient genomes. PLoS One 8(10):e78575

Shendure J, Ji H (2008) Next-generation DNA sequencing. Nat Biotechnol 26(10):1135–1145

Singleton MV, Guthery SL, Voelkerding KV, Chen K, Kennedy B, Margraf RL, Yandell M et al (2014) Phevor combines multiple biomedical ontologies for accurate identification of disease-causing alleles in single individuals and small nuclear families. Am J Hum Genet 94(4):599–610

Stewart-Ibarra AM, Ryan SJ, Kenneson A, King CA, Abbott M, Barbachano-Guerrero A, Endy TP et al (2018) The burden of dengue fever and chikungunya in southern coastal Ecuador: epidemiology, clinical presentation, and phylogenetics from the first two years of a prospective study. Am J Trop Med Hyg 98(5):1444

Toledo-Rueda W, Rosas-Murrieta NH, Muñoz-Medina JE, González-Bonilla CR, Reyes-Leyva J, Santos-López G (2018) Antiviral resistance markers in influenza virus sequences in Mexico, 2000–2017. Infect Drug Resist 11:1751

Turnbaugh PJ, Ley RE, Hamady M, Fraser-Liggett CM, Knight R, Gordon JI (2007) The human microbiome project. Nature 449(7164):804–810

Van Dijk EL, Jaszczyszyn Y, Thermes C (2014) Library preparation methods for next-generation sequencing: tone down the bias. Exp Cell Res 322(1):12–20

Voelkerding KV, Dames SA, Durtschi JD (2009) Next-generation sequencing: from basic research to diagnostics. Clin Chem 55(4):641–658

Wang RL, Biales A, Bencic D, Lattier D, Kostich M, Villeneuve D, Toth G et al (2008) DNA microarray application in ecotoxicology: experimental design, microarray scanning, and factors affecting transcriptional profiles in a small fish species. Environ Toxicol Chem 27(3):652–663

Wang L, Feng Z, Wang X, Wang X, Zhang X (2010) DEGseq: an R package for identifying differentially expressed genes from RNA-seq data. Bioinformatics 26(1):136–138

Wang J, Fan HC, Behr B, Quake SR (2012) Genome-wide single-cell analysis of recombination activity and de novo mutation rates in human sperm. Cell 150(2):402–412

Wang X, Wu X, Hong N, Jin W (2023) Progress in single-cell multimodal sequencing and multi-omics data integration. Biophys Rev:1–16

Weirather JL, Duggal P, Nascimento EL, Monteiro GR, Martins DR, Lacerda HG, Fakiola M, Blackwell JM, Jeronimo SM, Wilson ME (2016) Fine mapping under linkage peaks for symptomatic or asymptomatic outcomes of Leishmania infantum infection in Brazil. Infect Genet Evol 43:1–5

Willner D, Furlan M, Schmieder R, Grasis JA, Pride DT, Relman DA, Haynes M et al (2011) Metagenomic detection of phage-encoded platelet-binding factors in the human oral cavity. Proc Natl Acad Sci U S A 108(suppl_1):4547–4553

Xu L, Seki M (2020) Recent advances in the detection of base modifications using the nanopore sequencer. J Hum Genet 65(1):25–33

Zhang G, Zhang F, Ding G, Li J, Guo X, Zhu J, Dong X et al (2012) Acyl homoserine lactone-based quorum sensing in a methanogenic archaeon. ISME J 6(7):1336–1344

Zhang W, Xia W, Lv Z, Ni C, Xin Y, Yang L (2017a) Liquid biopsy for cancer: circulating tumor cells, circulating free DNA or exosomes? Cell Physiol Biochem 41(2):755–768

Zhang W, Chien J, Yong J, Kuang R (2017b) Network-based machine learning and graph theory algorithms for precision oncology. npj Precis Oncol 1(1):25

Zong C, Lu S, Chapman AR, Xie XS (2012) Genome-wide detection of single-nucleotide and copy-number variations of a single human cell. Science 338(6114):1622–1626

Advances in Structural Bioinformatics 2

Juveriya Israr, Shabroz Alam, Sahabjada Siddiqui, Sankalp Misra, Indrajeet Singh, and Ajay Kumar

Abstract

Structural bioinformatics is a captivating discipline that delves into the intricate realm related to proteins, RNA, and DNA, the macromolecules of life. Its primary focus lies in comprehending and foreseeing the enigmatic three-dimensional (3D) architecture of these fundamental entities. By employing cutting-edge computational techniques and advanced algorithms, structural bioinformatics unravels the complex interplay between structure and function, shedding light on the inner workings of life's molecular machinery. Bioinformatics is an interdisciplinary field that combines experimental and computational approaches to investigate various aspects of macromolecular 3D structure. By utilizing experimentally determined structures and computational models, bioinformatics aims to explore diverse inquiries related to macromolecules. These inquiries encompass understanding the distinctions and similarities between macro and micro structures, understanding the rules of molecular interaction, evolution, and folding, and revealing the complexity of structure-function correlations. Structural

J. Israr
Faculty of Biosciences, Institute of Biosciences and Technology, Shri Ramswaroop Memorial University, Barabanki, Uttar Pradesh, India

Department of Biotechnology, Era University, Lucknow, Uttar Pradesh, India

S. Alam · S. Siddiqui
Department of Biotechnology, Era University, Lucknow, Uttar Pradesh, India

S. Misra
Faculty of Biosciences, Institute of Biosciences and Technology, Shri Ramswaroop Memorial University, Barabanki, Uttar Pradesh, India

I. Singh · A. Kumar (✉)
Department of Biotechnology, Rama University, Kanpur, Uttar Pradesh, India
e-mail: drajay.fet@ramauniversity.ac.in

V. Singh, A. Kumar (eds.), *Advances in Bioinformatics*,
https://doi.org/10.1007/978-981-99-8401-5_2

bioinformatics, a specialized domain within the realm of computational structural biology, encompasses the study and analysis of biological structures. The term "structural" in this context aligns with its definition in the field of structural biology. The field of structural bioinformatics is dedicated to addressing biological challenges and unveiling novel insights through the development of innovative methodologies for the analysis of data pertaining to biological macromolecules.

Keywords

Structural biology · Protein data bank · Drug discovery · Molecular modeling

2.1 Introduction

The growing domain of structural bioinformatics encompasses the dynamic realm of prognosticating and scrutinizing protein architectures. Utilizing state-of-the-art methodologies and computational algorithms, bioinformatics researchers have achieved remarkable advancements in the realm of protein modeling. By employing these cutting-edge techniques, they can effectively simulate intricate protein structures, forecast protein-protein interactions, and discern critical binding sites crucial in order to create innovative medicines. This document aims to delve into the latest breakthroughs in the realm of bioinformatics. Important biological macromolecules including proteins, RNA, and DNA are the subject of structural bioinformatics, a subfield of bioinformatics that analyzes and predicts their three-dimensional structures (Patel et al. 2019). Bioinformatics is a multidisciplinary field that encompasses the study of macromolecular 3D structures. Analyzing and generalizing about these structures includes doing things like comparing overall folds and local motifs, learning the principles of molecular folding, learning about evolutionary relationships, learning about binding interactions, and learning how these structures are put together to perform specific tasks. (Gauthier et al. 2019). This comprehensive approach utilizes both experimentally determined structures and computational models to gain insights into the intricate world of macromolecules. In structural bioinformatics, the word "structural" has the same meaning as it does in structural biology. As a subfield of computational structural biology, structural bioinformatics is essentially an essential part of the field. Structural bioinformatics, as a field, is primarily focused on the development and implementation of innovative methodologies for the analysis and manipulation of biological macromolecular data. By harnessing these advanced techniques, researchers aim to address complex biological challenges and uncover novel insights into the intricate workings of living systems. The overarching goal is to not only expand our understanding of biology but also pave the way for the generation of transformative knowledge that can revolutionize various facets of scientific inquiry (Gu and Bourne 2011; Wei et al. 2014).

2.2 Protein Structure

In bioinformatics, understanding how a protein's structure contributes to its function is crucial. The intricate three-dimensional arrangement of amino acids within a protein dictates its ability to carry out specific biological tasks. This fundamental principle, known as structure-function relationship, forms the cornerstone of our understanding of protein biology (Rigden 2009). By deciphering the structural characteristics of proteins, bioinformaticians can unravel the underlying mechanisms that govern their diverse functions, ranging from enzymatic catalysis to molecular recognition. Consequently, the Proteins, through the strategic arrangement of distinct chemical moieties, exhibit enzymatic properties that facilitate the acceleration of diverse biochemical reactions. Primary, secondary, tertiary, and quaternary structures are the usual divisions into a protein's four stages of organization explained in Fig. 2.1 (Kocincová et al. 2017).

Structural bioinformatics is a field that primarily focuses on the analysis and understanding of interactions between biomolecular structures, with a particular emphasis on their spatial coordinates. The analysis of the primary structure is commonly undertaken within the purview of conventional bioinformatics disciplines. Specific constraints in the underlying genetic code of the supplied sequence allow for the formation of conserved regional topologies within the polypeptide chain, such as alpha-helices, beta-sheets, and loops. In the realm of bioinformatics, it is worth noting that the protein fold is fortified by a series of feeble interactions, including but not limited to hydrogen bonds. The stability of the protein

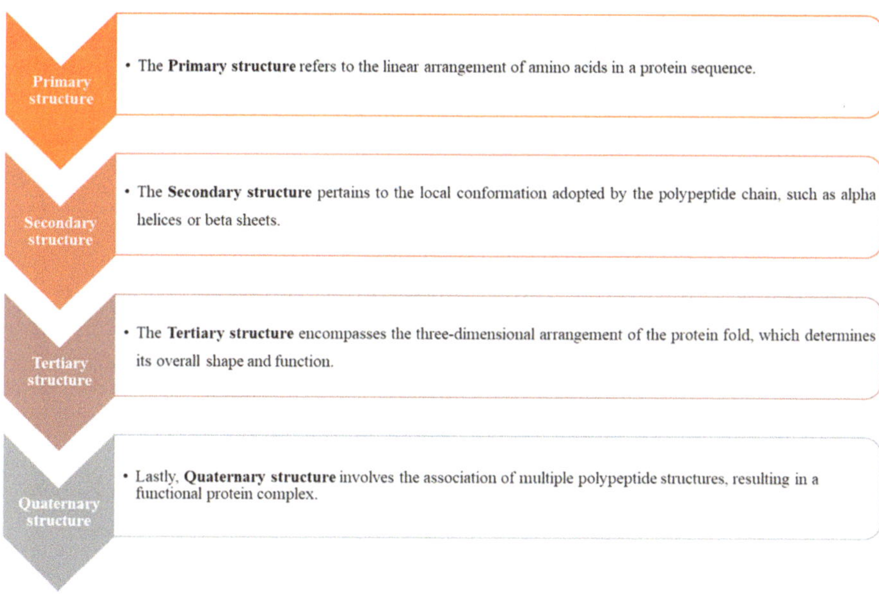

Fig. 2.1 Types of protein structure

structure depends on these interactions. In the realm of bioinformatics, interactions can manifest in two distinct ways: intrachain and interchain. Intrachain interactions transpire within the confines of a single protein monomer, specifically within its tertiary structure. On the other hand, interchain interactions take place between diverse structures, commonly referred to as the quaternary structure. In the realm of structural bioinformatics, researchers are currently engaged in the comprehensive examination of the intricate organization of interactions, encompassing both robust and delicate connections, as well as the interwoven complexities. This captivating domain employs cutting-edge methodologies, including circuit topology frameworks, to unravel the profound mysteries underlying the topological configuration of biological systems.

2.3 Structure Visualization

The visualization of protein structures holds significant importance within the field of structural bioinformatics. The user's text is not provided. The platform facilitates the visualization of both static and dynamic molecular representations, enabling the identification of molecular interactions that can be leveraged for the inference of underlying molecular mechanisms. In the field of bioinformatics, a plethora of visualization techniques have emerged to aid when it comes to deciphering and analyzing complex biological data. Among the most prevalent and widely utilized types of visualization are (Fig. 2.2) (Shi et al. 2017):

Cartoon: The utilization of cartoon representations in protein visualization serves as a valuable tool for highlighting variations in secondary structure. In the realm of bioinformatics, the α-helix, a fundamental structural motif, is often symbolically depicted as a helical screw, embodying its characteristic spiral conformation. Similarly, β-strands, another crucial element of protein structure, are commonly represented as arrows, symbolizing their linear arrangement. Furthermore, the flexible regions known as loops are typically denoted by straight lines, capturing their

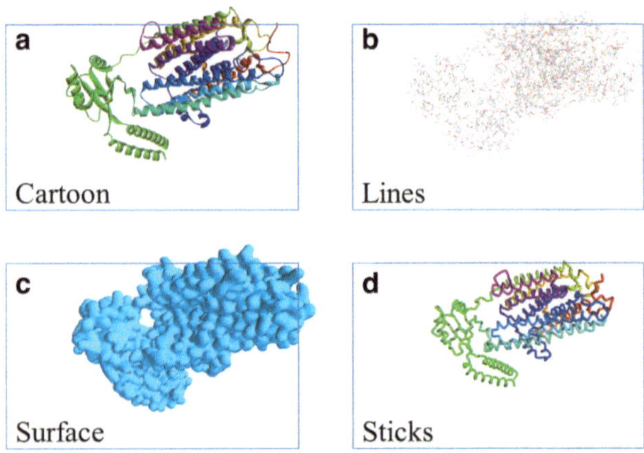

Fig. 2.2 Structural visualization of transmembrane protein 63B (PDB ID: 8EHX). (**a**) Cartoon; (**b**) lines; (**c**) surface; (**d**) sticks

inherent flexibility and variability. The study and analysis of protein structures can be aided by these visual representations, enabling researchers to comprehend and interpret the intricate three-dimensional arrangements of these vital biomolecules.

Lines: In this bioinformatics representation, the amino acid residues are elegantly depicted as slender lines, enabling efficient and cost-effective graphic rendering.

Surface: The visualization depicts the molecular structure's external shape, providing insights into its surface characteristics.

Sticks: Stick diagrams are a common way for bioinformaticians to visualize the complex world of molecular structures. The framework's slender sticks representing the covalent links between amino acid atoms perfectly capture the essence of this chemical connection. This method of visualization is typically used to better understand and portray the complex web of interactions that takes place between amino acids.

2.4 DNA Structure Background

The seminal elucidation of the DNA duplex structure was first expounded by the esteemed scientific duo of Watson and Crick, with notable contributions from the esteemed researcher Rosalind Franklin. The DNA molecule is a complex construction made composed of a phosphate group, a pentose sugar, and a nitrogenous base. These constituents, when combined, form the fundamental building blocks of DNA. The phosphate group serves as a crucial backbone, providing structural stability to the molecule. The pentose sugar, a five-carbon sugar, acts as a central framework, connecting the various components of DNA. Lastly, the nitrogenous bases, which include adenine, thymine, cytosine, and guanine, play a pivotal role in encoding genetic information within the DNA molecule (Travers and Muskhelishvili 2015). Together, these three substances harmoniously unite to form the remarkable DNA molecule, the cornerstone of life's genetic blueprint. Hydrogen bonding between complementary base pairs are responsible for maintaining the DNA double helix's structural integrity. Adenine and thymine (A–T) and cytosine and guanine (C–G) form hydrogen bonds in DNA. These hydrogen bonds play a crucial role in stabilizing the overall structure of DNA. Structural bioinformatics research endeavors have predominantly concentrated on elucidating the intricate interplay between deoxyribonucleic acid (DNA) and diminutive chemical entities. This captivating area of investigation has garnered considerable attention in the realm of drug design, with numerous studies dedicated to unravelling the underlying mechanisms governing these interactions (Fig. 2.3).

2.5 Interactions

Interactions encompass the intricate network of contacts that are established between various components of molecules operating at distinct hierarchical levels. As a vital component of the intricate world of bioinformatics, they assume the crucial role of

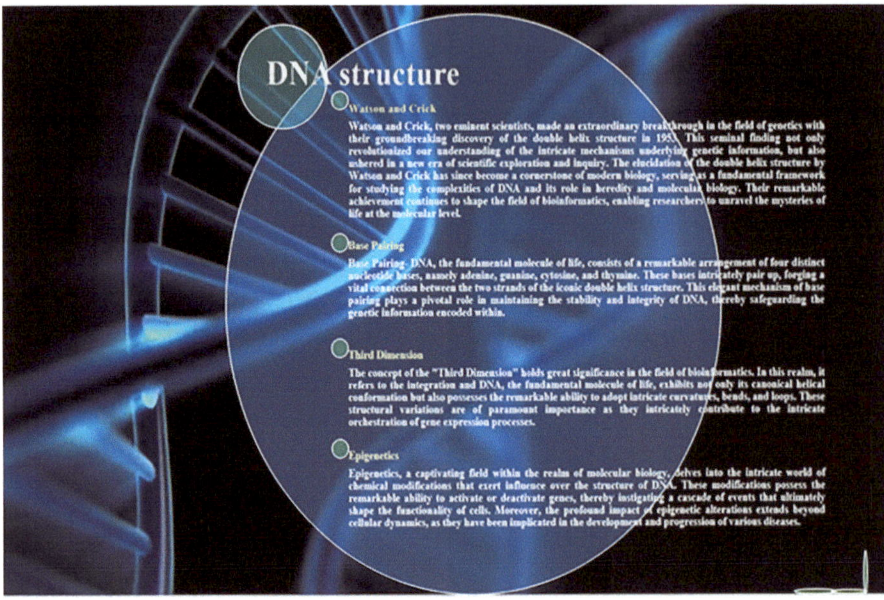

Fig. 2.3 DNA structure

ensuring the stability of protein structures while engaging in a diverse array of functional activities. In bioinformatics, the study of molecular interactions entails the identification, classification, and evaluation of sets of atoms or molecular areas that have an effect on one another. The hydrophobic effect, hydrogen bonding, and electrostatic forces are all possible explanations for these phenomena. Proteins engage in a wide variety of interactions, such as those between themselves and other proteins, between themselves and peptides, between themselves and ligands, and between themselves and DNA (Stanfield and Wilson 1995; Klebe 2015).

2.6 Calculating Contacts

The computation of contacts holds significant significance within the realm of structural bioinformatics. Docking and molecular dynamics analyses are made much easier using it, and it also helps predict protein structure and folding, thermodynamic stability, protein-protein and protein-ligand interactions, and more. In the realm of bioinformatics, conventional approaches have relied upon computational methodologies that leverage the concept of threshold distance, commonly referred to as cut-off, in order to identify potential interactions among atoms. The detection methodology employed in this study relies on the calculation of Euclidean distances and angles between atoms of specific types. In the realm of bioinformatics, it has been observed that a majority of the methodologies relying on the straightforward Euclidean distance principle tend to fall short in effectively identifying occluded

Fig. 2.4 Distance criteria for contact definition

Distance criteria for contact definition

Type	Max distance criteria
Hydrogen bond	3,9 Å
Hydrophobic interaction	5 Å
Ionic interaction	6 Å
Aromatic Stacking	6 Å

contacts. In recent years, the utilization of cut-off-free methodologies, such as Delaunay triangulation, has emerged as a prominent approach in the field of bioinformatics. Furthermore, the integration of a diverse range of parameters, such as physicochemical attributes, spatial proximity, molecular structure, and bond orientations, has been leveraged to enhance the accuracy of contact identification. Distance criteria for contact definition are explained in Fig. 2.4 (Martins et al. 2018; da Silveira et al. 2009).

2.7 Protein Data Bank (PDB)

Proteins, DNA, and RNA are just a few examples of the complex biomolecules that are represented in PDB, which is a repository for three-dimensional structural information relevant to macromolecules of biological significance. PDB is achieved and maintained by the acclaimed multinational cooperation known as the Worldwide Protein Data Bank (wwPDB). The PDB is managed and curated by a collaboration that includes PDBe, PDBj, the RCSB, and BMRB, to name a few of the regional organizations involved. The individual plays a crucial role in ensuring that online copies of PDB (Protein Data Bank) data are publicly available to anyone who wants to use them. Bioinformatics has come a long way, as seen by the exponential increase in the quantity of structural data stored in the Protein Data Bank (PDB). These invaluable databases have been growing steadily each year as cutting-edge techniques like X-ray crystallography, NMR spectroscopy, and cryo-electron microscopy have been used to collect them (Berman et al. 2000).

2.8 Data Format

The PDB format, also known as the PDB format, is a textual file format that has been widely adopted in the field of bioinformatics. It serves as a repository for storing crucial information pertaining to the three-dimensional structures of macromolecules. This format has been extensively utilized by the esteemed PDB, which is a renowned source for researchers and scientists worldwide. The PDB format, due to inherent limitations in its structural framework, imposes restrictions on the representation of expansive molecular structures. More specifically, it can't handle structures with more than 99,999 atoms or more than 62 chains. The macromolecular Crystallographic Information File (mmCIF), or Protein Data Bank exchange (PDBx), is a widely adopted and standardized text file format that serves as a comprehensive representation of crystallographic information. This format is specifically designed to capture and convey essential data pertaining to macromolecular structures obtained through crystallography techniques. By adhering to a consistent and well-defined structure, the PDBx/mmCIF format enables efficient storage, exchange, and analysis of crystallographic information, facilitating seamless collaboration and interoperability within the bioinformatics community. PDBx/mmCIF (Protein Data Bank exchange/macromolecular Crystallographic Information File) has been the primary way of exchanging data within the PDB archive since its introduction in 2014. This transition has allowed for enhanced data representation and improved interoperability within the field of bioinformatics. The PDBx/mmCIF file format, denoted by (.cif) extension, offers a more comprehensive and structured approach to storing and exchanging macromolecular structural information. By adopting this standardized format, researchers and scientists can seamlessly access and analyze protein structure data, facilitating advancements in various areas of bioinformatics research. The PDB (Protein Data Bank) format is a standardized file format that comprises a collection of records denoted by a keyword consisting of a maximum of 6 characters. In contrast, the PDBx/mmCIF format employs a distinct structure that relies on a key-value system. In this system, the key represents a specific name that serves to identify a particular attribute, while the value corresponds to the variable information associated with that attribute.

2.9 Other Structural Databases

Alongside the PDB, a multitude of databases housing comprehensive repositories of protein structures and diverse macromolecules have been established. In the realm of bioinformatics, a multitude of examples can be found that showcase the application of computational techniques to biological data analysis. These instances include:

The **Molecular Modeling Database (MMDB)** is an extensive repository that contains experimentally derived three-dimensional structures of biomolecules. These structures are derived from the Protein Data Bank (PDB), a repository of protein and nucleic acid structures. MMDB provides a valuable platform for researchers in the field of bioinformatics to access and analyze these structures, enabling a deeper understanding of the molecular architecture and function of biomolecules. By leveraging the wealth of information contained within MMDB, scientists can unravel the intricate relationships between structure and function, paving the way for advancements in fields such as drug discovery,

The **Nucleic Acid Database (NDB)** is a comprehensive depository of experimentally derived information pertaining to nucleic acids, including DNA and RNA.

The **Structural Classification of Proteins (SCOP)** is a comprehensive and intricate framework that elucidates the intricate structural and evolutionary connections among proteins that have been experimentally determined. By meticulously analyzing the three-dimensional structures of proteins, SCOP provides a detailed and systematic classification system that allows researchers to comprehend the intricate relationships and evolutionary patterns within the protein universe. The advancement of our knowledge of the complicated molecular machinery that underpins life has been greatly aided by this wonderful resource, which has become a cornerstone in the science of bioinformatics.

TOPOFIT-DB is a comprehensive database that specializes in protein structural alignments utilizing the cutting-edge TOPOFIT methodology. This innovative approach enables the accurate comparison and analysis of protein structures, facilitating the identification of conserved regions and functional motifs. By leveraging TOPOFIT-DB, researchers can gain valuable insights into the structural relationships between proteins, unravelling intricate molecular mechanisms and aiding in the discovery of novel therapeutic targets (Ilyin et al. 2004).

The **Electron Density Server (EDS)** is a cutting-edge bioinformatics tool that provides researchers with invaluable insights into the electron-density maps and statistical analyses pertaining to the fit of crystal structures and their corresponding maps. By leveraging advanced computational algorithms, EDS provides a robust environment for the study and interpretation of electron-density data, helping researchers gain insight into the structural features and molecular interactions of biological macromolecules. EDS's intuitive design and powerful features allow scientists to make sound judgments and propel innovative discoveries in structural biology.

The **Critical Assessment of Protein Structure Prediction (CASP)** is an internationally acclaimed program that encourages scientists everywhere to work together. It's a hub for massive-scale studies to determine how proteins fold in three dimensions. CASP's global reach allows it to convene specialists from a wide range of disciplines to develop protein structure prediction. The Critical Assessment of Protein Structure Prediction (CASP) is the industry benchmark for evaluating protein structure prediction methods. It allows scientists to evaluate the efficacy of their methods for predicting the three-dimensional structure of proteins. Redundancy in protein databases is a problem for protein structure analysis in bioinformatics. To access the PISCES server, the bioinformatics platform used here may compile a vetted set of Protein Data Bank (PDB) entries according to predefined sequence identity and structural quality thresholds. This program swiftly sifts through the millions of PDB structures in the database, using sophisticated algorithms and data mining techniques to zero in on entries that fulfil the user-specified sequence identity criterion and display excellent structural quality. This method guarantees that the resulting list contains only the most pertinent and trustworthy PDB items.

The **Structural Biology Knowledgebase (SBK)** is a comprehensive platform that offers a wide range of sophisticated tools and resources specifically tailored to facilitate and enhance protein research design. With a focus on structural biology, SBK provides invaluable assistance to scientists and researchers in their pursuit of unravelling the intricate details of protein structures and functions. Equipped with cutting-edge computational algorithms and advanced data analysis techniques, SBK empowers users to explore, analyze, and manipulate protein structures with utmost precision and efficiency. Its extensive repertoire.

ProtCID is an invaluable resource in the field of bioinformatics, specifically designed to cater to the needs of researchers studying protein-protein interactions. The Protein Common Interface Database, or ProtCID, is an extensive database of crystal structures that include homologous proteins that have comparable protein-protein interfaces. By meticulously curating and organizing a vast collection of crystal structures, ProtCID enables scientists to explore and analyze the intricate details of protein-protein interactions. This database offers a unique opportunity to investigate the similarities and differences in the interfaces of homologous proteins, shedding light on the underlying principles governing these crucial interactions. With its user-friendly interface and powerful search capabilities, ProtCID empowers researchers to delve into the wealth of information.

Alpha Fold is a cutting-edge bioinformatics tool that has revolutionized the field of protein structure prediction. Developed by DeepMind, Alpha Fold utilizes advanced machine learning algorithms to accurately protein three-dimensional structure prediction using only amino acid sequences by leveraging vast amounts of genomic and proteomic data.

2.10 Structure Comparison

2.10.1 Structural Alignment

Structural alignment, a fundamental technique in bioinformatics, facilitates the comparison of three-dimensional (3D) structures by evaluating their shape and conformation. The user, identified by the numerical identifier, has expressed a desire for their text to be Bioinformatics has revolutionized the field of evolutionary biology by enabling the inference of intricate evolutionary relationships among proteins, even in cases where their sequence similarity is relatively low. Through the application of sophisticated computational algorithms and statistical models, bioinformatics tools have unlocked the potential to unravel the evolutionary history of protein sequences, shedding light on their shared ancestry and divergent paths. By harnessing the power of bioinformatics, researchers can delve into the intricate tapestry of protein evolution, uncovering hidden connections and gaining valuable insights into the complex dynamics that shape the molecular world. Structural alignment is a fundamental technique in bioinformatics that involves the precise alignment of two or more protein structures in three-dimensional space. The atoms in identical places are rotated and translated so that one structure can be superimposed on top of another. Alpha carbon atoms or the backbone heavy atoms (carbon, nitrogen, oxygen) are typically used for the alignment. Protein structures can be compared and analyzed with this alignment method, revealing previously hidden information on evolutionary links, functional domains, and conserved areas. The root-mean-square deviation (RMSD) of atomic locations is commonly used in bioinformatics to evaluate the quality of an alignment. This index calculates the typical separation between stacked atoms.

The distance between atom i and either a reference atom that corresponds to the same atom in another structure or the mean coordinate of N similar atoms is denoted by the variable i in structural bioinformatics. This distance measurement is crucial for analyzing and comparing the spatial arrangement of atoms in different structures, aiding in the identification of structural similarities and differences. By quantifying the spatial relationships between atoms, δ_i provides valuable insights into the structural characteristics and dynamics of biomolecules. The measurement of the root mean square deviation (RMSD) outcome is typically expressed in the Ångström (Å) unit, a widely used scale in bioinformatics and structural biology. This unit is equivalent to 10^{-10} m, providing a precise and standardized representation of the structural differences between biomolecules. In bioinformatics, the root mean square deviation (RMSD) is a widely used metric to quantify the similarity between protein or nucleic acid structures. The higher the similarity between the two structures, the smaller the RMSD number will be.

$$\mathrm{RMSD} = \sqrt{\frac{1}{N}\sum_{i=1}^{N}\delta_i^2}$$

2.10.2 Graph-Based Structural Signatures

Structural signatures, commonly referred to as fingerprints, serve as representations of macromolecule patterns, enabling the inference of similarities and distinctions. Comparing a large number of proteins using the Root Mean Square Deviation (RMSD) approach is complicated by the high computing expense of creating structural alignments. The growth of structural signatures in bioinformatics has been aided by the use of graph distance patterns among atom pairs. These signatures have proven to be instrumental in the identification of protein vectors and the detection of non-trivial information within protein structures. In addition, the integration of linear algebra and machine learning techniques has proven to be invaluable in the field of bioinformatics. Protein signature clustering, ligand identification, free energy prediction, and Euclidean distance-based mutation recommendation are just some of the many applications that have benefited from the use of these effective technologies (Pires et al. 2011; Mariano et al. 2019).

2.10.3 Structure Prediction

Bioinformatics work relies heavily on the determination of molecular structures. Methods including X-ray crystallography (XRC), nuclear magnetic resonance

(NMR) spectroscopy, and three-dimensional electron microscopy are used for this purpose. These techniques enable scientists to gain insights into the intricate arrangements of atoms within molecules. However, it is important to note that these methods can be resource-intensive, both in terms of financial costs and experimental efforts. Additionally, certain molecular structures, particularly those of membrane proteins, pose unique challenges and may require specialized approaches for successful establishment. Therefore, the utilization of computational methodologies becomes imperative in the determination of three-dimensional (3D) structures of macromolecules. In the field of bioinformatics, the study of protein structure prediction encompasses various methodologies, primarily categorized into two main approaches: comparative modeling and de novo modeling.

2.10.4 Comparative Modeling

Protein 3D structure prediction is a common use of comparative modeling, also known as homology modeling, in the field of bioinformatics. To do this, they compare the target protein's amino acid sequence to that of a known-structure template protein. By comparing the similarities between the target and template sequences, computational algorithms can predict the three-dimensional structure of the target protein. Comparative modeling is a valuable tool in the field of bioinformatics, enabling researchers to gain insights into protein structure and function. The scientific literature extensively documents the phenomenon wherein proteins that share evolutionary ancestry exhibit a remarkably preserved three-dimensional conformation. The user, identified as, is seeking assistance in rewriting their text to align with the field Furthermore, it is worth noting that protein sequences exhibiting a degree of dissimilarity greater than 20% may exhibit distinct structural conformations (Kaczanowski and Zielenkiewicz 2010; Chothia and Lesk 1986).

2.10.5 De Novo Modeling

In the realm of structural bioinformatics, the concept of de novo modeling, or ab initio modeling, pertains to methodologies employed to derive three-dimensional structures from sequences, obviating the requirement for a pre-existing homologous 3D structure. The field of de novo protein structure prediction continues to captivate the scientific community, as it grapples with the persistent challenge of unravelling protein structures through the application of novel algorithms and methodologies (Meyers et al. 2021). Despite the notable advancements witnessed in recent years, this pursuit remains an unresolved frontier in the realm of modern science.

2.10.6 Structure Validation

Following the process of structure modeling, it becomes imperative to incorporate an essential subsequent phase of structure validation. This is due to the fact that numerous algorithms and tools employed in both comparative and "de novo" modeling rely on heuristic approaches to assemble the three-dimensional structure, thereby leading to the potential generation of a multitude of errors. In the realm of bioinformatics, a plethora of validation strategies are employed to ascertain the accuracy and reliability of computational models. One such approach involves the calculation of energy scores, which are subsequently juxtaposed against experimentally determined structures. This comparative analysis serves as a means to evaluate the fidelity and plausibility of the computational predictions, enabling researchers to gain valuable insights into the structural integrity and functional characteristics of biomolecules. The DOPE score, a widely employed energy score in bioinformatics, plays a pivotal role within the MODELLER tool. Its primary function revolves around the assessment and selection of optimal models. One additional validation strategy involves the computation of φ and ψ backbone dihedral angles for each residue, followed by the construction of a Ramachandran plot. The conformational space of amino acids is influenced by the side-chain properties and the interplay of interactions within the backbone. Consequently, the Ramachandran plot serves as a valuable tool for visualizing the permissible conformations by constraining these two angles. The presence of a substantial abundance of amino acids positioned in non-permissive locations within the chart serves as an indicative characteristic of a modeling outcome of inferior quality (Webb and Sali 2014).

2.10.7 Prediction Tools

The compendium of protein structure prediction software encompasses a comprehensive array of frequently employed computational tools. De novo protein structure prediction, protein threading, comparison modeling, and secondary structure prediction are only few of the methods included in this collection.

2.10.8 Molecular Docking

Among the most often used computational methods in bioinformatics is molecular docking, is employed to forecast the precise spatial arrangement and coordinates of a ligand molecule upon its binding to a receptor or target molecule. The binding phenomenon predominantly occurs via non-covalent interactions, although investigations into covalently linked binding are also conducted. Molecular docking is a computational approach utilized in bioinformatics to forecast potential conformations, also known as binding modes, of a ligand as it engages with distinct regions on a receptor. Bioinformatics software applications employ molecular docking algorithms that leverage the principles of force fields to ascertain an objective score, thereby facilitating the ranking of optimal molecular conformations. This scoring mechanism is designed to prioritize poses that exhibit enhanced intermolecular interactions between the two biomolecules under investigation.

Docking protocols are widely employed in the field of bioinformatics to computationally forecast the intricate interplay between minute chemical compounds and proteins. Protein docking, peptide docking, DNA/RNA docking, lipid docking, and carbohydrate docking are just a few examples of how docking is used as a powerful computational tool in bioinformatics to decipher complex connections and binding patterns in macromolecules.

2.10.9 Virtual Screening

Virtual screening is an indispensable computational methodology employed in the realm of bioinformatics to expedite the screening process of vast compound libraries, thereby facilitating the discovery of potential drug candidates. In the realm of bioinformatics, virtual screening is a widely employed technique that leverages the power of docking algorithms to prioritize small molecules based on their affinity towards a specific target receptor.

In the realm of contemporary scientific inquiry, a multitude of cutting-edge computational tools have been harnessed to assess the efficacy and potential of virtual screening methodologies within the realm of pharmaceutical drug discovery. In the realm of bioinformatics, the docking process encounters various challenges that impede its efficacy. These obstacles encompass issues such as incomplete data, flawed comprehension of drug-like molecular characteristics, suboptimal scoring functions, and inadequate docking strategies. The current body of literature indicates that the technology in question is not yet regarded as fully developed or mature within the field of bioinformatics (Dhasmana et al. 2019; Wermuth et al. 2015).

2.10.10 Molecular Dynamics

Molecular dynamics (MD) is a widely employed computational technique in the field of bioinformatics that enables the simulation and analysis of molecular systems by studying the dynamic behavior of molecules and their constituent atoms over a specified time frame (Pagadala et al. 2017). Through MD simulations, intricate details of molecular interactions and their effects can be elucidated, providing valuable insights into the underlying mechanisms governing various biological processes. By leveraging the principles of classical mechanics and statistical physics, MD offers a powerful tool for investigating the structural and functional properties of biomolecules, facilitating the exploration of complex biological phenomena at the atomic level. This bioinformatics approach enables the comprehensive examination of molecular dynamics and intermolecular interactions within a holistic system. In the realm of bioinformatics, the elucidation of system behavior and trajectory determination is achieved through the utilization of molecular dynamics (MD). This powerful computational technique relies on the application of Newton's equation of motion, coupled with molecular mechanic's methodologies, to estimate the interparticle forces, commonly referred to as force fields. By harnessing these fundamental principles, MD enables the comprehensive analysis of complex biological systems, shedding light on their dynamic behavior and facilitating a deeper understanding of their intricate molecular interactions (Costa et al. 2019; Alder and Wainwright 1959; Yousif 2020).

2.11 Applications

To better comprehend the three-dimensional structures of biological macromolecules like proteins and nucleic acids, structural bioinformatics employs methods from biology, computer science, and mathematics. In this field, informatics approaches play a crucial role in the analysis and interpretation of structural data. One of the key information, the process of target selection in bioinformatics involves the identification of potential targets through a comprehensive comparison with databases containing both structural and sequence information. By leveraging these databases, researchers can effectively evaluate the suitability of various targets for further investigation and analysis. The determination of a target's significance can be predicated upon an extensive analysis of the existing body of published scientific literature. Target selection can be guided by the identification of protein domains present within the target. Protein domains, the fundamental units of protein structure, possess the remarkable ability to undergo rearrangements, thereby facilitating the generation of novel proteins with diverse functionalities. Isolation of these entities can be undertaken as an initial step in their study (Gong et al. 2011).

The utilization of X-ray crystallography in the field of bioinformatics enables the elucidation of the intricate three-dimensional architecture of proteins. In the realm of bioinformatics, the utilization of X-ray technology for the examination of protein crystals necessitates the prior formation of highly refined and unadulterated protein crystals. This intricate process often entails a substantial number of experimental iterations to achieve the desired outcome. The imperative to monitor the conditions and outcomes of experiments arises from the inherent complexity of scientific investigations, necessitating a comprehensive approach to data management and analysis. Moreover, the utilization of supervised machine learning algorithms enables the analysis of the accumulated data to discern potential factors that could enhance the production of pristine crystals.

The investigation and interpretation of X-ray crystallographic data is a fundamental aspect of bioinformatics research. X-ray crystallography is used to discover the three-dimensional structures of biological macromolecules like proteins and nucleic acids, and then scientists use cutting-edge computer methods and tools to study these structures. The Fourier transform of the electron density distribution can be used to make sense of the diffraction pattern generated by shining X-rays on electrons. The demand for bioinformatics algorithms capable of deconvolving

Fourier transforms in the presence of partial information is evident. This arises from the inherent limitations in phase information, as detectors are only able to measure the amplitude of diffracted X-rays, while the phase shifts remain elusive. The utilization of advanced computational methods, such as the Multiwavelength Anomalous Dispersion technique, holds immense potential in the field of bioinformatics. By employing this technique, researchers can effectively generate electron density maps, which serve as invaluable tools for deciphering the intricate structures of biological macromolecules. In particular, the precise positioning of selenium atoms within these maps serves as a crucial reference point, enabling the accurate determination of the remaining components of the molecular architecture. The generation of the standard Ball-and-stick model involves the utilization of the electron density map.

Nuclear magnetic resonance (NMR) spectroscopy data is a key component of the inquiry and will be analyzed and interpreted. NMR spectroscopy experiments yield multi-dimensional datasets, wherein each discernible peak represents a distinct chemical moiety present within the analyzed sample. The application of optimization techniques is pivotal in the transformation of spectral data into intricate three-dimensional molecular structures.

The integration of structural and functional information is a fundamental aspect of bioinformatics research. By correlating structural data with functional insights, researchers can gain valuable insights into the intricate relationship between a biomolecule's structure and its biological activity. Structural studies serve as powerful tools to probe and elucidate the structural-functional relationship, shedding light on the underlying mechanisms governing molecular function. Through this interdisciplinary approach, bioinformatics researchers strive to unravel the complex interplay between structure and function, ultimately advancing our understanding of the intricate workings of biological systems.

2.12 Tools

2.12.1 List of Structural Bioinformatics Tools

2.12.1.1 I-TASSER (https://zhanggroup.org/I-TASSER)

Protein structure prediction involves building a three-dimensional model using the protein's amino acid sequence. Protein 3D models are created from amino acid sequences using the bioinformatics method I-TASSER (Iterative Threading ASSEmbly Refinement). The Protein Data Bank's structure templates are located using a technique called fold recognition (or threading). Replica exchange Monte Carlo simulations are used to reconstruct structural parts derived from threading templates into models of the whole structures. Among the best protein structure prediction algorithms, I-TASSER has shown itself to be in the large-scale community-wide CASP trials. Structural matching of target protein models to the known proteins in protein function databases has allowed researchers to get annotations for ligand binding site, gene ontology, and enzyme commission (Yang et al. 2015). To

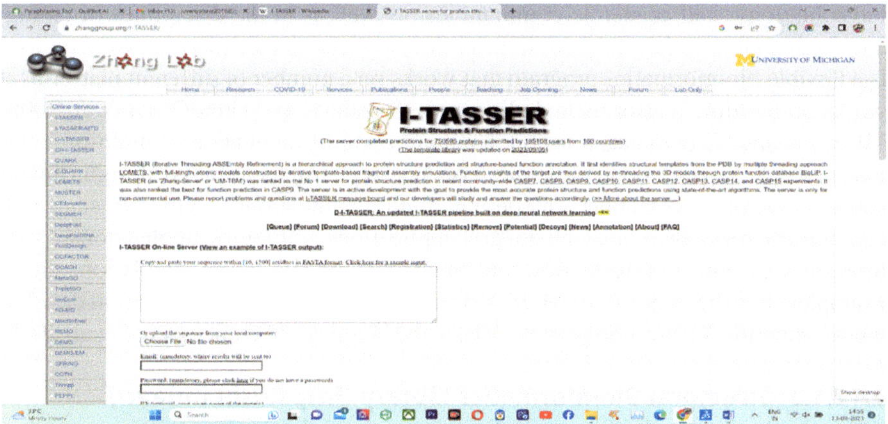

Fig. 2.5 I-TASSER webpage

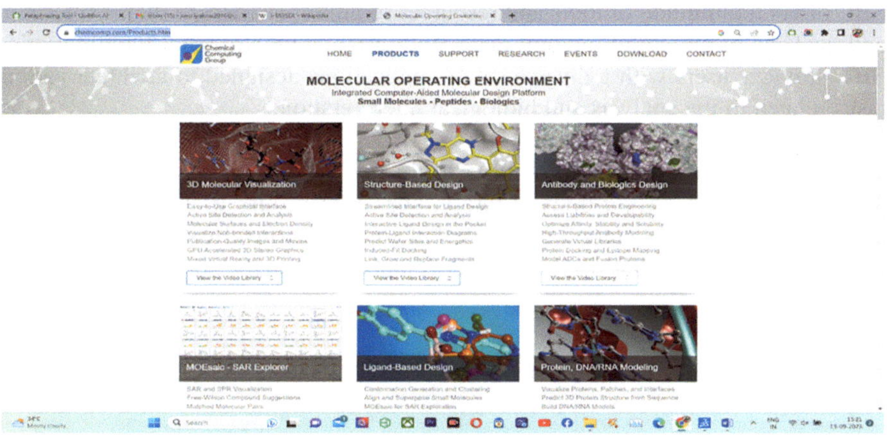

Fig. 2.6 Molecular Operating Environment webpage

help forecast the structures and roles of sequences, the Yang Zhang lab at the University of Michigan in Ann Arbor built a web service. You can download the full version of I-TASSER from the website (Fig. 2.5).

2.12.1.2 Molecular Operating Environment (https://www.chemcomp.com/Products.htm)

Modern drug discovery laboratories often use the state-of-the-art Molecular Operating Environment (MOE) platform (Fig. 2.6). Visualization, modeling, simulation, and the creation of new methodologies are only few of the features that are effortlessly included into the whole. MOE allows scientists and researchers to quickly investigate and evaluate molecular structures, predict their behavior, and quicken the pace at which new medicines are developed. In the pharmaceutical, biotechnology, and academic communities, as well as in the disciplines of biology and medicinal chemistry, the MOE scientific applications are put to good use. MOE is a flexible bioinformatics program that works on a number of different platforms. It works on multiple platforms including macOS, Linux, and Unix (Vilar et al. 2008). MOE's adaptability makes it useful in a number of different areas of bioinformatics. Pharmacophore discovery, structure-based design, fragment-based design, ligand-based design, molecular modeling and simulations, virtual screening, cheminformatics, medicinal chemistry applications, biologics applications, structural biology and bioinformatics, and molecular modeling and simulations are all examples. For the majority of MOE's command, scripting, and application development, Scientific Vector Language (SVL) is the language of choice (Vilar et al. 2008).

2.12.1.3 Structural Bioinformatics Library (https://sbl.inria.fr/)

SBL, also known as the Structural Bioinformatics Library (Fig. 2.7), is a comprehensive software package that encompasses a wide range of end-user applications and advanced algorithms. Developed specifically for the field of bioinformatics, SBL offers a multitude of tools and resources to aid researchers in their exploration and analysis of structural biology data. With a focus on structural bioinformatics, SBL provides users with a diverse set of applications designed to facilitate various tasks related to the analysis and biological interpretation.

2.12.1.4 BALLView (https://ball-project.org/ballview/)

The Biochemical Algorithms Library, or BALL, is a sophisticated and all-encompassing C++ class framework that contains a broad variety of algorithms and data structures that have been developed especially for the purposes of

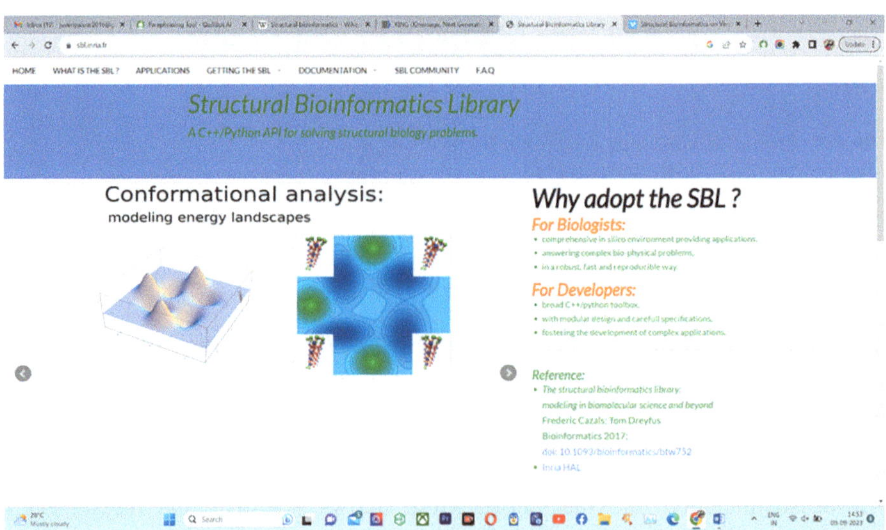

Fig. 2.7 Structural Bioinformatics Library webpage

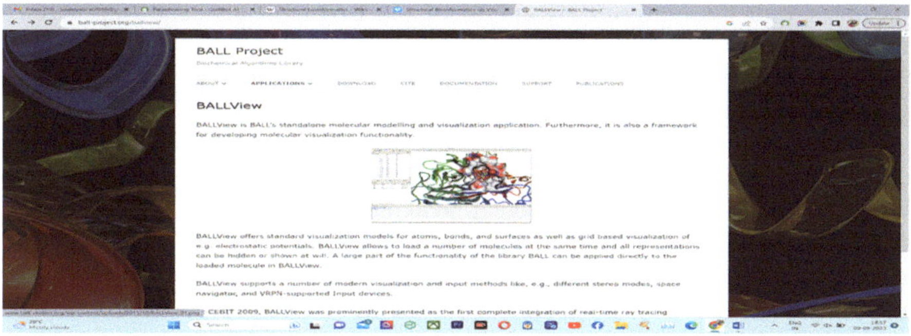

Fig. 2.8 BALLView webpage

molecular modeling and computational structural bioinformatics (Fig. 2.8). This powerful library also offers a Python interface, enabling seamless integration with the Python programming language. Additionally, BALL provides a visually appealing and user-friendly molecule viewer called BALLView, which serves as a graphical user interface for the library (Hildebrandt et al. 2010). With its extensive capabilities and versatile features, BALL is an indispensable tool for researchers and scientists in the field of bioinformatics.

The BALL software has undergone a remarkable transformation, transitioning from a proprietary commercial product to a freely available open-source solution. With this change, the software's license has also been updated; the GNU Lesser General Public License (LGPL) is now in effect. BALLView, an exceptional software tool, proudly operates under the esteemed GNU General Public License (GPL) license. This license, renowned for its commitment to promoting freedom and collaboration, ensures that BALLView remains accessible to all, fostering a vibrant community of bioinformatics enthusiasts. With its powerful features and user-friendly interface, BALLView empowers researchers to visualize and analyze complex biological data with unparalleled precision and efficiency. By embracing the GPL license, BALLView exemplifies the spirit of open-source bioinformatics, enabling scientists.

The widely utilized bioinformatics software tools, BALL and BALLView, have been successfully adapted and made compatible with various operating systems including Linux, macOS, Solaris, and Windows. This extensive porting effort ensures that researchers and scientists across different computational environments can seamlessly leverage the functionalities and capabilities offered by these powerful tools (Nickels et al. 2013).

BALLView is an advanced molecular visualization tool that has been meticulously crafted by the esteemed BALL project team. This cutting-edge software is implemented in C++ and leverages the power of Qt and OpenGL, while employing the remarkable real-time ray tracer RTFact as its rendering back-ends. BALLView is a cutting-edge software tool that provides advanced capabilities for three-dimensional and stereoscopic visualization in various modes. It seamlessly integrates with the powerful algorithms of the BALL library, allowing users to leverage its functionalities through an intuitive graphical user interface. With BALLView, researchers and bioinformaticians can effortlessly explore complex molecular structures and gain valuable insights into their data.

Acclaimed research groups from the illustrious Saarland University, Mainz University, and University of Tübingen have painstakingly developed and diligently maintained the BALL project, a ground-breaking endeavor in the field of bioinformatics. In the fields of learning and discovery, the library and the viewer are vital resources that make the gathering and processing of knowledge possible. Users now have easy access to this powerful bioinformatics tool thanks to the Debian project's incorporation of BALL packages into its repository.

2.12.1.5 PyMoL (https://pymol.org/2/)

Warren Lyford DeLano's cutting-edge molecular visualization system, PyMOL, uses open source technology and his proprietary expertise. PyMOL helps scientists visualize and analyze complex molecular structures with its advanced features and user-friendly interface. PyMOL helps users understand biomolecules, leading to bioinformatics breakthroughs. Pioneering private software company DeLano Scientific LLC commercialized this innovative technology. Developing cutting-edge tools with broad accessibility for scientific and educational communities, DeLano Scientific LLC helped advance this advancement (Yuan et al. 2017). Schrödinger, Inc., a bioinformatics giant, commercializes this technology. The permissive software license was removed. Later software is distributed under a custom license instead of the Python license. This custom license grants extensive use, redistribution, and modification rights while transferring copyright ownership to Schrodinger, LLC. Note that some source code is no longer available. PyMOL, a powerful bioinformatics tool, creates visually appealing three-dimensional representations of chemical compounds and complex biological macromolecules like proteins (Rosignoli and Paiardini 2022). The primary author claims that PyMOL, a popular software tool, gained popularity in bioinformatics by 2009. Nearly 25% of 3D protein structure images in scientific literature were created using PyMOL. PyMOL is a popular structural biology model visualization tool (Fig. 2.9). It stands out among the few open-source software options in this domain. The "Py" prefix indicates that the software is written in Python. PyMOL, a versatile molecular visualization software, uses GLEW and Free GLUT. PyMOL performs well in solving complex Poisson-Boltzmann equations using the Adaptive Poisson Boltzmann Solver. No user text is provided. PyMOL, a powerful bioinformatics tool, uses Tk for its GUI widgets. Schrödinger provided macOS Aqua binaries. With version 2.0, PyMOL switched to PyQt, ensuring a consistent experience across platforms. No user text is provided.

2.12.1.6 Visual Molecular Dynamics (https://www.ks.uiuc.edu/Research/vmd/)

Visual Molecular Dynamics (VMD) is a cutting-edge bioinformatics software application that serves as a powerful tool for molecular modeling and visualization (Fig. 2.10). Developed specifically for the purpose of analyzing complex molecular

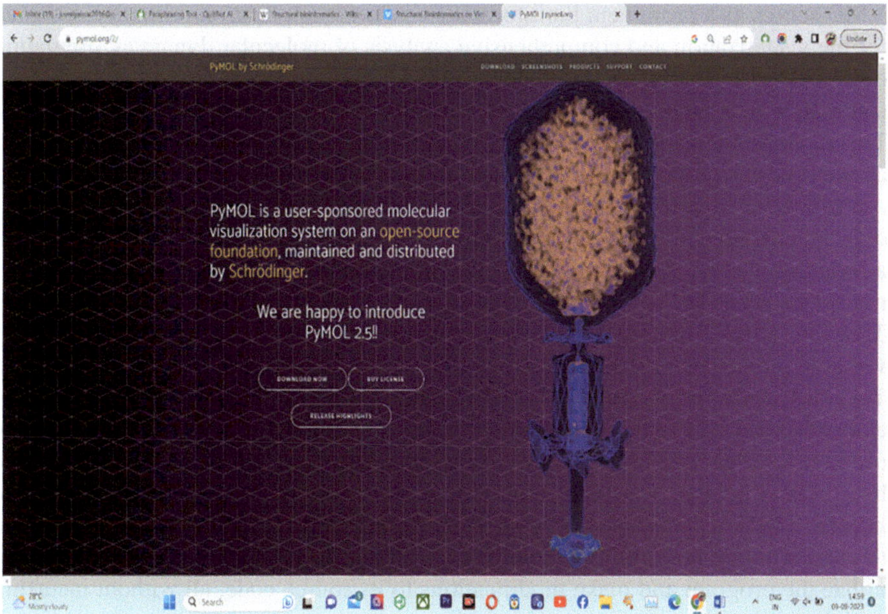

Fig. 2.9 PyMoL webpage

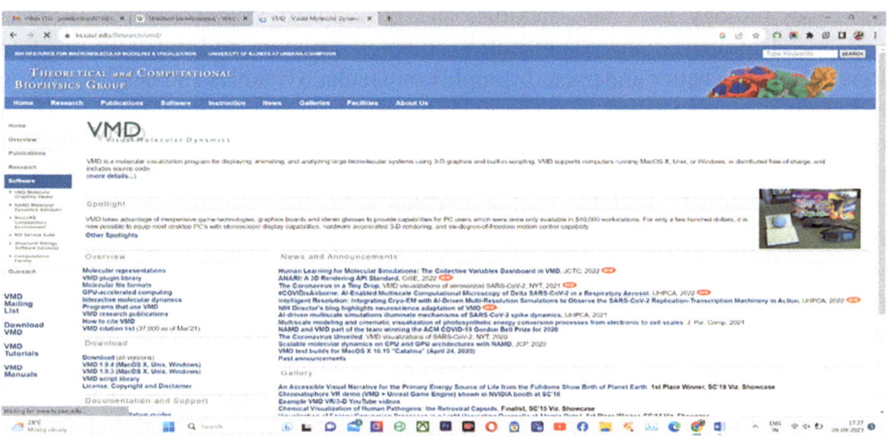

Fig. 2.10 Visual Molecular Dynamics webpage

systems, VMD enables researchers to gain valuable insights into the intricate structures and dynamic behaviors of biomolecules (Mackoy et al. 2021). By employing advanced computational algorithms and sophisticated graphical rendering techniques, VMD empowers scientists to explore and manipulate molecular data with unparalleled precision and clarity. With its user-friendly interface and extensive range of features, VMD has become an indispensable resource in the field of bio

VMD serves as a prominent bioinformatics software application primarily designed for the visualization and comprehensive analysis of molecular dynamics simulation outcomes. The bioinformatics field encompasses a wide range of tools and techniques that facilitate the analysis and manipulation of various types of data. This includes the handling of volumetric data, such as three-dimensional structures, as well as the processing of sequence data, such as DNA or protein sequences. Additionally, bioinformatics tools are designed to work with arbitrary graphics objects, enabling researchers to visualize and interpret complex biological data in a meaningful way. It is common practice in the bioinformatics profession to export molecular sceneries to external rendering programs like POV-Ray, Render Man, Tachyon, Virtual Reality Modeling Language (VRML), and many more. VMD, a versatile molecular visualization software, empowers users to execute personalized Tcl and Python scripts seamlessly. This capability is facilitated by the inclusion of embedded Tcl and Python interpreters within the VMD framework. VMD may be used on a variety of platforms, including Unix, macOS, and Windows, making it a highly flexible tool for molecular visualization. Through a distribution-specific license, VMD, a potent piece of molecular visualization software, is made available to users who are not profiting from its use. This license grants users the freedom to utilize the program and make modifications to its source code, all without incurring any charges.

2.12.1.7 KiNG (http://kinemage.biochem.duke.edu/software/king/)

Structural biology uses fast, flexible, and customized visualization software to understand biological macromolecules' complex structure and dynamic function. Researchers can better understand complex molecular dynamics and behaviors using bioinformatics visualizations. These software applications must display three-dimensional annotations of model errors or significant interaction sites alongside the structural depiction to be effective.

The Java-based, modular, and extensible scientific visualization tool KiNG (Kinemage, Next Generation) focuses on macromolecular visualization. KiNG, a versatile molecular visualization software, is similar to PyMOL, SwissPdbViewer, Chimera, RasMol, and JMol (Fig. 2.11). These programs provide the means for real-time manipulation and exploration of molecular structures in three dimensions. KiNG's dynamic 3D rotation, translation, cropping, and zooming aids in comprehending molecular depth perception and spatial interactions (Chen et al. 2009). KiNG's molecule-agnostic kinemage graphics format stands out with its versatile color palette, advanced depth cueing, and extensive tools and features. This distinguishes KiNG from other bioinformatics software.

KiNG is state-of-the-art software that builds on the past three decades of progress in molecular graphics, especially in the area of protein ribbon diagrams. It proudly stands on the shoulders of Mage, the pioneering kinemage graphics program that inspired it. Mage, a front-line bioinformatics tool, was carefully designed to help create stunning and accurate molecular illustrations. Its flexible functionality lets researchers and educators easily add captivating visuals to journal articles and classroom materials. Mage's innovative features and user-friendly interface enable

Fig. 2.11 KiNG webpage

users to visually communicate complex molecular concepts, improving scientific communication. Mage quickly became essential to the lab's research program due to its adaptability. Reimagining kinemage functionality from scratch in the KiNG framework produced a modern user interface that resembles its predecessor. It has also simplified the data structure, making maintenance and expansion easier. KiNG, an essential part of the lab's research, leads their scientific efforts.

Mage and KiNG's bioinformatics collaborations have led to different development paths. The simultaneous development of two kinemage viewers has improved both software applications. As mentioned in the "Results and Discussion" section, the KiNG and Mage software platforms initially integrated high-dimensional visualization techniques for specific purposes. We created a robust, versatile, and enhanced functionality through collaboration and synergistic integration of diverse implementations.

Bioinformatics becomes more versatile and adaptable by decoupling molecular information, such as PDB files, from its visual representation using KiNG and Mage. In bioinformatics, "7" provides no context or information to rewrite. Secondary annotations such as helix axis, local validation outliers, and interface contact dots can be seamlessly incorporated into main structural data such as models, ribbons, electron density, and NMR using the proposed method. Additionally, this strategy allows for fully non-molecular visualizations in the same computational tool.

Bioinformatics enthusiasts can find many examples and format documentation at http://kinemage.biochem.duke.edu. The "Materials and Methods" section can help interested parties understand the topic. Kinemage is a versatile and efficient plain text format for bioinformatics manual editing and program generation.

KiNG's adaptability is especially useful when the kinemage format is rigid. The runtime-loaded Java plug-in modules enhance existing features. The flexible graphics engine can be used in a new computational framework. To enable high-dimensional analysis, this study uses protein reconstruction plug-ins and molecular visualization tools while improving the core software. KiNG, a novel bioinformatics tool, can quickly develop and integrate new modules, increasing its flexibility.

2.12.1.8 STRIDE (Algorithm) (https://webclu.bio.wzw.tum.de/cgi-bin/stride/stridecgi.py)

Structural Identification (STRIDE), a sophisticated bioinformatics tool, uses atomic coordinates from cutting-edge methods like X-ray crystallography, protein NMR, and other protein structure determination methods (Fig. 2.12). STRIDE accurately assigns secondary structure elements to proteins using a robust algorithm, revealing their structural organization. STRIDE's ability to decipher a protein's intricate atom arrangement helps us understand protein structure-function relationships and facilitate bioinformatics analyses (Matarazzo and Pakzad 2014). In bioinformatics, dihedral angle potentials in hydrogen bond criteria improve the DSSP algorithm in the STRIDE framework. This method uses more complicated secondary structure definition criteria than the popular DSSP algorithm. The STRIDE energy function includes a distance-dependent 8–6 potential for the hydrogen-bond term, which is inspired by the work of Lennard-Jones. The optimal planarity of the hydrogen bond geometry is captured by integrating two angular dependency components Like DSSP, this method relies on empirical examinations of solved structures from the Protein Data Bank that have had their secondary structure elements visually assigned, and then uses statistical likelihood factors to identify these elements. One of the first and most used bioinformatics tools is the Dictionary of Secondary Structure of Proteins (DSSP). DSSP remains the most popular structural assignment method despite its age. However, the original definition of STRIDE, another popular method, claimed to outperform DSSP in at least 70% of structural assignments. In the DSSP method, shorter secondary structures are often assigned than by expert crystallographers. The STRIDE algorithm has been improved to address this issue. Minor local structural variations near secondary structure element termini cause this discrepancy. To address this issue, STRIDE has been improved to predict secondary structures more accurately. The bioinformatics-popular STRIDE and DSSP algorithms agree 95.4% of the time. A sliding-window technique reduces single-terminal residue assignment discrepancies, achieving this agreement. No user text is provided. STRIDE and DSSP may underestimate secondary structure elements like pi helices.

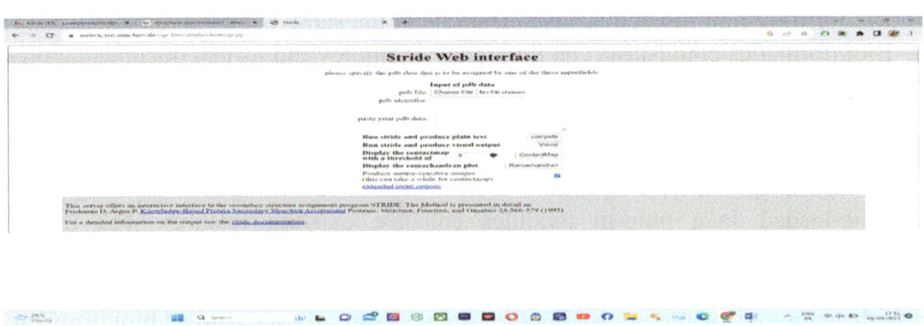

Fig. 2.12 STRIDE (algorithm) webpage

2.12.1.9 DSSP (Algorithm) (https://www.blopig.com/blog/2014/08/dssp/)

The DSSP algorithm (Fig. 2.13), widely recognized as the gold standard in the field of bioinformatics, serves as the primary tool for the precise determination of secondary structure elements within protein sequences. This system efficiently assigns secondary structure annotations to individual amino acids using atomic-resolution protein coordinates, providing researchers with a wealth of information about the structural organization and functional features of proteins (Sekihara et al. 2016). The acronym, mentioned singularly within the confines of the 1983 publication, delineates the nomenclature assigned to the Pascal software application responsible for executing the algorithm denoted as Define Secondary Structure of Proteins.

2.12.1.10 MolProbity (http://molprobity.manchester.ac.uk/)

MolProbity is an essential web-based tool that validates the quality of complicated 3D structures, such as those of proteins, nucleic acids, and complexes. The software provides an exhaustive examination of all atom interactions, which aids in the detection of steric hindrances in molecular structures (Fig. 2.14). In addition, it can calculate and display hydrogen bond and van der Waals interactions between molecules at their interfaces. Polar and non-polar hydrogen atoms must be included and refined thoroughly as part of the aforementioned procedure (Williams et al. 2018). The KiNG viewer presents the results in a number of formats, including numerical scores, lists, downloadable PDB and graphics files, and, most crucially, online interactive 3D kinemage images. The aforementioned service is provided at no cost to users and can be accessed at http://kinemage.biochem.duke.edu.

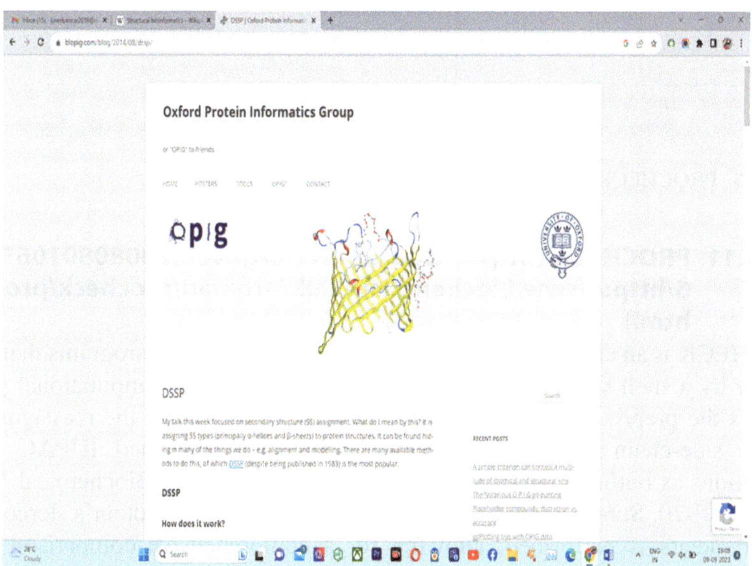

Fig. 2.13 DSSP (algorithm) webpage

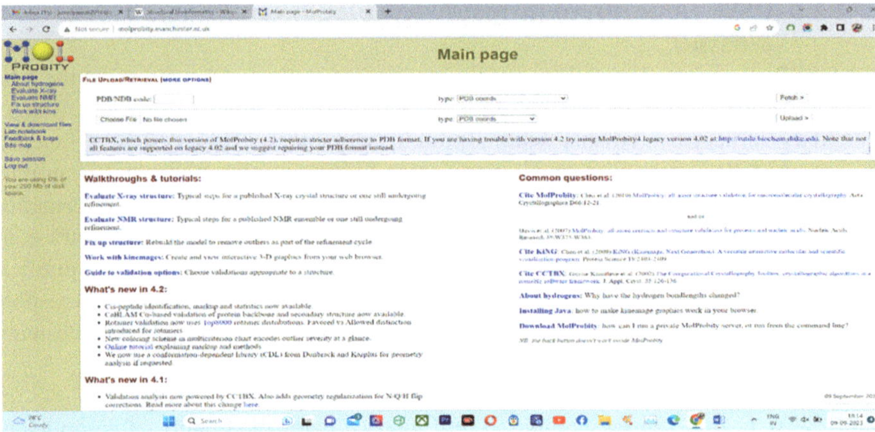

Fig. 2.14 MolProbity (algorithm) webpage

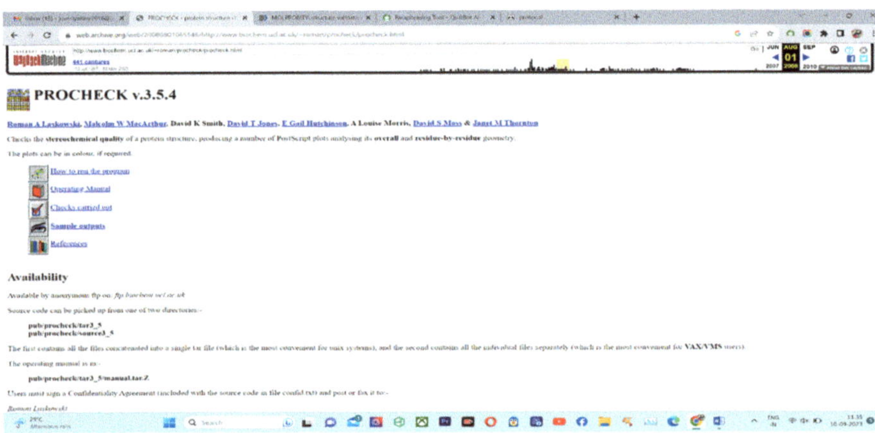

Fig. 2.15 PROCHECK webpage

2.12.1.11 PROCHECK (https://web.archive.org/web/20080801065546/http://www.biochem.ucl.ac.uk/~roman/procheck/procheck.html)

PROCHECK is an all-inclusive suite of separate Fortran and C programs that are run in order by a shell script (Fig. 2.15). The initial step of the computational pipeline involves the preprocessing of the input PDB file. This entails the reassignment of specific side-chain atoms in accordance with the established IUPAC naming conventions as outlined by the IUPAC–IUB Commission on Biochemical Nomenclature in 1970. Subsequently, an exhaustive analysis of the protein's stereo chemical parameters is performed, allowing for a comprehensive comparison against established norms. At the end of the process, the pipeline produces a visually beautiful PostScript output in addition to a meticulously detailed summary of the protein's structural features, residue by residue. The exclusion of hydrogen atoms and atoms with zero occupancy is a standard practice in bioinformatics analyses. When atoms can take on multiple shapes, only the one with the highest occupancy rate is taken into account (Yao and Cao 2023).

The comprehensive collection of program source codes can be accessed at the esteemed web address http://www.biochem.ucl.ac.uk/roman/procheck/procheck.html. The software in question, which has been integrated into the CCP4 suite of programs, was developed as part of the Collaborative Computational Project, Number 4 in 1994. More information about the CCP4 suite can be found at http://www.dl.ac.uk/CCP/CCP4/main.html. Additionally, users have the option to directly access and utilize the software through the Biotech Validation Server, which is available at http://biotech.embl-ebi.ac.uk:8400/.

2.12.1.12 CheShift (http://www.cheshift.com/)

The innovative bioinformatics tool CheShift-2 revolutionizes the computation of 13Cα and 13Cβ protein chemical shifts and offers vital insights into protein structure validation (Fig. 2.16). CheShift-2's cutting-edge algorithms and methods help researchers and scientists understand protein chemistry's complicated intricacies. The study analyzes 13Cα and 13Cβ chemical shift patterns in connection to the torsional angles (φ, ψ, ω, and χ1, χ2) of 20 amino acids using quantum mechanics simulations (Vila et al. 2009).

Bioinformatics tool CheShift-2 analyzes PDB protein structures to gain insights. CheShift-2 generates a complete collection of theoretical chemical shift values using advanced algorithms. Researchers use this knowledge to understand protein behavior and structural and functional qualities. CheShift-2 helps bioinformaticists analyze and interpret protein structures quickly and accurately with its correct predictions (Martin et al. 2012). A supplied PDB file and chemical shift values enable three-dimensional protein model display in the software. The 3D protein

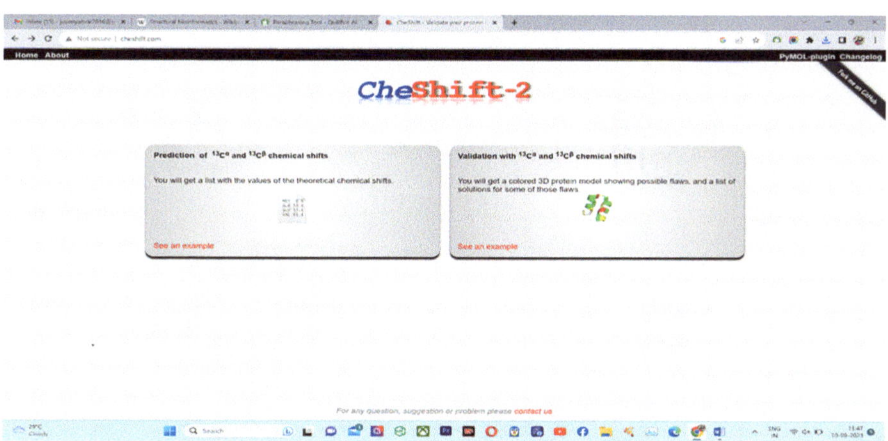

Fig. 2.16 CheShift webpage

model's five-color code shows the differences between anticipated and experimental chemical shift values. Using the discrepancies between experimentally measured and projected 13Cα and 13Cβ chemical shifts can reveal probable abnormalities in protein structures. A strong bioinformatics method, CheShift-2, uses 13Cα and 13Cβ chemical shifts to identify alternate $\chi 1$ and $\chi 2$ side-chain torsional angles. CheShift-2 generates a complete list of probable torsional angles by minimizing chemical shift discrepancies. These insights can improve protein structure quality and accuracy by fixing flaws.

The official website http://www.cheshift.com makes CheShift-2, a cutting-edge bioinformatics tool, accessible online. This powerful program can also be smoothly incorporated into PyMOL, a popular molecular visualization platform, via a plugin.

2.12.1.13 3Dmol.js (https://3dmol.csb.pitt.edu/)

3Dmol.js uses WebGL (Fig. 2.17), a cutting-edge technology, to create spectacular and interactive molecular images on web platforms. This JavaScript module makes real-time molecular structure exploration easy via hardware acceleration. 3Dmol.js helps bioinformatics researchers understand molecular biology with its powerful features (Rego and Koes 2015). Many different types of molecular data files and presentation formats are supported by the software. These include volumetric data like cube files and simulation data like AMBER or GROMACS data (Shkurti et al. 2016). A rich JavaScript API lets users alter molecular structures with 3Dmol.js, a sophisticated bioinformatics application. Additionally, its embedding API lets molecular views be seamlessly integrated into web pages using a concise div declaration. A hosted viewer API from 3Dmol.js uses URLs to easily retrieve and visualize molecular data. The introduced observer, http://3dmol.csb.pitt.edu/viewer.html, has been carefully designed to include all the data needed for molecular visualization in the URL. In bioinformatics, a captivating scenario can be created

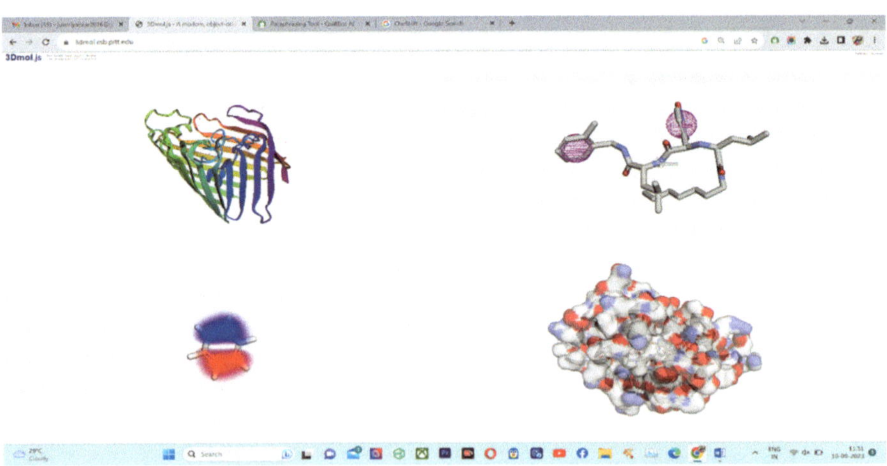

Fig. 2.17 3Dmol.js webpage

by carefully importing molecular data and applying styles using the edit panel. Next, pupils can be given the URL of this carefully designed scene, perhaps as a QR code. This allows students to actively participate in the same scene, making learning lively and participatory. The viewer can load molecular data using a PubChem compound id to ensure a 3D structure. An externally hosted file or PDB identification can likewise be used. By extending the functionality of the 3Dmol.js API and hosted viewer, a dynamic learning environment has been established. In this interactive whiteboard, students react to questions by selecting molecules in three dimensions.

2.12.1.14 PROPKA (https://web.archive.org/web/20070113065659/http://propka.ki.ku.dk/)

PROPKA predicts protein ionizable residue pK(a) values by using non-proteinaceous ligands and their ionizable group pK(a) values (Fig. 2.18) (Saoudi et al. 2011). PROPKA 2.0 extensively uses 1.0's empirical criteria for ligand functional groups. Due to its speed, PROPKA can calculate the pK (a) values of all ionizable groups in seconds for most proteins. Several protein-ligand complexes are explored, comparing PROPKA 2.0 predictions to experimental results. This complex contains trypsin, thrombin, three pepsins, HIV-1 protease, chymotrypsin, xylanase, hydroxynitrile lyase, and dihydrofolate reductase. Four of the 14 trypsin-thrombin ligand complexes have considerable protonation state changes ($|n| > 0.5$). PROPKA 2.0 and Klebe's PEOE method show a 0.4-unit protonation shift at pH 6.5 and 7.0 when plasmin II, cathepsin D, and endothiapepsin bind to pepstatin. PROPKA 2.0 data shows that ligand binding alters structure effect proton uptake/release and residues away from the binding site. The residues' surroundings and hydrogen bonding network have changed, generating these alterations. PROPKA 2.0 can quickly and correctly forecast the protonation states of critical residues and ligand functional groups at a protein's binding or active site

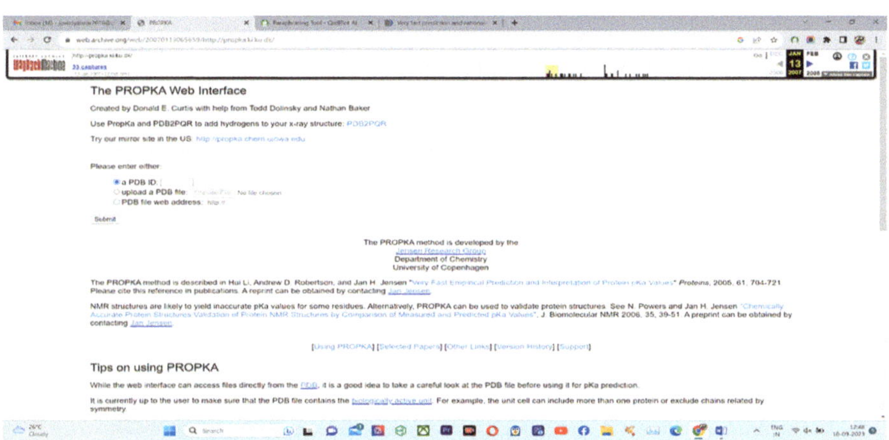

Fig. 2.18 PROPKA webpage

by describing protein-ligand interactions that modify titratable groups' pK(a) values (Saoudi et al. 2011).

2.12.1.15 CARA (http://cara.nmr.ch/doku.php)

CARA (Computer Aided Resonance Assignment) (Fig. 2.19), a cutting-edge bioinformatics tool for structural biology resonance assignment. CARA transforms nuclear magnetic resonance (NMR) data analysis and interpretation with its powerful algorithms and user-friendly interface. Visit our website to learn about CARA's remarkable features and join the expanding community of scientists who use it for structural elucidation. Advanced bioinformatics software CARA is powerful. Its main use is NMR spectrum analysis and computer-aided resonance assignment (Bosso et al. 2017). CARA is helpful in molecular and structural biology due to its concentration on bio macromolecules. Researchers and scientists use NMR spectroscopy to understand biomacromolecular structures, and its powerful capabilities and specific features make it indispensable. Structure determination is aided by dedicated software for identifying backbones, assigning side chains, and integrating peaks. These devoted tools simplify biomolecule structure elucidation steps.

2.12.1.16 Docking Server (https://www.dockingserver.com/web)

Docking Server handles ligand and protein setup and molecular docking with a web-based, easy-to-use interface (Fig. 2.20). Docking Server's user-friendly interface lets researchers from all biochemistry fields calculate and evaluate docking results, but advanced users can set ligand and protein parameters and docking calculations (Yu et al. 2016). The app can dock and analyze single ligands and dock ligand libraries to target proteins at fast speed. Parameters for optimizing ligand shape, minimizing energy, calculating charges, docking molecules, and representing protein-ligand complexes are all precisely calculated using Docking Server's computational chemistry program. Thus, Docking Server combines a number of popular in silico chemistry products into a single, all-encompassing web service, allowing for extremely fast and accurate docking computations.

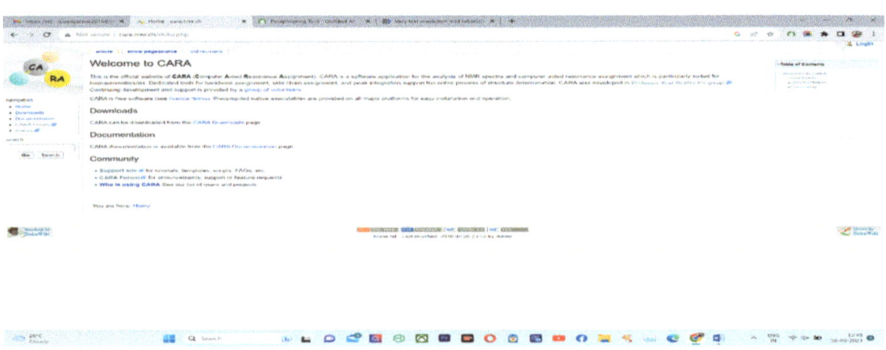

Fig. 2.19 CARA webpage

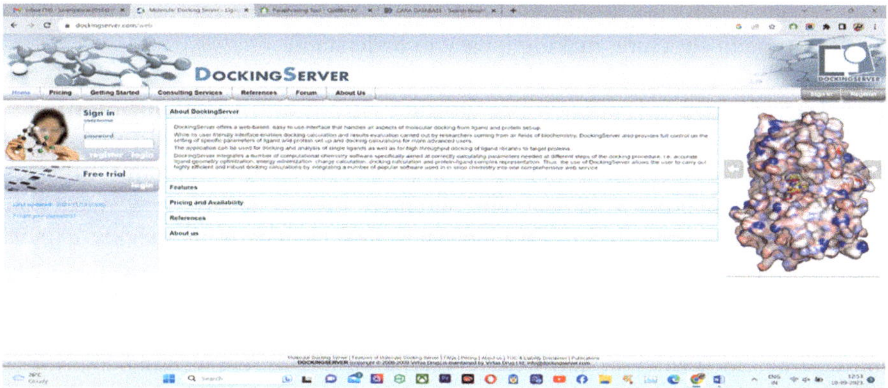

Fig. 2.20 Docking server webpage

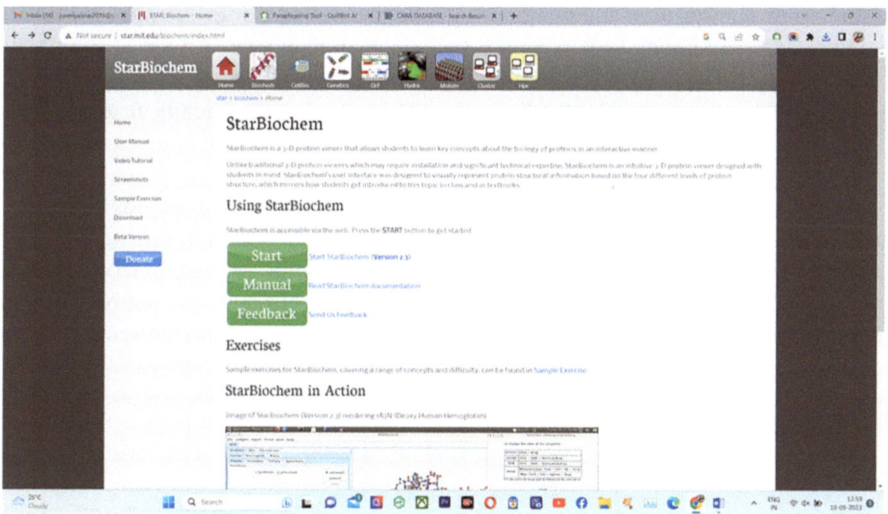

Fig. 2.21 StarBiochem webpage

2.12.1.17 StarBiochem (http://star.mit.edu/biochem/)

Star Biochem (Fig. 2.21), a forefront bioinformatics tool, helps researchers understand protein fundamentals. With three-dimensional visualization, this cutting-edge protein viewer lets students interact and immerse themselves in biological topics. Star Biochem helps students grasp protein structures and functions with its intuitive UI and powerful capabilities.

The creative and user-friendly Star Biochem 3-D protein viewer serves students. Star Biochem unlike ordinary viewers requires no complex setups or technical expertise. Its user-friendly interface makes protein structure visualization easy. The Star Biochem user interface was carefully designed to present protein structural data in accordance with classroom and textbook pedagogy. It smoothly unifies the four levels of protein structure, making it easy to understand.

2.12.1.18 SPADE (Structural Proteomics Application Development Environment) (https://sites.google.com/view/spade)

SPADE visualizes and studies molecular structures on multiple platforms. SPADE's various functionalities aid protein structure research by academics and bioinformaticians (Fig. 2.22). This advanced tool enables users quickly explore, alter, and analyze structural data to comprehend complex biological systems. SPADE shows protein three-dimensional architecture and dynamic behavior to help bioinformaticists address molecular challenges (Manjasetty et al. 2012). I enjoy Structure Prediction and Design Engine, a protein engineering innovation. Its innovative design may help structural biology algorithm developers.

Bioinformatics tool SPADE has a simple UI and apps. The powerful evolutionary calculating visualization tool Sequence Pad is one. Researchers can precisely visualize protein-protein interactions with SequencePad (Li et al. 2017). SPADE helps researchers understand complex molecular interactions. RAVE is a sophisticated bioinformatics tool for chemically probing projected structure models for experimental validation. Researchers can swiftly and methodically test computational predictions with RAVE's numerous functions. RAVE studies molecular structures at unprecedented depth utilizing cutting-edge algorithms and data processing. RAVE uses experimental data and computational simulations to understand complex biological systems and uncover new treatment targets faster. SPADE is a powerful computer platform for Molnir, a genetic algorithm-driven hybrid protein structure modeling tool. Programmers in bioinformatics use many computational tools. These technologies improve accuracy and efficiency through numerous means. Calculators assess biomolecular solvent and surface accessibility. Multi-feature dynamic programming easily aligns and compares large biological

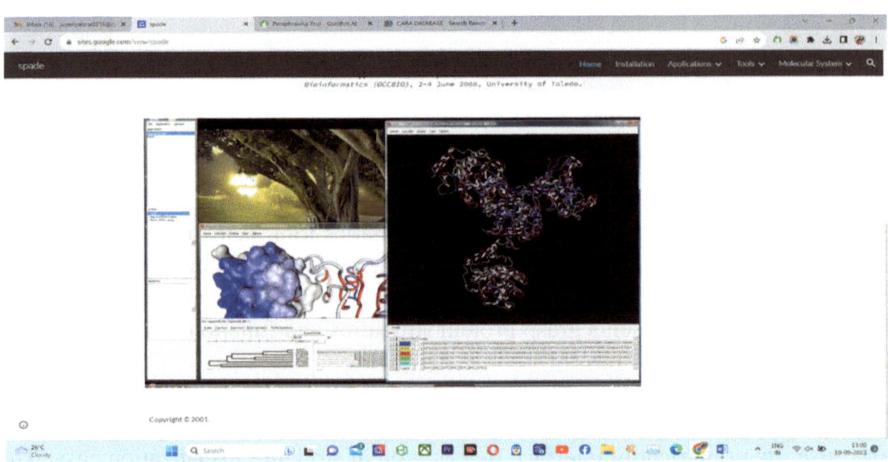

Fig. 2.22 SPADE webpage

sequences. Hydrogen bond calculators let programmers measure hydrogen atom–molecular interactions. Bioinformatics programmers can clearly comprehend biological systems with these essential components. Using algorithms, bioinformatics interprets biological data, including molecular structures. Unconstrained multiple structure alignment that supports distantly linked structures is a big development in this field. Different structures are aligned by this innovative method. It may freely align structurally diverse molecules, showing conserved areas and functional patterns in many biological units. The algorithm's remote structure handling. The massive biological data set awaits discovery.

2.13 Conclusion, Future prospective and challenges

Structural bioinformatics uses computer methods to study biological molecules' complex arrangements and functions, making it dynamic and progressive. The study of structural bioinformatics has led to several important findings and exciting prospects. Computational and experimental methods have revealed biomolecules' three-dimensional structures, improving our understanding of their functions and interactions. These advances have enabled new drug discovery, protein engineering, and molecular design methods. Machine learning algorithms and artificial intelligence have changed protein structure prediction, making it highly accurate. As structural bioinformatics advances, biological complexity can be unraveled.

Numerous bioinformatics triumphs have marked successes enabled over the past decade, structural bioinformatics has advanced greatly. These advances have helped predict protein structures, understand complex protein–protein interactions, and produce new drugs. The tremendous growth in processing capability, smart algorithms, and growing data have made these astonishing discoveries possible.

Several bioinformatics fields are set for significant advances. These include improving protein structure prognostication, using machine learning to analyze large datasets, and combining structural and functional data to better understand complex biological systems.

While structural bioinformatics has made significant progress, it still faces several obstacles in understanding biomolecular structures. In bioinformatics, protein structures are difficult to predict, especially for large, complex proteins. Researchers in bioinformatics must constantly focus on many issues. Integrating data from several sources is a difficult task. Harmonizing information allows a holistic perspective of biological phenomena. To improve molecular simulations, more accurate force fields are needed. Simulating molecular activity and interactions with these force fields illuminates their complex dynamics and functions. Big data presents another challenge: efficient algorithms. The exponential rise of datasets requires new computational methods that can quickly and effectively analyze massive amounts of data. The search for more efficient algorithms aims to reveal patterns and insights in these massive datasets, improving our understanding of biological systems. In conclusion, bioinformatics must integrate varied data sources, optimize molecular simulations with precise force fields, and build efficient algorithms to evaluate enormous information. These issues motivate bioinformatics research and innovation.

Structured bioinformatics offers several opportunities for biological researchers to make significant contributions. Combining computational and experimental methods helps scientists understand biological macromolecules' complex architecture and dynamic activity. This synergistic strategy has great potential for uncovering their mechanisms of action, enabling the discovery and design of novel pharmacological drugs and therapeutic interventions for a variety of diseases.

In conclusion, structural bioinformatics has advanced biological molecule investigation to new heights, with good prospects for future growth. Bioinformatics researchers must constantly develop new methods and algorithms to overcome several hurdles. Despite these challenges, structural bioinformatics offers many opportunities for researchers to make biological science advances.

References

Alder BJ, Wainwright TE (1959) Studies in molecular dynamics. I. General method. J Chem Phys 31(2):459–466. Bibcode:1959JChPh, 31.459A. ISSN 0021-9606. https://doi.org/10.1063/1.1730376

Berman HM, Westbrook J, Feng Z, Gilliland G, Bhat TN, Weissig H, Shindyalov IN, Bourne PE (2000) The Protein Data Bank. Nucleic Acids Res 28(1):235–242. https://doi.org/10.1093/nar/28.1.235. PMID: 10592235; PMCID: PMC102472

Bosso A, Pirone L, Gaglione R, Pane K, Del Gatto A, Zaccaro L, Di Gaetano S, Diana D, Fattorusso R, Pedone E, Cafaro V (2017) A new cryptic host defense peptide identified in human 11-hydroxysteroid dehydrogenase-1 β-like: from *in silico* identification to experimental evidence. Biochim Biophys Acta Gen Subj 1861(9):2342–2353

Chen VB, Davis IW, Richardson DC (2009) KING (Kinemage, next generation): a versatile interactive molecular and scientific visualization program. Protein Sci 18(11):2403–2409

Chothia C, Lesk AM (1986) The relation between the divergence of sequence and structure in proteins. EMBO J 5(4):823–826. https://doi.org/10.1002/j.1460-2075.1986.tb04288.x. PMC 1166865. PMID 3709526

Costa LS, Mariano DC, Rocha RE, Kraml J, Silveira CH, Liedl KR et al (2019) Molecular dynamics gives new insights into the glucose tolerance and inhibition mechanisms on β-glucosidases. Molecules 24(18):3215. https://doi.org/10.3390/molecules24183215. PMC 6766793. PMID 31487855

da Silveira CH, Pires DEV, Minardi RC, Ribeiro C, Veloso CJM, Lopes JCD, Meira W, Neshich G, Ramos CHI, Habesch R, Santoro MM (2009) Protein cutoff scanning: a comparative analysis of cutoff dependent and cutoff free methods for prospecting contacts in proteins. Proteins Struct Funct Bioinform 74(3):727–743. https://doi.org/10.1002/prot.22187

Dhasmana A, Raza S, Jahan R, Lohani M, Arif JM (2019) Chapter 19—high-throughput virtual screening (HTVS) of natural compounds and exploration of their biomolecular mechanisms: an *in silico* approach. In: Ahmad Khan MS, Ahmad I, Chattopadhyay D (eds.) New look to phytomedicine. Academic, pp 523–548. https://doi.org/10.1016/b978-0-12-814619-4.00020-3. isbn:978-0-12-814619-4. S2CID 69534557

Gauthier J, Vincent AT, Charette SJ, Derome N (2019) A brief history of bioinformatics. Brief Bioinformatics 20(6):1981–1996

Gong S, Worth CL, Cheng TM, Blundell TL (2011) Meet me halfway: when genomics meets structural bioinformatics. J Cardiovasc Transl Res 4:281–303

Gu J, Bourne PE (2011) Structural bioinformatics. Wiley. Gu J, Bourne PE (2009-03-16). Structural bioinformatics. Wiley. 978-0-470:18105-8

Hildebrandt A, Dehof AK, Rurainski A, Bertsch A, Schumann M, Toussaint NC, Moll A, Stöckel D, Nickels S, Mueller SC, Lenhof HP (2010) BALL-biochemical algorithms library 1.3. BMC Bioinformatics 11(1):1–5

Ilyin VA, Abyzov A, Leslin CM (2004) Structural alignment of proteins by a novel TOPOFIT method, as a superimposition of common volumes at a topomax point. Protein Sci 13(7): 1865–1874. https://doi.org/10.1110/ps.04672604. PMC 2279929. PMID 15215530

Kaczanowski S, Zielenkiewicz P (2010) Why similar protein sequences encode similar three-dimensional structures? Theor Chem Accounts 125(3–6):643–650. https://doi.org/10.1007/s00214-009-0656-3. issn:1432-881X. S2CID 95593331

Klebe G (2015) Protein-ligand interactions as the basis for drug action. In: Scapin G, Patel D, Arnold E (eds) Multifaceted roles of crystallography in modern drug discovery. NATO science for peace and security series a: chemistry and biology. Springer, Dordrecht, pp 83–92. https://doi.org/10.1007/978-3-642-17907-5_4. isbn:978-3-642-17906-8

Kocincová L, Jarešová M, Byška J, Parulek J, Hauser H, Kozlíková B (2017) Comparative visualization of protein secondary structures. BMC Bioinformatics 18:23. https://doi.org/10.1186/s12859-016-1449-z. PMID: 28251875; PMCID: PMC5333176

Li H, Chang YY, Lee JY, Bahar I, Yang LW (2017) DynOmics: dynamics of structural proteome and beyond. Nucleic Acids Res 45(W1):W374–W380

Mackoy T, Kale B, Papka ME, Wheeler RA (2021) View Sq, a visual molecular dynamics (VMD) module for calculating, analyzing, and visualizing X-ray and neutron structure factors from atomistic simulations. Comput Phys Commun 264:107881

Manjasetty BA, Büssow K, Panjikar S, Turnbull AP (2012) Current methods in structural proteomics and its applications in biological sciences. 3 Biotech 2:89–113

Mariano DC, Santos LH, Machado KD, Werhli AV, de Lima LH, de Melo-Minardi RC (2019) A computational method to propose mutations in enzymes based on structural signature variation (SSV). Int J Mol Sci 20(2):333. https://doi.org/10.3390/ijms20020333. PMC 6359350. PMID 30650542.

Martin OA, Vila JA, Scheraga HA (2012) Che Shift-2: graphic validation of protein structures. Bioinformatics 28(11):1538–1539

Martins PM, Mayrink VD, de Silveira S, da Silveira CH, de Lima LH, de Melo-Minardi RC (2018) How to compute protein residue contacts more accurately? Proceedings of the 33rd annual ACM symposium on applied computing. Pau: ACM Press, pp 60–67. isbn:978-1-4503-5191-1. S2CID 49562347 https://doi.org/10.1145/3167132.3167136

Matarazzo TJ, Pakzad SN (2014) Modal identification of golden gate bridge using pseudo mobile sensing data with STRIDE. In: Dynamics of civil structures, vol. 4: proceedings of the 32nd IMAC, a conference and exposition on structural dynamics. Springer International Publishing, pp 293–298

Meyers J, Fabian B, Brown N (2021) De novo molecular design and generative models. Drug Discov Today 26(11):2707–2715

Nickels S, Stöckel D, Mueller SC, Lenhof HP, Hildebrandt A, Dehof AK (2013) Presenta BALL—A powerful package for presentations and lessons in structural biology. In: 2013 IEEE symposium on biological data visualization (BioVis) 2013 Oct 13. IEEE, pp 33–40

Pagadala NS, Syed K, Tuszynski J (2017) Software for molecular docking: a review. Biophys Rev 9:91–102

Patel B, Singh V, Patel D (2019) Structural bioinformatics. In: Essentials of bioinformatics, vol I: Understanding bioinformatics: genes to proteins, pp 169–199

Pires DE, de Melo-Minardi RC, dos Santos MA, da Silveira CH, Santoro MM, Meira W (2011) Cutoff scanning matrix (CSM): structural classification and function prediction by protein

inter-residue distance patterns. BMC Genomics 12 Suppl 4(S4):S12. https://doi.org/10.1186/1471-2164-12-S4-S12. PMC 3287581. PMID 22369665
Rego N, Koes D (2015) 3Dmol. js: molecular visualization with WebGL. Bioinformatics 31(8): 1322–1324
Rigden DJ (2009) From protein structure to function with bioinformatics. In: Rigden DJ (ed) Springer, Berlin
Rosignoli S, Paiardini A (2022) Boosting the full potential of PyMOL with structural biology plugins. Biomolecules 12(12):1764
Saoudi N, Latcu DG, Rinaldi JP, Ricard P (2011) Graphical analysis of pH-dependent properties of proteins predicted using PROPKA. Bull Acad Natl Med 192:1029–1041
Sekihara K, Kawabata Y, Ushio S, Sumiya S, Kawabata S, Adachi Y, Nagarajan SS (2016) Dual signal subspace projection (DSSP): a novel algorithm for removing large interference in bio magnetic measurements. J Neural Eng 13(3):036007
Shi M, Gao J, Zhang MQ (2017) Web3DMol: interactive protein structure visualization based on WebGL. Nucleic Acids Res 45(W1):W523–W527. https://doi.org/10.1093/nar/gkx383. PMID: 28482028; PMCID: PMC5570197
Shkurti A, Goni R, Andrio P, Breitmoser E, Bethune I, Orozco M, Laughton CA (2016) pyPcazip: a PCA-based toolkit for compression and analysis of molecular simulation data. SoftwareX 1(5): 44–50
Stanfield RL, Wilson IA (1995) Protein-peptide interactions. Curr Opin Struct Biol 5(1):103–113. https://doi.org/10.1016/0959-440X(95)80015-S. PMID: 7773739
Travers A, Muskhelishvili G (2015) DNA structure and function. FEBS J 282(12):2279–2295
Vila JA, Arnautova YA, Martin OA, Scheraga HA (2009) Quantum-mechanics-derived 13Cα chemical shift server (Che shift) for protein structure validation. Proc Natl Acad Sci U S A 106(40):16972–16977
Vilar S, Cozza G, Moro S (2008) Medicinal chemistry and the molecular operating environment (MOE): application of QSAR and molecular docking to drug discovery. Curr Top Med Chem 8(18):1555–1572
Webb B, Sali A (2014) Comparative protein structure modeling using Modeller. Curr Protoc Bioinformatics 47(1):5.6.1–32. PMC: 4186674. PMID: 25199792 https://doi.org/10.1002/0471250953.bi0506s47
Wei D, Xu Q, Zhao T, Dai H (2014) Advance in structural bioinformatics. Springer
Wermuth CG, Villoutreix B, Grisoni S, Olivier A, Rocher JP (2015) Strategies in the search for new lead compounds or original working hypotheses. In: Wermuth CG, Aldous D, Raboisson P, Rognan D (eds) The practice of medicinal chemistry. Academic, pp 73–99. https://doi.org/10.1016/B978-0-12-417205-0.00004-3. isbn:978-0-12-417205-0
Williams CJ, Headd JJ, Moriarty NW, Prisant MG, Videau LL, Deis LN, Verma V, Keedy DA, Hintze BJ, Chen VB, Jain S (2018) MolProbity: more and better reference data for improved all-atom structure validation. Protein Sci 27(1):293–315
Yang J, Yan R, Roy A, Xu D, Poisson J, Zhang Y (2015) The I-TASSER suite: protein structure and function prediction. Nat Methods 12(1):7–8
Yao Z, Cao D (2023) 4.3.9 Prochek module. PyBioMed Documentation 10:53
Yousif RH (2020) Exploring the molecular interactions between neoculin and the human sweet taste receptors through computational approaches. Sains Malays 49(3):517–525. https://doi.org/10.17576/jsm-2020-4903-06
Yu J, Vavrusa M, Andreani J, Rey J, Tufféry P, Guerois R (2016) InterEvDock: a docking server to predict the structure of protein–protein interactions using evolutionary information. Nucleic Acids Res 44(W1):W542–W549
Yuan S, Chan HS, Hu Z (2017) Using PyMOL as a platform for computational drug design. Wiley Interdiscipl Rev Comput Mol Sci 7(2):e1298

Functional Genomics and Network Biology 3

Amit Joshi, Ajay Kumar, and Vikas Kaushik

Abstract

Functional genomics and network biology have emerged as interdisciplinary fields that provide a comprehensive understanding of biological systems. This discussion explored various aspects of these fields, including genomic approaches for studying gene function, transcriptomics and gene expression analysis, proteomics and protein function, epigenomics and epigenetic regulation, metabolomics and metabolic networks, integrative omics analysis, network biology, network inference and reconstruction, network analysis and visualization tools, systems biology and network modeling, and the application of functional genomics and network biology in disease research and drug discovery. The potential of functional genomics and network biology was highlighted in unraveling the complexities of biological systems, identifying disease-associated genes and pathways, and developing personalized treatment strategies. Integration of multi-omics data, single-cell technologies, machine learning and artificial intelligence, and consideration of dynamic and temporal aspects were discussed as emerging trends in the field. These trends offer promising opportunities to advance our understanding of biological systems and accelerate discoveries in biomedicine and personalized medicine. The conclusion emphasized the transformative impact of functional genomics and network biology, with their potential to drive discoveries in disease research, drug discovery, and systems-level

A. Joshi
Department of Biochemistry, Kalinga University, Naya Raipur, Chhattisgarh, India

A. Kumar (✉)
Department of Biotechnology, Rama University, Kanpur, Uttar Pradesh, India
e-mail: drajay.fet@ramauniversity.ac.in

V. Kaushik
Science Habitat, Markham, ON, Canada

V. Singh, A. Kumar (eds.), *Advances in Bioinformatics*,
https://doi.org/10.1007/978-981-99-8401-5_3

biology. The integration of high-throughput experimental techniques, computational modeling, and network-based analyses has enabled a holistic and systems-level understanding of biological processes. The future perspectives and emerging trends discussed suggest that functional genomics and network biology will continue to play a pivotal role in unraveling the complexities of life and driving advancements in biomedicine and personalized medicine. Functional genomics and network biology provide valuable tools and approaches to understand the molecular basis of diseases, identify potential therapeutic targets, and develop personalized treatment strategies. The ongoing advancements in these fields, coupled with the integration of multi-omics data, single-cell technologies, and computational tools, hold great promise for accelerating discoveries and improving human health in the future.

Keywords

Functional genomics · Network biology · Omics integration · Personalized medicine · Network analysis

3.1 Introduction

Functional genomics and network biology are interdisciplinary fields that have revolutionized our understanding of biological systems at a molecular level. By integrating high-throughput experimental techniques with computational approaches, these fields provide valuable insights into the complex interplay of genes, proteins, and other molecular components within a cell or organism (Geschwind and Konopka 2009). Functional genomics focuses on deciphering the function of genes and their products, such as proteins and RNA molecules, within the context of the entire genome (see Table 3.1). Traditional genetics and molecular biology approaches often study individual genes or proteins in isolation, but functional genomics takes a holistic approach, aiming to understand the coordinated interactions among various components of the genome (Davidsen et al. 2016). One of the key tools in functional genomics is transcriptomics, which involves studying the expression patterns of all genes within a given tissue or organism. This can be done using microarray technology or more recently, next-generation sequencing techniques such as RNA-Seq. By comparing gene expression profiles under different conditions or in different cell types, researchers can identify genes that are specifically activated or repressed, providing insights into their roles in different biological processes. Proteomics is another important component of functional genomics, focusing on the study of the entire complement of proteins present in a cell or organism. Mass spectrometry-based techniques enable the identification and quantification of proteins, allowing researchers to analyze protein abundance, modifications, and interactions (van Bergen et al. 2022). By integrating proteomic data with other genomic information, researchers can gain a more comprehensive understanding of cellular processes and signaling pathways. In epigenomics,

Table 3.1 Various aspects of functional genomics and network biology

Aspect	Functional genomics	Network biology
Definition	Study of gene function and activity on a global scale	Study of biological systems through networks of interactions
Focus	Individual genes and their functions	Interactions between genes, proteins, and other molecules
Techniques	High-throughput sequencing, microarrays, RNA interference	Network construction, data integration, computational models
Data analysis	Gene expression profiling, pathway enrichment analysis	Network visualization, topological analysis, clustering
Biological insight	Identification of differentially expressed genes	Identification of key network components and modules
Application	Understanding gene function, disease mechanisms	Systems-level analysis, drug target discovery
Challenges	Data interpretation, noise reduction, functional annotation	Network complexity, data integration, dynamic modeling
Examples of studies	Gene expression profiling, functional annotation studies	Protein-protein interaction networks, gene regulatory networks
Related disciplines	Molecular biology, genetics, bioinformatics	Systems biology, computational biology, data science

alterations to gene transcription patterns that are inherited but not brought on by changes in the underlying DNA sequence are investigated. It entails the investigation of dynamically regulating epigenetic alterations, such as DNA methylation and histone modifications. Epigenetic changes play a crucial role in development, aging, and diseases, and understanding their impact on gene function is a key area of research in functional genomics. Metabolomics complements genomics and proteomics by focusing on the comprehensive analysis of small molecules, or metabolites, within a biological system. By profiling metabolites, researchers can gain insights into the metabolic pathways and networks that underlie various physiological processes. Epigenomics, an interdisciplinary field at the intersection of genomics and epigenetics, has revolutionized our understanding of gene expression and heritable modifications beyond DNA sequence changes. It explores the complex mechanisms by which cells interpret and utilize genetic information, shedding light on the dynamic nature of our genome and its influence on human health and disease. At its core, epigenomics investigates the modifications that occur to our DNA and its associated proteins, known as histones, which can impact gene activity without altering the underlying genetic code. These modifications, such as DNA methylation and histone acetylation, act as switches, turning genes on or off in response to various signals from the environment or developmental cues. Through epigenetic modifications, cells can maintain their specialized functions and respond to changing conditions throughout an organism's lifespan. Epigenomic research has uncovered a myriad of exciting findings with far-reaching implications. It has revealed how epigenetic alterations play a crucial role in embryonic development, ensuring the appropriate activation or silencing of genes as cells differentiate into specialized tissues. Moreover, epigenetic modifications have been linked to various diseases, including cancer, neurological disorders, and autoimmune conditions, highlighting their significance in human health. One of the key tools in epigenomics is high-throughput sequencing technology, which enables the comprehensive profiling of epigenetic marks across the entire genome. Researchers can map DNA methylation patterns or analyze histone modifications on a global scale, generating vast amounts of data. These datasets provide valuable insights into epigenetic landscapes, identifying regulatory elements, enhancers, and regions prone to genetic instability. Epigenomic studies have also paved the way for personalized medicine approaches. By understanding the epigenetic alterations associated with specific diseases, researchers can develop targeted therapies that restore normal gene expression patterns and potentially reverse aberrant epigenetic states. Epigenomics holds immense promise in precision medicine, enabling tailored treatments that take into account an individual's unique epigenetic profile. However, despite the remarkable progress in epigenomic research, many questions remain unanswered. Scientists are still unraveling the intricate mechanisms underlying epigenetic modifications and their interplay with genetic and environmental factors. Additionally, ethical considerations and potential long-term consequences of manipulating epigenetic marks require careful evaluation to ensure the responsible application of this knowledge. Epigenomics is a cutting-edge field that has revolutionized our understanding of gene regulation and inheritance. By studying the dynamic modifications that occur on top of our DNA, researchers are uncovering new insights into development, disease, and personalized medicine. As the field continues to advance, epigenomics holds immense potential to transform healthcare and improve human well-being. Metabolomics data can be integrated with other omics data to unravel the complex interactions between genes, proteins, and metabolites within a biological system (Fraunhoffer et al. 2022). Network biology is a powerful framework that leverages the vast amount of molecular data generated by functional genomics approaches. It involves the construction and analysis of biological networks, which represent the interactions between genes, proteins, and other molecules. Network biology provides a systems-level view of biological processes, highlighting the interconnectedness and emergent properties of molecular components. By analyzing these networks, researchers can uncover key regulatory hubs, identify novel gene functions, and gain insights into disease mechanisms (Yan and Hu 2022). In this era of big data, functional genomics and network biology are revolutionizing our ability to decipher the complexity of living systems (see Table 3.1). By integrating various omics data and computational modeling, these fields hold great promise for advancing our understanding of biological processes, improving disease diagnosis and treatment, and facilitating the development of personalized medicine. The following sections will delve deeper into the specific approaches and applications within functional genomics and network biology, highlighting their significance in contemporary research and discovery.

3.2 Genomic Approaches for Studying Gene Function

Understanding the function of genes is a fundamental goal in molecular biology and genetics. Genomic approaches have revolutionized the study of gene function by providing comprehensive insights into the structure, organization, and regulation of genes within an organism's genome. These approaches employ high-throughput technologies and computational analyses to unravel the complexity of gene function and regulation on a genome-wide scale (Kustatscher et al. 2022; Khawaja et al. 2023). One of the key genomic approaches for studying gene function is gene knockout or gene silencing. In gene knockout experiments, specific genes are intentionally inactivated or disrupted, either in cell lines or in model organisms. This allows researchers to observe the phenotypic consequences of gene loss and infer the gene's function based on the resulting changes in cellular or organismal behavior. Gene knockout experiments have been particularly valuable in identifying essential genes, deciphering gene networks, and uncovering the underlying molecular mechanisms of diseases. Another powerful genomic approach is gene expression profiling, which provides a comprehensive snapshot of the genes that are active or turned on in a specific tissue or under specific conditions. This can be accomplished using techniques such as microarrays or RNA sequencing (RNA-Seq). By comparing gene expression patterns across different tissues, developmental stages, or disease states, researchers can gain insights into the specific functions of genes and the biological processes they regulate. Functional genomics also encompasses the study of non-coding regions of the genome, which make up a significant portion of the genome but do not code for proteins. These regions include regulatory elements such as enhancers and promoters, as well as non-coding RNA molecules. Techniques such as chromatin immunoprecipitation sequencing (ChIP-Seq) and assay for transposase-accessible chromatin using sequencing (ATAC-Seq) allow researchers to map and characterize these non-coding regions, providing insights into their role in gene regulation and cellular function (Zou et al. 2022). In recent years, CRISPR-Cas9 technology has emerged as a revolutionary tool for studying gene function. CRISPR-Cas9 allows for precise and efficient genome editing by enabling researchers to introduce specific mutations or modifications at desired locations within the genome (Wang et al. 2022; Sahel et al. 2023). This technology has significantly accelerated the study of gene function, enabling the rapid generation of gene knockout or knock-in models in a wide range of organisms. CRISPR-based screens such as CRISPR-Cas9 knockout screens or CRISPR activation/repression screens have also been developed to systematically interrogate gene function on a genome-wide scale. Furthermore, comparative genomics is an essential approach for studying gene function by examining the similarities and differences in gene content and organization across different species. By comparing genomes, researchers can identify conserved genes and regulatory elements that play critical roles in various biological processes. Comparative genomics also enables the identification of gene families, which consist of related genes that have diversified in function through evolution. Genomic approaches have revolutionized the study of gene function by providing comprehensive and systematic insights into the structure, organization, and regulation of genes within genomes (Hernández-Plaza et al. 2023). Techniques such as gene knockout, gene expression profiling, analysis of non-coding regions, and the use of CRISPR-Cas9 technology have greatly advanced our understanding of gene function and its role in development, physiology, and disease. The continued advancement of genomic approaches will undoubtedly lead to further discoveries and breakthroughs in our understanding of gene function and its implications in various biological processes.

3.3 Transcriptomics and Gene Expression Analysis

Transcriptomics is a branch of functional genomics that focuses on the study of the complete set of RNA molecules produced by a cell or organism, known as the transcriptome. It provides valuable insights into gene expression patterns, alternative splicing events, and the regulation of gene expression at a global level. Transcriptomics has revolutionized our understanding of gene function and the dynamic nature of gene expression in various biological processes (Ahuja et al. 2022; D'Agostino et al. 2022). Gene expression analysis is a key component of transcriptomics, involving the measurement and quantification of RNA molecules to determine which genes are active and to what extent. The advent of high-throughput technologies, such as microarrays and next-generation sequencing (RNA-Seq), has greatly facilitated gene expression analysis by enabling the simultaneous profiling of thousands to millions of transcripts in a single experiment (Erhard et al. 2022). Microarray technology allows for the measurement of gene expression levels by hybridizing labeled RNA samples to a microarray chip containing thousands of gene-specific probes (Cathryn et al. 2022). By comparing the fluorescence intensities of the labeled samples, researchers can determine the relative abundance of RNA transcripts and identify genes that are differentially expressed across different conditions or tissues. Microarrays have been widely used in transcriptomics research and they have contributed to significant discoveries in various fields including developmental biology, cancer research, and drug discovery. RNA-Seq, on the other hand, is a more recent and powerful technique for gene expression analysis. It involves sequencing the cDNA libraries generated from RNA samples, allowing for the direct quantification and profiling of RNA molecules. RNA-Seq provides several advantages over microarrays, including the ability to detect novel transcripts, quantify low-abundance transcripts with greater accuracy, and provide information on alternative splicing events and RNA editing (Negi et al. 2022). The advent of RNA-Seq has greatly expanded our understanding of the transcriptome and its complexity. In addition to measuring gene expression levels, transcriptomics also encompasses the study of non-coding RNA molecules, which do not code for proteins but have critical regulatory functions. Non-coding RNAs, such as microRNAs and long non-coding RNAs, play key roles in gene regulation, cellular processes, and disease development. Transcriptomics approaches, such as small RNA sequencing or total RNA sequencing, enable the identification and profiling of these non-coding RNA molecules, providing insights into their functions and mechanisms of action. The analysis of transcriptomics data requires advanced computational methods and bioinformatics tools (Ekiz Kanik et al. 2022; Xie et al. 2022). Data preprocessing, normalization, and differential expression analysis are essential steps in transcriptomics data analysis. Various statistical algorithms and machine learning approaches have been developed to identify differentially expressed genes and uncover gene regulatory networks. Gene set enrichment analysis (GSEA) is a popular method used to determine whether predefined sets of genes, such as those belonging to specific pathways or gene ontology categories, are overrepresented in the transcriptomics data. Transcriptomics and gene expression analysis have transformed our understanding of gene function, cellular processes, and disease mechanisms (Fang et al. 2023). By providing a comprehensive view of gene expression patterns, alternative splicing events, and non-coding RNA molecules, transcriptomics has become an invaluable tool in biological and medical research. The continued advancement of high-throughput technologies, computational methods, and integrative analyses will undoubtedly lead to further discoveries and insights into the complexity of gene expression and its regulation in diverse biological systems.

3.4 Proteomics and Protein Function

Proteomics is a field of study that focuses on the comprehensive analysis of proteins, their structures, functions, and interactions within a biological system. It complements genomics and transcriptomics by providing insights into the functional aspects of gene expression and the intricate roles that proteins play in various biological processes. Proteins are the workhorses of the cell, carrying out diverse functions essential for life (Chafran et al. 2022). Proteomics aims to identify and characterize all the proteins present in a cell, tissue, or organism, collectively known as the proteome. By studying the proteome, researchers can gain a deeper understanding of protein function, post-translational modifications, protein-protein interactions, and their involvement in disease mechanisms (Pandy et al. 2023). One of the key techniques in proteomics is mass spectrometry (MS), which allows for the identification and quantification of proteins. In MS-based proteomics, proteins are enzymatically digested into smaller peptides, which are then separated and ionized before being analyzed by a mass spectrometer. The mass spectra obtained are compared against protein sequence databases to identify the proteins present in the sample. Quantitative proteomics techniques, such as label-free quantification or stable isotope labeling, can further provide information about protein abundance changes under different conditions. Another important aspect of proteomics is the study of post-translational modifications (PTMs) that occur on proteins. PTMs play critical roles in regulating protein function, localization, and interactions. Examples of PTMs include phosphorylation, acetylation, glycosylation, and ubiquitination. Proteomics techniques, such as phosphoproteomics or glycoproteomics, aim to identify and quantify specific PTMs, providing insights into their functional relevance and potential implications in disease (Solari et al.

2023). Protein-protein interactions (PPIs) are essential for cellular processes, and proteomics plays a crucial role in mapping and characterizing these interactions. Various approaches, such as affinity purification coupled with mass spectrometry (AP-MS) or yeast two-hybrid (Y2H) assays, can be employed to identify and validate PPIs (Mani et al. 2022; Benz et al. 2022). The resulting protein interaction networks can shed light on complex biological pathways and help elucidate the molecular mechanisms underlying cellular functions. Functional proteomics involves studying protein function on a global scale. It encompasses methods such as protein expression profiling, where changes in protein abundance are correlated with specific biological conditions or disease states. Functional proteomics can also involve examining the subcellular localization of proteins, determining their enzymatic activities, or investigating protein complexes and their dynamics (Shoko et al. 2023; Qu et al. 2022). The integration of proteomics with other omics data, such as genomics, transcriptomics, and metabolomics, enables a more comprehensive understanding of biological systems. These integrative approaches, known as multi-omics or systems biology, can unravel the complexity and interconnectedness of molecules within cells and organisms. Proteomics has numerous applications in biomedical research and clinical diagnostics. It has been instrumental in biomarker discovery, where specific proteins or protein patterns are identified as indicators of disease or therapeutic response (Sempionatto et al. 2022). Proteomics also plays a significant role in drug discovery, aiding in target identification and evaluation of drug efficacy. Proteomics is a powerful tool for studying protein function and unraveling the complexities of cellular processes. By analyzing the proteome and protein-protein interactions, researchers can gain insights into protein function, post-translational modifications, and their roles in health and disease. The continued advancements in proteomic technologies and data analysis methods hold immense potential for further understanding the intricate workings of proteins and their impact on biological systems.

3.5 Epigenomics and Epigenetic Regulation

Epigenomics is a rapidly evolving field of research that focuses on studying the epigenetic modifications and mechanisms that influence gene expression and cellular function. Epigenetics refers to changes in gene expression patterns that occur without alterations in the underlying DNA sequence. These changes can be heritable and reversible, providing a mechanism for cells to adapt and respond to their environment. Epigenetic modifications play a crucial role in various biological processes, including development, aging, and disease. Epigenomics aims to map and characterize these modifications across the genome, providing insights into their functional significance and their impact on gene regulation (Kreitmaier et al. 2023). One of the key epigenetic modifications is DNA methylation, which involves the addition of a methyl group to the DNA molecule. DNA methylation typically occurs at cytosine residues in the context of CpG dinucleotides. It plays a critical role in gene regulation by influencing the accessibility of DNA to transcription factors and other regulatory proteins. Aberrant DNA methylation patterns have been implicated in numerous diseases, including cancer, neurodegenerative disorders, and cardiovascular diseases (Vachher et al. 2022). Histone modifications are another important aspect of epigenetic regulation. Histones are proteins around which DNA is wrapped, forming a structure called chromatin. Chemical modifications, such as methylation, acetylation, phosphorylation, and ubiquitination, can occur on histone tails and influence the compaction and accessibility of DNA. These modifications can either activate or repress gene expression, and their dysregulation has been associated with various diseases. Epigenomics techniques enable the profiling and characterization of epigenetic modifications on a genome-wide scale. These techniques include chromatin immunoprecipitation sequencing (ChIP-Seq), which allows for the identification and mapping of histone modifications and transcription factor binding sites, and bisulfite sequencing, which can determine the DNA methylation patterns at single-nucleotide resolution (Hino et al. 2022). These methods provide a comprehensive view of the epigenome and allow researchers to study the interplay between different epigenetic marks and their impact on gene expression. Epigenomics also encompasses the study of non-coding RNA molecules, such as microRNAs and long non-coding RNAs, which play important roles in gene regulation and epigenetic processes. These non-coding RNAs can interact with DNA, RNA, and proteins, influencing chromatin structure and gene expression patterns. The dysregulation of non-coding RNAs has been implicated in various diseases, and their study is an active area of research in epigenomics. Understanding epigenetic regulation has significant implications in both basic biological research and clinical applications. Epigenetic modifications and mechanisms have been linked to various diseases, and targeting epigenetic processes has emerged as a promising avenue for therapeutic interventions. Epigenomics research has led to the development of epigenetic drugs, such as DNA methyltransferase inhibitors and histone deacetylase inhibitors, which can modulate gene expression patterns and potentially reverse aberrant epigenetic marks associated with diseases (Szczepanek et al. 2023). Epigenomics is a field that investigates the epigenetic modifications and mechanisms that control gene expression and cellular function. It provides valuable insights into the regulation of genes and their impact on biological processes. The study of epigenomics has the potential to revolutionize our understanding of development, disease, and therapeutic interventions by uncovering the intricate interplay between the genome and its epigenetic modifications.

Mapping a network based on genomic information using bioinformatics typically involves analyzing gene expression data to identify regulatory relationships and interactions among genes (see Fig. 3.1). Here's a general workflow for mapping a gene regulatory network using bioinformatics:

Data collection: Obtain gene expression data, such as RNA-seq or microarray data, from the relevant experimental conditions or tissues. This data will provide information on the expression levels of genes across different samples.

Data preprocessing: Clean and preprocess the gene expression data to remove noise and normalize the expression values. This step ensures that the data is suitable for downstream analysis.

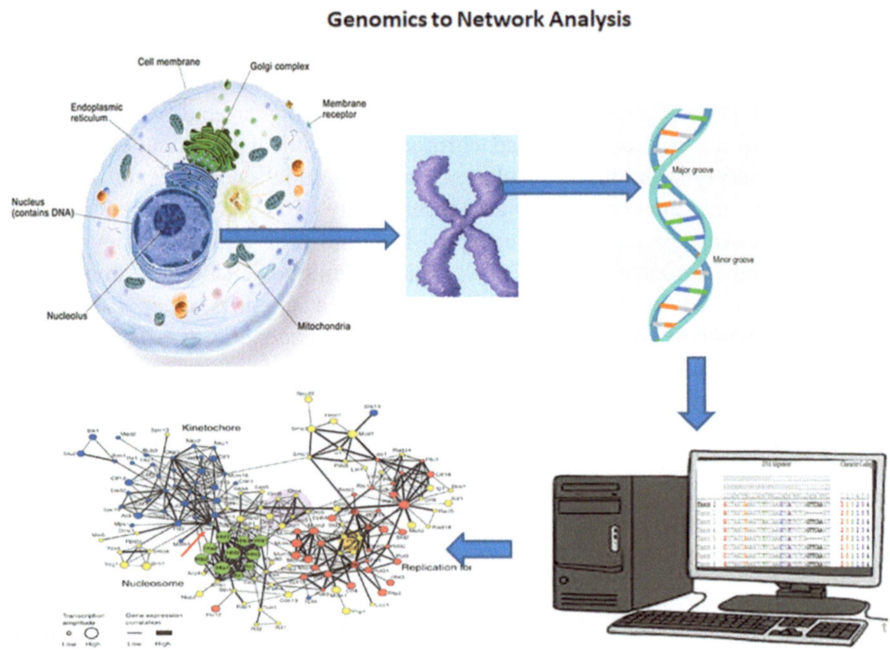

Fig. 3.1 Mapping network on the basis of genomic information by using bioinformatics

Differential gene expression analysis: Identify differentially expressed genes (DEGs) by comparing gene expression levels between different conditions or groups. DEG analysis helps identify genes that are significantly upregulated or downregulated under specific conditions.

Gene co-expression analysis: Perform gene co-expression analysis to identify groups of genes that show similar expression patterns across samples. This analysis helps identify potential regulatory relationships between genes that are co-expressed.

Gene regulatory network construction: Use computational algorithms and tools to construct a gene regulatory network based on the co-expression data and known interactions between genes. Various methods, such as correlation-based networks, mutual information-based networks, or machine learning approaches, can be employed for network construction.

Network visualization and analysis: Visualize the gene regulatory network using network visualization software or programming libraries. Analyze the network to identify key hub genes, gene modules, or regulatory motifs within the network.

Functional enrichment analysis: Perform functional enrichment analysis on the genes within the network to understand the biological processes, pathways, or functions associated with the network. This analysis helps interpret the functional implications of the gene regulatory network.

Validation and experimental verification: Validate the predicted gene regulatory network by comparing it with existing knowledge or by conducting experimental validations, such as knockdown or overexpression experiments, chromatin immunoprecipitation (ChIP), or gene perturbation studies.

Iterative refinement: Refine the gene regulatory network by incorporating additional data, such as transcription factor binding data, epigenetic data, or protein-protein interaction data. This iterative process helps improve the accuracy and completeness of the network.

Bioinformatics tools and software, such as R/Bioconductor packages (e.g., limma, WGCNA), Cytoscape, or network analysis platforms like GeneMANIA or STRING, can assist in various steps of this workflow.

3.6 Metabolomics and Metabolic Networks

Systems biology is rapidly expanding discipline of metabolomics focuses on the thorough examination of tiny molecules, or metabolites, in a biological system. It provides valuable insights into the metabolic pathways, biochemical reactions, and the overall metabolic state of cells, tissues, and organisms. Metabolomics plays a crucial role in understanding the intricate interplay between genes, proteins, and metabolites in various biological processes (Bassareo and McMahon 2022). Metabolites are the end products of cellular processes and represent the chemical fingerprints of cellular activities. By studying the metabolome, researchers can gain a deeper understanding of cellular metabolism, identify metabolic biomarkers, and elucidate the mechanisms underlying complex biological phenotypes and diseases. Metabolomics encompasses a range of analytical techniques, including mass spectrometry (MS) and nuclear magnetic resonance (NMR) spectroscopy, to profile and quantify metabolites in biological samples (Rey-Stolle et al. 2022). These techniques allow for the identification and quantification of hundreds to thousands of metabolites, providing a snapshot of the metabolic profile at a given time. Metabolic networks represent the interconnected pathways and reactions that occur within a cell or organism. Metabolomics plays a pivotal role in mapping and characterizing these networks by providing information about the substrates, intermediates, and end products of metabolic reactions. By integrating metabolomics data with other omics data, such as genomics and proteomics, researchers can construct comprehensive models of metabolic networks, unraveling the complexity of cellular metabolism. Metabolic profiling, one of the key applications of metabolomics, involves comparing the metabolic profiles of different samples to identify metabolites that are differentially expressed or show altered abundance under specific conditions. This can provide insights into metabolic dysregulation associated with diseases, drug responses, and environmental changes. Metabolic profiling has been particularly useful in cancer research, where it has helped identify metabolic signatures and potential therapeutic targets. Metabolic flux analysis is another important application of metabolomics, which aims to quantify the rates of metabolic reactions and fluxes within a metabolic network. By tracing the flow of isotopically labeled substrates through metabolic pathways, researchers can gain insights into the dynamics and regulation of metabolic processes (Harrieder et al. 2022). This information is crucial for understanding metabolic adaptations and for optimizing metabolic engineering strategies in biotechnology and bioengineering. Metabolomics also plays a vital role in understanding the impact of diet, nutrition, and gut microbiota on metabolism. It enables the identification of dietary biomarkers and the study of host-microbiota interactions. Metabolomics data can be integrated with other clinical and phenotypic data to unravel the intricate relationships between metabolism, health, and disease. The analysis of metabolomics data requires advanced computational methods and bioinformatics tools. Data preprocessing, metabolite identification, statistical analysis, and pathway analysis are essential steps in metabolomics data analysis. Multivariate statistical approaches, such as principal component analysis (PCA) and partial least squares-discriminant analysis (PLS-DA), are commonly used for pattern recognition and biomarker discovery (Shahzad et al. 2022). Metabolomics is a powerful tool for studying metabolism and unraveling the complexity of cellular processes. By analyzing the metabolome, researchers can gain insights into metabolic pathways, identify biomarkers, and understand the metabolic changes associated with various biological phenomena. The continued advancements in analytical techniques, computational tools, and integrative analyses will further enhance our understanding of metabolic networks and their implications in health, disease, and biotechnology applications.

3.7 Integrative Omics Analysis

Integrative omics analysis is a multidisciplinary approach that combines data from various omics disciplines, such as genomics, transcriptomics, proteomics, metabolomics, and epigenomics, to gain a holistic understanding of biological systems. By integrating multiple layers of molecular information, researchers can uncover complex interactions, identify key regulatory mechanisms, and unravel the underlying molecular processes that drive biological phenomena (Belay and Caleb 2022). High-throughput technologies have caused a boom in omics data, producing enormous volumes of molecular data. Integrative omics analysis offers a powerful framework to make sense of this wealth of data and extract meaningful insights that would be difficult to obtain by analyzing individual omics datasets in isolation. One of the primary goals of integrative omics analysis is to elucidate the relationship between different molecular layers. For example, integrating genomics and transcriptomics data can help identify genetic variants or mutations that drive changes in gene expression patterns. Similarly, combining transcriptomics and proteomics data can provide insights into the relationship between mRNA levels and protein abundance, shedding light on post-transcriptional and translational regulation (Chen et al. 2023). Integrative omics analysis can also facilitate the identification of regulatory networks and pathways. By integrating multiple omics datasets, researchers can infer gene regulatory networks, protein-protein interaction networks, and metabolic pathways. This integrative approach allows for a more comprehensive understanding of how genes, proteins, and metabolites work together to perform specific biological functions. Furthermore, integrative omics analysis enables the identification of biomarkers and potential therapeutic targets. By integrating clinical data with omics data, researchers can identify molecular signatures associated with specific diseases or treatment responses. This can lead to the development of personalized medicine approaches, where treatment strategies can be tailored to an individual's molecular profile. Integrative omics analysis relies on advanced computational methods and bioinformatics tools to integrate, analyze, and interpret complex datasets. Statistical approaches, such as correlation analysis, clustering algorithms, and machine learning techniques, are commonly employed to identify patterns, classify samples, and prioritize key features. Network-based approaches, including pathway analysis and module identification, are used to understand the functional relationships between molecules and uncover biological processes. The challenges in integrative omics analysis include data integration, normalization, batch effects, and the interpretation of complex results. Standardization of data formats, the development of robust computational algorithms, and the establishment of community standards are ongoing efforts to overcome these challenges and ensure the reproducibility and reliability of integrative omics studies. Integrative omics analysis has wide-ranging applications in biomedical research, including disease classification, drug discovery, biomarker identification, and systems biology modeling. It has the potential to revolutionize our understanding of complex biological systems, providing a more comprehensive view of molecular interactions and their implications in health and disease (Rohner et al. 2022). Integrative omics analysis is a powerful approach that combines data from multiple omics disciplines to gain a deeper understanding of biological systems. By integrating different molecular layers, researchers can uncover complex interactions, identify regulatory mechanisms, and derive meaningful insights that transcend individual omics datasets. The continued advancements in technology, data analysis methods, and collaborative efforts are driving the field of integrative omics analysis and opening new avenues for discoveries in biology and medicine.

3.8 Network Biology: Understanding Biological Networks

The study of and comprehension of biological systems as linked networks are the main goal of the multidisciplinary area of network biology. It leverages concepts and techniques from graph theory, systems biology, and computational biology to analyze the complex interactions between biological entities, such as genes, proteins, metabolites, and even whole organisms. By representing biological systems as networks, researchers can gain valuable insights into their structure, dynamics, and functional properties. Biological networks are made up of nodes, which stand in for individual biological units, and edges, which show their relationships to one another. These connections can be physical interactions, such as protein-protein interactions, or functional relationships, such as gene regulatory interactions. Biological networks can span different levels of organization, from molecular interactions within cells to ecological interactions between organisms in ecosystems. Network analysis provides a framework to study the structure and topology of biological networks

(Gosak et al. 2022). By analyzing network properties such as degree distribution, clustering coefficient, and network motifs, researchers can uncover patterns and organizational principles that govern the behavior of biological systems. For example, the scale-free nature of many biological networks, characterized by a few highly connected nodes (hubs) and many poorly connected nodes, is a common feature observed in various biological contexts. Biological networks also allow for the identification of central or essential nodes that play critical roles in maintaining network integrity and function. These nodes, known as network hubs, often correspond to key genes, proteins, or metabolites that regulate essential cellular processes or control information flow within the network. Disruption of these hubs can have significant consequences on the overall network stability and function, making them potential targets for therapeutic interventions. Network biology enables the study of network dynamics, including how networks change over time or respond to perturbations (Kumar et al. 2022). Dynamic network analysis can uncover important information about the resilience, robustness, and adaptability of biological systems. It can also reveal the underlying mechanisms driving disease progression, drug responses, and other biological phenomena. One of the key applications of network biology is in the identification of functional modules or communities within biological networks. Modules are subsets of nodes that exhibit higher levels of connectivity among themselves compared to the rest of the network. These modules often correspond to groups of genes or proteins that work together to perform specific biological functions or participate in common pathways. Identifying and characterizing these modules can provide insights into the functional organization of biological systems and help elucidate the molecular mechanisms underlying complex phenotypes and diseases. Network biology is instrumental in the integration of diverse omics data. By overlaying omics data onto biological networks, researchers can integrate and contextualize large-scale molecular datasets. This integrative approach allows for the identification of candidate genes, proteins, or metabolites associated with specific biological processes or diseases. It also facilitates the interpretation of omics data in a systems biology context, where the interactions and relationships between molecules are taken into account. Network biology provides a powerful framework for understanding biological systems as complex networks of interactions. By analyzing the structure, dynamics, and functional properties of biological networks, researchers can gain valuable insights into the organization and behavior of living systems. We can better understand biology, disease causes, and therapeutic approaches by combining network biology with other fields including genomics, systems biology, and computational biology.

3.9 Network Inference and Reconstruction

Network inference and reconstruction are fundamental processes in network biology that involve the identification and construction of biological networks from experimental data (Hasman et al. 2023). Biological networks, which represent the interactions and relationships between biological entities, such as genes, proteins, or metabolites, provide valuable insights into the structure, dynamics, and functionality of biological systems. Network inference refers to the computational or statistical methods used to infer the connections or interactions between biological entities based on experimental data. These methods leverage various data types, such as gene expression data, protein-protein interaction data, or genetic variation data, to uncover the underlying relationships within a biological system. Network inference can be performed using different approaches, including correlation-based methods, Bayesian approaches, and machine learning algorithms. Correlation-based methods, such as correlation coefficients or mutual information, are commonly used to identify pairwise relationships between variables in large-scale datasets. These methods measure the statistical dependencies between variables and can be applied to gene expression data, proteomics data, or other omics datasets (Sequeira et al. 2022). Correlation-based methods provide a simple and intuitive way to infer associations between biological entities but may not capture causal relationships. Bayesian approaches, on the other hand, aim to infer the underlying network structure by incorporating prior knowledge or assumptions about the relationships between variables. These methods use probabilistic models to estimate the likelihood of different network structures given the data. Bayesian network inference, for example, allows for the representation of causal relationships between variables and provides a framework for understanding the directed dependencies within a network. Machine learning algorithms, such as neural networks or support vector machines, have also been applied to network inference tasks. These algorithms can learn patterns and relationships from large-scale datasets and make predictions about the presence or absence of interactions between biological entities. Machine learning-based network inference approaches are particularly useful when dealing with high-dimensional data or complex network structures. Network reconstruction, on the other hand, involves the assembly or construction of a complete network from partial or incomplete data. This process aims to fill in missing connections or edges in the network, refine network topology, and improve the accuracy and completeness of the inferred network. Network reconstruction methods often combine network inference techniques with additional information, such as prior knowledge from databases or biological pathways, to guide the construction process. Validation and evaluation of inferred networks are critical steps in network inference and reconstruction. Various metrics and techniques are used to assess the quality and reliability of inferred networks, such as network topology measures, predictive performance, or comparison with known networks or experimental data. Validation helps ensure that the inferred networks accurately represent the underlying biological relationships and can provide meaningful insights into the system under study. Network inference and reconstruction have wide-ranging applications in biology and medicine. They are used to uncover regulatory networks, protein-protein interaction networks, metabolic networks, and other types of biological networks. Inference and reconstruction of disease-specific networks can help identify key genes, proteins, or pathways associated with diseases, aiding in the understanding of disease mechanisms and the development of targeted therapeutic interventions. Network inference and reconstruction are essential processes in network biology that enable the identification and construction of biological networks from experimental data. These methods leverage computational and statistical techniques to infer relationships and connections within a system. Network inference and reconstruction have broad applications in understanding biological systems, unraveling disease mechanisms, and facilitating personalized medicine approaches. The continued development of advanced algorithms, integration of diverse data types, and validation strategies will further enhance the accuracy and utility of network inference and reconstruction methods.

3.10 Network Analysis and Visualization Tools

Network analysis and visualization tools play a crucial role in the study of biological networks and their interpretation. These tools provide researchers with the ability to explore, analyze, and visualize complex network structures, uncovering hidden patterns and extracting meaningful insights from the data (Sagulkoo et al. 2022). With the ever-increasing availability of large-scale biological datasets, network analysis and visualization tools have become indispensable in understanding the organization and behavior of biological systems. Network analysis tools offer a wide range of functionalities to investigate the structure and properties of biological networks. These tools can compute various network metrics such as node degree, clustering coefficient, betweenness centrality, and network motifs, to quantify the characteristics of the network. By calculating these metrics, researchers can identify central nodes, network hubs, and densely connected regions within the network. Additionally, network analysis tools can help in community detection, revealing functional modules or groups of nodes with similar connectivity patterns. One popular network analysis tool is Cytoscape, an open-source platform that provides a comprehensive suite of features for network analysis and visualization (Chiliński et al. 2022). Cytoscape allows users to import network data from various file formats, perform network analysis algorithms, and visualize networks with customizable layouts and styles. It also offers a wide range of plugins and extensions that enhance its capabilities for specific biological applications, such as gene expression analysis, protein-protein interaction networks, and pathway analysis. Other network analysis tools include Gephi, a powerful software for interactive visualization and exploration of networks. Gephi provides an intuitive interface for network data import, manipulation, and analysis. It offers a wide range of layout algorithms, including force-directed layouts, which position nodes based on attractive and repulsive forces, allowing for the visualization of complex network structures. Gephi also supports dynamic network visualization, enabling the exploration of network evolution over time. Network visualization tools play a crucial role in representing complex network structures in a visually appealing and intuitive manner. These tools allow researchers to explore the network's connectivity, identify important nodes or clusters, and communicate their findings effectively. Visualization tools often provide various layout algorithms to position nodes and edges, allowing for better visualization of network topology. They also offer customization options, such as node and edge color-coding, size scaling, and label placement, to convey additional information and improve the interpretability of the network. Cytoscape and Gephi, mentioned earlier as network analysis tools, also offer powerful visualization capabilities. These platforms allow users to customize the visual appearance of networks, apply different layouts, and integrate additional data attributes, such as gene expression levels or functional annotations, for more informative visualizations. Other popular network visualization tools include GigaGalaxy, VisANT, and NetworkX, each with their own unique features and strengths (Chaudhary et al. 2022). In recent years, web-based network analysis and visualization tools have gained popularity due to their accessibility and ease of use. Tools like Cytoscape Web, NetworkAnalyst, and NAViGaTOR provide web-based interfaces that allow users to perform network analysis and visualize networks directly in a web browser, eliminating the need for installation and setup. These web-based tools often offer interactive features, such as zooming, panning, and filtering, which enhance the exploration and analysis of networks. Network analysis and visualization tools are essential for studying and interpreting biological networks. These tools enable researchers to analyze network properties, detect functional modules, and visualize complex network structures in an intuitive manner. With the rapid advancements in technology and the availability of large-scale biological datasets, network analysis and visualization tools continue to evolve, providing researchers with powerful resources to uncover the hidden intricacies of biological systems.

3.11 Systems Biology and Network Modeling

Systems biology is an interdisciplinary field that aims to understand biological systems by studying their components and their interactions in a holistic and integrative manner. It combines experimental techniques, computational modeling, and mathematical analysis to gain a comprehensive understanding of the complex behavior and dynamics of biological systems. At the core of systems biology lies network modeling which involves the construction and analysis of mathematical models that capture the interactions and relationships between components within a biological system (Bhatt et al. 2022). Network modeling provides a framework to represent and study the structure, dynamics, and function of biological networks. These networks can represent various biological systems, such as gene regulatory networks, protein-protein interaction networks, signaling networks, or metabolic networks. By quantitatively describing the interactions between components, network models enable researchers to simulate and predict the behavior of biological systems under different conditions. Network modeling typically involves the use of mathematical equations, such as differential equations, Boolean logic, or stochastic models, to capture the dynamics of biological processes. These models incorporate parameters that represent the rates of biochemical reactions, the strengths of interactions, or the probabilities of molecular events. By simulating the model equations, researchers can explore the behavior of the system over time and make predictions about its response to external stimuli or perturbations. Network modeling can be used to study a wide range of biological phenomena. For example, in gene regulatory network modeling, researchers aim to understand how genes interact and regulate each other's expression. By simulating the dynamics of gene regulatory networks, it becomes possible to identify key regulatory elements, predict the effects of genetic mutations, and gain insights into the mechanisms underlying cellular processes and diseases (Demirjian et al. 2023). Similarly, protein-protein interaction network modeling allows researchers to investigate the complex interactions between proteins and understand how these interactions drive cellular functions. By integrating experimental data with network models, researchers can identify critical protein nodes, predict protein functions, and uncover functional modules within the network. Metabolic network modeling focuses on understanding the flow of metabolites and the interconnectedness of metabolic pathways within a cell or organism. These models enable researchers to simulate metabolic fluxes, predict the effects of genetic or environmental perturbations on metabolic processes, and optimize metabolic engineering strategies for biotechnological applications. Systems biology and network modeling also play a significant role in drug discovery and personalized medicine. By constructing models that capture the interactions between drugs, target proteins, and disease pathways, researchers can simulate the effects of drug treatments, predict drug responses, and identify potential drug targets for specific diseases or patient populations. The success of network modeling in systems biology relies on the availability of high-quality experimental data, as well as computational methods for model construction, parameter estimation, and model validation. Model calibration and validation are critical steps to ensure that the model accurately represents the observed behavior of the biological system. Systems biology and network modeling provide powerful tools to understand the behavior and dynamics of biological systems. By constructing mathematical models that capture the interactions between components within a network, researchers can simulate and predict the behavior of biological systems, uncover underlying mechanisms, and make predictions that guide experimental investigations. As technology and computational methods continue to advance, systems biology and network modeling will continue to contribute to our understanding of complex biological phenomena and pave the way for new discoveries in biology and medicine.

3.12 Application of Functional Genomics and Network Biology in Disease Research

Functional genomics and network biology have revolutionized disease research by providing powerful tools and approaches to understand the molecular basis of diseases. These fields integrate high-throughput experimental techniques, computational analyses, and network-based approaches to unravel the complexities of diseases at the molecular level. By studying the functional relationships between genes, proteins, and other biological entities, researchers can gain valuable insights into disease mechanisms, identify potential therapeutic targets, and develop personalized treatment strategies. One of the key applications of functional genomics in disease research is the identification of disease-associated genes and genetic variants. Genome-wide association studies (GWAS) have been instrumental in identifying genetic variations associated with various diseases (Li et al. 2023). By analyzing the genomes of large cohorts of individuals, researchers can pinpoint genetic variations that are more prevalent in individuals with a specific disease compared to healthy controls. Functional genomics techniques, such as gene expression profiling, chromatin accessibility assays, or DNA methylation analyses, can provide further insights into how these genetic variants impact gene function and contribute to disease development. Network biology approaches complement functional genomics by providing a framework to understand the complex interactions and relationships between disease-associated genes and proteins. By constructing disease-specific networks, researchers can identify key nodes or hub proteins that play critical roles in disease pathways. These hub proteins often serve as potential therapeutic targets, as their manipulation can have significant effects on the overall network and disease progression. Network-based analysis also allows for the identification of functional modules or subnetworks that are dysregulated in specific diseases, providing a systems-level understanding of disease processes. Functional genomics and network biology are particularly valuable in studying complex diseases that involve multiple genetic and environmental factors. By integrating genomics data with other omics data, such as transcriptomics, proteomics, or metabolomics, researchers can generate multi-layered datasets that capture the molecular changes associated with disease. These integrative omics analyses enable the identification of disease-specific molecular signatures, the discovery of novel biomarkers for disease diagnosis and prognosis, and the identification of potential drug targets. Another important application of functional genomics and network biology in disease research is the identification of drug targets and the development of personalized medicine approaches. Network-based analyses can identify key nodes or network modules that are dysregulated in specific patient populations or disease subtypes. This information can guide the development of targeted therapies that specifically modulate these dysregulated components, leading to more effective and personalized treatments. Furthermore, functional genomics and network biology provide a valuable framework for understanding drug response and resistance mechanisms. By integrating drug perturbation data with molecular profiles of cells or tissues, researchers can identify molecular features that predict drug response or resistance. This information can aid in the development of biomarkers for patient stratification and the identification of combination therapies that can overcome drug resistance. The application of functional genomics and network biology in disease research has transformed our understanding of the molecular basis of diseases. By integrating high-throughput experimental techniques, computational analyses, and network-based approaches, researchers can unravel the complexities of diseases, identify disease-associated genes and pathways, and develop personalized treatment strategies. The continued advancements in these fields hold great promise for accelerating the discovery of novel therapeutic targets and improving patient outcomes in various diseases.

3.13 Drug Discovery and Target Identification Using Network Approaches

Network approaches have revolutionized the field of drug discovery and target identification by providing a powerful framework to understand the complex interactions and relationships within biological systems. These approaches leverage the knowledge of molecular networks, such as protein-protein interaction networks, gene regulatory networks, or signaling networks, to identify potential drug targets and accelerate the development of new therapeutic interventions. One of the primary applications of network approaches in drug discovery is target identification. Traditionally, drug discovery has relied on a target-centric approach, focusing on individual molecules or proteins believed to be directly involved in the disease process. However, this approach often overlooks the intricate network of interactions underlying disease pathways (Koivisto et al. 2022). Network approaches provide a more comprehensive view by considering the interactions and dependencies between multiple components within the network. By analyzing network topology and identifying critical nodes or hubs within the network, network-based target identification can uncover key proteins or genes that play pivotal roles in disease pathways. These proteins or genes, known as network centrality nodes, are often essential for maintaining the integrity and function of the network. Targeting these central nodes can have a significant impact on the overall network and disease progression. Network-based target identification also allows for the identification of functional modules or network clusters that are dysregulated in specific diseases, providing insights into potential therapeutic targets (Vincent et al. 2022). Network approaches also aid in the repurposing of existing drugs for new indications. By integrating drug-target interaction data with network information, researchers can identify potential off-target effects of existing drugs or uncover new targets that may be modulated by the drug. This approach has the advantage of leveraging existing knowledge about the safety and pharmacokinetics of the drug, potentially reducing the time and cost associated with traditional drug discovery. Furthermore, network-based approaches can provide insights into the mechanisms of drug action and resistance. By analyzing the effects of drugs on network dynamics and identifying changes in network structure or activity upon drug treatment, researchers can gain a better understanding of how drugs modulate disease pathways. This information can guide the development of combination therapies or drug repurposing strategies to overcome drug resistance. Network-based approaches also facilitate the exploration of drug combinations and polypharmacology, where multiple drugs are used in combination to target multiple components within a network simultaneously. By considering the interactions and dependencies within the network, researchers can identify drug combinations that have synergistic effects or can modulate multiple disease pathways simultaneously. This approach has the potential to improve treatment efficacy, overcome drug resistance, and reduce adverse effects. In addition to target identification and drug combination strategies, network approaches also aid in the optimization of drug development processes. By integrating network-based models with pharmacokinetic and pharmacodynamic data, researchers can simulate and predict the effects of drug candidates in a more comprehensive and holistic manner. This approach can help in prioritizing drug candidates, optimizing dosing regimens, and predicting potential side effects. Network approaches have revolutionized drug discovery and target identification by providing a systems-level understanding of disease pathways. By leveraging the knowledge of molecular networks, researchers can identify potential drug targets, repurpose existing drugs, explore drug combinations, and optimize drug development processes. The continued advancements in network-based approaches, combined with high-throughput experimental techniques and computational tools, hold great promise for accelerating the discovery and development of effective therapeutic interventions for a wide range of diseases.

3.14 Future Perspectives and Emerging Trends in Functional Genomics and Network Biology

Functional genomics and network biology have made significant contributions to our understanding of complex biological systems, and their potential for further advancements and applications is vast. As technology continues to evolve and our knowledge expands, several future perspectives and emerging trends are expected to shape the field of functional genomics and network biology (Yocca and Edger 2022; Mittler and Shulaev 2013). One of the emerging trends is the integration of multi-omics data. While functional genomics has traditionally focused on a single omics layer, such as gene expression or protein-protein interactions, the integration of multiple omics data, including genomics, transcriptomics, proteomics, metabolomics, and epigenomics, offers a more comprehensive view of biological systems. Integrating these diverse datasets enables a deeper understanding of the interactions and dependencies between different molecular components, allowing for a more accurate modeling of complex biological processes. Another promising trend is the development of single-cell functional genomics and network biology approaches. Traditional bulk assays have provided valuable insights into average cellular behavior, but they often mask cellular heterogeneity. Single-cell technologies, such as single-cell RNA sequencing and mass cytometry, allow for the profiling of individual cells, enabling the study of cellular diversity within tissues and the identification of rare cell populations. Integrating single-cell data with network analysis can uncover cell-specific interactions, cellular communication networks, and cell-state transitions, providing a more nuanced understanding of biological systems. The application of machine learning and artificial intelligence (AI) techniques is also poised to have a transformative impact on functional genomics and network biology. Machine learning algorithms can handle large-scale data, identify patterns, and make predictions, thus enabling the discovery of novel biological insights and the development of predictive models. AI can assist in the analysis of complex networks, enabling the identification of important nodes, the prediction of network dynamics, and the inference of missing interactions. Integrating AI with functional genomics and network biology approaches has the potential to uncover hidden relationships, accelerate data interpretation, and facilitate the discovery of novel biomarkers and therapeutic targets. As the field progresses, there is also a growing focus on understanding the dynamics and temporal aspects of biological systems. Dynamic modeling approaches, such as dynamic network analysis and time-series analysis, enable the investigation of how molecular interactions change over time and in response to external stimuli. Incorporating time-resolved data into functional genomics and network biology analyses allows for the identification of causal relationships, the detection of transient network states, and the prediction of system behavior under different conditions (Pathak and Kim 2022). Furthermore, the field is increasingly shifting toward a more systems-level and holistic perspective. Instead of studying individual components in isolation, researchers are embracing the study of entire systems and their emergent properties. This involves considering the interplay between different levels of biological organization, such as molecular networks, cellular processes, and organismal phenotypes. Systems biology approaches, which integrate functional genomics, network biology, and computational modeling, are becoming more prevalent in understanding complex biological phenomena, disease mechanisms, and drug responses. In terms of technology, the development of novel experimental techniques and tools will continue to drive progress in functional genomics and network biology. Advances in high-throughput sequencing, proteomics, imaging, and genome editing technologies will provide increasingly detailed and comprehensive data, enabling the construction of more accurate and dynamic network models. Additionally, the development of advanced computational algorithms, data integration methods, and visualization tools will facilitate the interpretation and exploration of complex networks. The future of functional genomics and network biology is promising, with emerging trends and perspectives shaping the field. The integration of multi-omics data, the utilization of single-cell technologies, the application of machine learning and AI techniques, the consideration of dynamic and temporal aspects, and the embrace of a systems-level perspective will all contribute to the advancement of our understanding of biological systems (Joshi et al. 2021). With the continued development of technology and the collaboration between experimentalists and computational biologists, functional genomics and network biology will play a pivotal role in unraveling the complexities of life and in driving discoveries in biomedicine, personalized medicine, and systems-level biology.

3.15 Conclusion and Future Perspective

In conclusion, functional genomics and network biology have emerged as powerful interdisciplinary fields that offer a comprehensive and systems-level understanding of biological systems. Through the integration of high-throughput experimental techniques, computational modeling, and network-based analyses, researchers have been able to unravel complex molecular interactions, identify key components within biological networks, and gain valuable insights into disease mechanisms, drug discovery, and personalized medicine. The discussed topics have highlighted the diverse applications of functional genomics and network biology. Transcriptomics and gene expression analysis have provided insights into gene function and regulation, while proteomics has allowed for the study of protein function and interactions. Epigenomics has shed light on the role of epigenetic modifications in gene regulation and disease development, and metabolomics has provided a comprehensive understanding of cellular metabolism. Integrative omics analysis has emerged as a powerful approach to integrate multiple omics data sets, enabling researchers to uncover complex relationships between different molecular layers and identify novel biomarkers and therapeutic targets. Network biology has played a vital role in understanding biological networks and their dynamic behavior, aiding in the identification of key nodes, functional modules, and drug targets. The future perspectives and emerging trends discussed in the field offer exciting possibilities for further advancements. The integration of multi-omics data, the utilization of single-cell technologies, the application of machine learning and AI techniques, the consideration of dynamic and temporal aspects, and the embrace of a systems-level perspective will all contribute to the continued growth of functional genomics and network biology. As technology continues to advance and computational tools become more sophisticated, functional genomics and network biology will continue to drive discoveries in various fields, including disease research, drug discovery, and personalized medicine. These approaches will provide a deeper understanding of complex biological processes, facilitate the development of targeted therapies, and pave the way for precision medicine approaches tailored to individual patients. In conclusion, functional genomics and network biology have transformed our understanding of biological systems, and their ongoing development and application hold immense promise for future breakthroughs in biomedical research and healthcare. By unraveling the complexities of molecular interactions and networks, functional genomics and network biology are paving the way for a deeper understanding of life and opening new avenues for improving human health.

References

Ahuja AK, Pontiggia L, Moehrlen U, Biedermann T (2022) The dynamic nature of human dermal fibroblasts is defined by marked variation in the gene expression of specific cytoskeletal markers. Life (Basel) 12(7):935

Bassareo PP, McMahon CJ (2022) Metabolomics: a new tool in our understanding of congenital heart disease. Children 9(12):1803

Belay ZA, Caleb OJ (2022) Role of integrated omics in unravelling fruit stress and defence responses during postharvest: a review. Food Chem (Oxf) 5:100118

Benz C, Ali M, Krystkowiak I, Simonetti L, Sayadi A, Mihalic F et al (2022) Proteome-scale mapping of binding sites in the unstructured regions of the human proteome. Mol Syst Biol 18(1):e10584

Bhatt P, Sethi K, Gangola S, Bhandari G, Verma A, Adnan M et al (2022) Modeling and simulation of atrazine biodegradation in bacteria and its effect in other living systems. J Biomol Struct Dyn 40(7):3285–3295

Cathryn RH, Kumar SU, Younes S, Zayed H, Doss CGP (2022) A review of bioinformatics tools and web servers in different microarray platforms used in cancer research. Adv Protein Chem Struct Biol 131:85–164

Chafran L, Carfagno A, Altalhi A, Bishop B (2022) Green hydrogel synthesis: emphasis on proteomics and polymer particle-protein interaction. Polymers 14(21):4755

Chaudhary A, Jain N, Kumar A (2022) Tools for social network analysis and mining. In: 2022 11th international conference on system modeling & advancement in research trends (SMART). IEEE, pp 1063–1067

Chen D, Liang J, Jiang C, Wu D, Huang B, Teng X, Tang Y (2023) Mitochondrion participated in effect mechanism of manganese poisoning on heat shock protein and ultrastructure of testes in chickens. Biol Trace Elem Res 201(3):1432–1441

Chiliński M, Sengupta K, Plewczynski D (2022) From DNA human sequence to the chromatin higher order organisation and its biological meaning: using biomolecular interaction networks to understand the influence of structural variation on spatial genome organisation and its functional effect. In: Seminars in cell & developmental biology, vol. 121. Academic, pp 171–185

D'Agostino N, Li W, Wang D (2022) High-throughput transcriptomics. Sci Rep 12(1):20313

Davidsen PK, Turan N, Egginton S, Falciani F (2016) Multilevel functional genomics data integration as a tool for understanding physiology: a network biology perspective. J Appl Physiol 120(3):297–309

Demirjian C, Vailleau F, Berthomé R, Roux F (2023) Genome-wide association studies in plant pathosystems: success or failure? Trends Plant Sci 28:471

Ekiz Kanik F, Celebi I, Sevenler D, Tanriverdi K, Lortlar Ünlü N, Freedman JE, Ünlü MS (2022) Attomolar sensitivity microRNA detection using real-time digital microarrays. Sci Rep 12(1): 16220

Erhard F, Saliba AE, Lusser A, Toussaint C, Hennig T, Prusty BK et al (2022) Time-resolved single-cell RNA-seq using metabolic RNA labelling. Nat Rev Methods Primers 2(1):77

Fang Z, Liu X, Peltz G (2023) GSEApy: a comprehensive package for performing gene set enrichment analysis in Python. Bioinformatics 39(1):btac757

Fraunhoffer NA, Abuelafia AM, Bigonnet M, Gayet O, Roques J, Nicolle R et al (2022) Multi-omics data integration and modeling unravels new mechanisms for pancreatic cancer and improves prognostic prediction. npj Precis Oncol 6(1):57

Geschwind DH, Konopka G (2009) Neuroscience in the era of functional genomics and systems biology. Nature 461(7266):908–915

Gosak M, Milojević M, Duh M, Skok K, Perc M (2022) Networks behind the morphology and structural design of living systems. Phys Life Rev 41:1–21

Harrieder EM, Kretschmer F, Böcker S, Witting M (2022) Current state-of-the-art of separation methods used in LC-MS based metabolomics and lipidomics. J Chromatogr B 1188:123069

Hasman M, Mayr M, Theofilatos K (2023) Uncovering protein networks in cardiovascular proteomics. Mol Cell Proteomics 22:100607

Hernández-Plaza A, Szklarczyk D, Botas J, Cantalapiedra CP, Giner-Lamia J, Mende DR et al (2023) eggNOG 6.0: enabling comparative genomics across 12 535 organisms. Nucleic Acids Res 51(D1):D389–D394

Hino S, Sato T, Nakao M (2022) Chromatin immunoprecipitation sequencing (ChIP-seq) for detecting histone modifications and modifiers. In: Epigenomics: methods and protocols. Springer US, New York, pp 55–64

Joshi A, Pathak DC, Mannan MAU, Kaushik V (2021) In-silico designing of epitope-based vaccine against the seven banded grouper nervous necrosis virus affecting fish species. Netw Model Anal Health Inform Bioinform 10(1):37

Khawaja M, Siddiqui R, Virani SS, Amos CI, Bandyopadhyay D, Virk HUH et al (2023) Integrative genetic approach facilitates precision strategies for acute myocardial infarction. Genes 14(7):1340

Koivisto AP, Belvisi MG, Gaudet R, Szallasi A (2022) Advances in TRP channel drug discovery: from target validation to clinical studies. Nat Rev Drug Discov 21(1):41–59

Kreitmaier P, Katsoula G, Zeggini E (2023) Insights from multi-omics integration in complex disease primary tissues. Trends Genet 39:46

Kumar S, Wang X, Strachan JP, Yang Y, Lu WD (2022) Dynamical memristors for higher-complexity neuromorphic computing. Nat Rev Mater 7(7):575–591

Kustatscher G, Collins T, Gingras AC, Guo T, Hermjakob H, Ideker T et al (2022) Understudied proteins: opportunities and challenges for functional proteomics. Nat Methods 19(7):774–779

Li JH, Brenner LN, Kaur V, Figueroa K, Schroeder P, Huerta-Chagoya A et al (2023) Genome-wide association analysis identifies ancestry-specific genetic variation associated with acute response to metformin and glipizide in SUGAR-MGH. Diabetologia 66(7):1260–1272

Mani DR, Krug K, Zhang B, Satpathy S, Clauser KR, Ding L et al (2022) Cancer proteogenomics: current impact and future prospects. Nat Rev Cancer 22(5):298–313

Mittler R, Shulaev V (2013) Functional genomics, challenges and perspectives for the future. Physiol Plant 148(3):317–321

Negi A, Shukla A, Jaiswar A, Shrinet J, Jasrotia RS (2022) Applications and challenges of microarray and RNA-sequencing. Bioinformatics:91–103

Pandy N, Franklin KA, Haynes KA, Rapé M, Cristea IM (2023) Adding post-translational modifications and protein–protein interactions to protein schematics. Trends Biochem Sci 48(5):407–409

Pathak RK, Kim JM (2022) Vetinformatics from functional genomics to drug discovery: insights into decoding complex molecular mechanisms of livestock systems in veterinary science. Front Vet Sci 9:1008728

Qu JH, Tarasov KV, Chakir K, Tarasova YS, Riordon DR, Lakatta EG (2022) Proteomic landscape and deduced functions of the cardiac 14-3-3 protein interactome. Cells 11(21):3496

Rey-Stolle F, Dudzik D, Gonzalez-Riano C, Fernández-García M, Alonso-Herranz V, Rojo D et al (2022) Low and high resolution gas chromatography-mass spectrometry for untargeted metabolomics: a tutorial. Anal Chim Acta 1210:339043

Rohner E, Yang R, Foo KS, Goedel A, Chien KR (2022) Unlocking the promise of mRNA therapeutics. Nat Biotechnol 40(11):1586–1600

Sagulkoo P, Chuntakaruk H, Rungrotmongkol T, Suratanee A, Plaimas K (2022) Multi-level biological network analysis and drug repurposing based on leukocyte transcriptomics in severe COVID-19: in silico systems biology to precision medicine. J Personal Med 12(7):1030

Sahel DK, Vora LK, Saraswat A, Sharma S, Monpara J, D'Souza AA et al (2023) CRISPR/Cas9 genome editing for tissue-specific in vivo targeting: nanomaterials and translational perspective. Adv Sci 10(19):e2207512

Sempionatto JR, Lasalde-Ramírez JA, Mahato K, Wang J, Gao W (2022) Wearable chemical sensors for biomarker discovery in the omics era. Nat Rev Chem 6(12):899–915

Sequeira JC, Rocha M, Alves MM, Salvador AF (2022) UPIMAPI, reCOGnizer and KEGGCharter: bioinformatics tools for functional annotation and visualization of (meta)-omics datasets. Comput Struct Biotechnol J 20:1798–1810

Shahzad K, Nawaz H, Majeed MI, Nazish R, Rashid N, Tariq A et al (2022) Classification of tuberculosis by surface-enhanced Raman spectroscopy (SERS) with principal component analysis (PCA) and partial least squares–discriminant analysis (PLS-DA). Anal Lett 55(11): 1731–1744

Shoko R, Magogo B, Pullen J, Mudziwapasi R, Ndlovu J (2023) Construction and analysis of protein–protein interaction networks based on nuclear proteomics data of the desiccation-tolerant Xerophyta schlechteri leaves subjected to dehydration stress. Commun Integr Biol 16(1):2193000

Solari FA, Krahn D, Swieringa F, Verhelst S, Rassaf T, Tasdogan A et al (2023) Multi-omics approaches to study platelet mechanisms. Curr Opin Chem Biol 73:102253

Szczepanek J, Skorupa M, Jarkiewicz-Tretyn J, Cybulski C, Tretyn A (2023) Harnessing epigenetics for breast cancer therapy: the role of DNA methylation, histone modifications, and MicroRNA. Int J Mol Sci 24(8):7235

Vachher M, Bansal S, Kumar B, Yadav S, Burman A (2022) Deciphering the role of aberrant DNA methylation in NAFLD and NASH, vol 8. Heliyon, p e11119

van Bergen W, Heck AJ, Baggelaar MP (2022) Recent advancements in mass spectrometry–based tools to investigate newly synthesized proteins. Curr Opin Chem Biol 66:102074

Vincent F, Nueda A, Lee J, Schenone M, Prunotto M, Mercola M (2022) Phenotypic drug discovery: recent successes, lessons learned and new directions. Nat Rev Drug Discov 21(12): 899–914

Wang SW, Gao C, Zheng YM, Yi L, Lu JC, Huang XY et al (2022) Current applications and future perspective of CRISPR/Cas9 gene editing in cancer. Mol Cancer 21(1):1–27

Xie W, Fang Y, Yu K, Min X, Li W (2022) MFRAG: multi fitness RankAggreg genetic algorithm for biomarker selection from microarray data. Chemom Intell Lab Syst 226:104573

Yan W, Hu G (2022) Structural biology meets biomolecular networks: the post-AlphaFold era. Curr Bioinform 17(6):493–497

Yocca AE, Edger PP (2022) Current status and future perspectives on the evolution of cis-regulatory elements in plants. Curr Opin Plant Biol 65:102139

Zou Z, Ohta T, Miura F, Oki S (2022) ChIP-atlas 2021 update: a data-mining suite for exploring epigenomic landscapes by fully integrating ChIP-seq, ATAC-seq and bisulfite-seq data. Nucleic Acids Res 50(W1):W175–W182

Bioinformatics in Gene and Genome Analysis

4

Nhat Le Bui, Van-Quy Do, and Dinh-Toi Chu

Abstract

Gene and genome analysis play important roles in molecular biology research and individualized medicine. Thanks to the development of sequencing techniques, sequencing data is getting more and more abundant, which requires bioinformatic tools to handle. As a combination of computational methods, statistics, and molecular biology, bioinformatics is a bridge between sequencing data and clinical interpretation. Via a half of decade development, bioinformatics has obtained novel achievements in data storage, assembly's speed and accuracy, variant identification, and friendly-to-user interfaces. In this chapter, we focus on the history and development of bioinformatics as well as introduced the principles and several popular computational tools for each step in the workflow of gene and genome analysis, including data generation, genome assembly, annotation, comparative analysis, variant calling, and finally interpretation. Since the genomes of prokaryotes are distinguished from eukaryotes, we also mentioned the differences in the data process between humans as well as animals and microorganisms.

Keywords

Bioinformatics · Computational tools · Gene and genome analysis · Comparative analysis

N. Le Bui · D.-T. Chu (✉)
Faculty of Applied Sciences, International School, Vietnam National University, Hanoi, Vietnam

Center for Biomedicine and Community Health, International School, Vietnam National University, Hanoi, Vietnam
e-mail: toicd@vnu.edu.vn

V.-Q. Do
Center for Biomedicine and Community Health, International School, Vietnam National University, Hanoi, Vietnam

V. Singh, A. Kumar (eds.), *Advances in Bioinformatics*,
https://doi.org/10.1007/978-981-99-8401-5_4

Abbreviations

BAM	Binary SAM
DNA	Deoxyribose nucleic acid
KEGG	Kyoto Encyclopedia of Genes and Genomes
NGS	Next generation sequencing
RNA	Ribosenucleic acid
SAM	Sequence Alignment/Map
SNPs	Single nucleotide polymorphisms
VCF	Variant call format

4.1 Introduction

The genetics field is currently of paramount importance to both research and clinical practice (Lazaridis et al. 2014; Bowdin et al. 2016). Thanks to the development of sequencing techniques, gene, and genome sequencing data have a great contribution to understanding the pathogenesis of both genetic and lifestyle-induced diseases at molecular levels (Bluestone et al. 2010; Abou Ziki and Mani 2016). Thus, it enhances the accuracy of both early diagnosis and prognosis. Since the therapeutic response depends on individuals, genetic testing provides more reliable evidence for doctors to adjust patients' treatment, which catches up with the trend toward personalized medicine (Crews et al. 2012). Besides, gene and genome analysis also provides information for the detection of biological markers for precise diagnosis and biological targets for drug design (Lazaridis et al. 2014).

Bioinformatics has been an important bridge between gene and genome data and the interpretation for clinical applications (Pereira et al. 2020). Bioinformatics, which main purpose is to develop special algorithms and tools to handle problems in interpreting a large amount of biological data into its clinical meaning, is considered an interdisciplinary field of the life sciences. The study subjects of bioinformatics are the sequence data of genetic materials including DNA, RNA, proteins, and other biological molecules such as metabolites (Akalın 2006). Thanks to the development of sequencing techniques, genome sequencing is now more accessible to worldwide laboratories and clinics with lower costs and shorter times. Consequently, the huge and arising amount of genome sequencing data that cannot be manually handled requires effective bioinformatic tools to manage and analyze (Hu et al. 2021). Since the first time the "Bioinformatics" term appeared in 1970 (Hesper and Hogeweg 1970), remarkable achievements in bioinformatics have been performed thanks to the increasing capacity of computation and the advanced software in dealing with these big data (Lelieveld et al. 2016). The amount of publications related to bioinformatics has dramatically increased in recent years (Fig. 4.1).

One of the first successful computer methods recorded is the protein sequence atlas created by Dayhoff et al., in which the proteins were classified into distinct

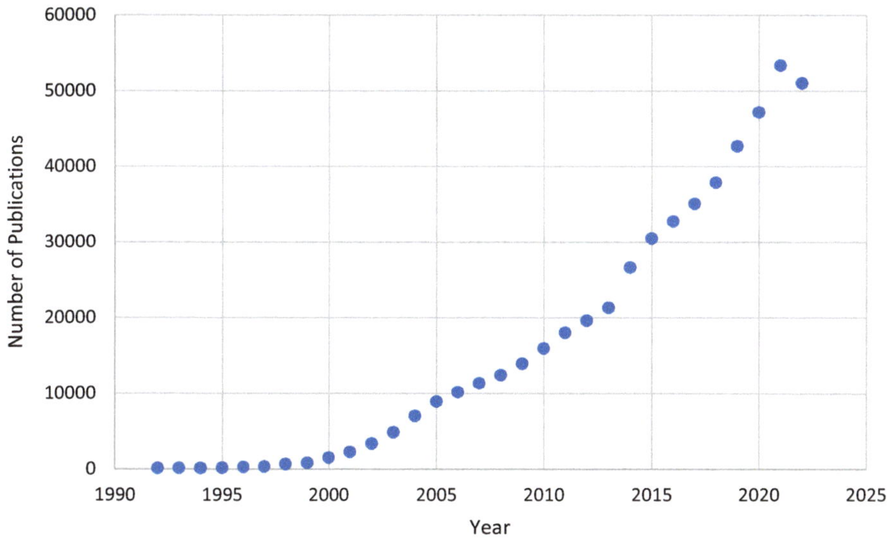

Fig. 4.1 The amount of publications related to bioinformatics in PubMed in the recent 20 years

groups and even subgroups based on the percent accepted mutation and sequence similarity (Dayhoff and Eck 1972). From then, new tools and algorithms of protein description including putative sequence, function, and structure based on the sequence of corresponding DNA were also developed. Moreover, to precisely predict the function of proteins, bioinformatics tools were designed to mimic and visualize the three-dimensional and folding structures of proteins. These applications are widely used in enzyme, ligand, or targeted drug design, as well as biological macromolecule structure prediction (Can 2014; Kumar et al. 2022). Regarding gene expression profiles, bioinformatics has resolved the problem of separating the main signal from background and noise and converting the signal into expression levels in high throughput studies like RNA-Seq, transcriptome profiling, and microarray. Data from the above applications including sequencing, annotation, expression, and protein prediction and simulation has contributed to organizing genetic pathways and networks of biological molecules and processes. Not only assessing the role of a single gene, integrated information from bioinformatics helps to assess the activity of millions of genes under the effects of regulation or co-expression of other factors (Daoud and Mayo 2019; Marco-Puche et al. 2019). From that, the complex relationship between genes has been constructed and visualized in the form of biological networks and pathways. Several examples of pathways databases are the Reactome Knowledgebase, the MetaCyc database, the Kyoto Encyclopedia of Genes and Genomes (KEGG), etc. (Caspi et al. 2020; Fabregat et al. 2016; Kanehisa 2002).

In the genomics field, bioinformatics is needed to design algorithms and tools to solve issues that cannot be handled manually. For example, raw reads resulting from sequencing techniques need to be assembled into longer chains called contigs, scaffolds, or even the completed genome. These sequences are then annotated into introns, exons, or the start, stop, or enhancer regions, which vary among different organisms (Cantacessi et al. 2010). In this chapter, we primarily focused on the application of bioinformatics in gene and genome analysis as well as updated novel developments and tools in this field.

4.2 Advances in Bioinformatics for Gene and Genome Analysis

Although recently becoming more popular, bioinformatics is still an immature scientific discipline that has only appeared in the past 70 years. Starting with the first sequence of insulin in the late 1950s, questions about the arrangements of genetic materials including DNA, RNA, and proteins were rising among scientists (Gauthier et al. 2019). It was not until Dayhoff's pioneering in applying computational methods to biochemistry, bioinformatics really got attention and flourished (Dayhoff and Eck 1972).

During the 1970–1980 period, the first computational tools to analyze Sanger sequences were published instead of manually extracting before. They could detect the overlap between sequence reads, rejoin the raw reads into contigs, and set the basement for annotation (Staden 1979). Phylogenetic algorithms like maximum parsimony and maximum likelihood were also first introduced during that period (Felsenstein 1981; Haeckel 1866).

From 1980 to 1990, thanks to milestones in molecular biology including DNA fractionation and DNA amplification by the polymerase chain reaction (PCR) (Schochetman et al. 1988), available sequencing reads were more abundant for bioinformatics processing. Consequently, software packages integrating command-line tools were developed rapidly with the first one being the GCG package (Womble 1999). The Free Software Foundation developed by Richard Stallman opened an era of free running, studying, distributing, and improving software among bioinformatics scientists (Stallman 2003). Since then, genetic databases like the GenBank, the European Molecular Biology Laboratory (EMBL), and the DNA Data Bank of Japan (DDBJ) were established, which is a huge contribution to the development of genome analysis till now (Benson et al. 2013; Kanz et al. 2005; Tateno et al. 2002).

Due to the development of shotgun sequencing, the demand for whole genome sequencing was rising. The Human Genome Project was launched in 1991 (Olson 1993), while the whole genome of living organisms was first sequenced and completed in 1995 (Fleischmann et al. 1995). The period 1990 to 2000 witnessed a revolution of the Web and Internet, which led to the worldwide accessibility of bioinformatics resources. The above genetic databases and the precise website of NCBI became available online at that time.

The period from 2000 to 2010 was the decade of high-throughput sequencing data with the evolution of next-generation sequencing (NGS), which allows scientists to sequence millions of DNA molecules per run (Sam and Patrick 2013). Dramatically increased tools designed to process bioinformatic data made it hard to select an appropriate one. Thus, a quality control problem was emerged with many indexes like largest contig length, L50 and N50 contig, and the minimum data required for a genomic sequence was defined (Endrullat et al. 2016). At that time, bioinformatic scientists also had to deal with storage volume and Big data issues. Typical FASTQ files from sequencing instruments seemed not to be suitable for long storage. Sequence Alignment/Map (SAM) file format, binary SAM (BAM) file format, and variants call format (VCF) file appeared with the sizes reduced 3 to 4 times compared to the original FASTQ file (Li et al. 2009; Peter et al. 2015; Danecek et al. 2011). However, greater infrastructure with special experts was still required, which then led to cooperation between worldwide laboratories.

Since 2010, the appearance of the term "Bioinformatician" indicated the role of Bioinformatics as a scientific field. Challenges in genomic sequence processing including quality control, data storage, and variant identification have been mostly resolved (Lelieveld et al. 2016). Bioinformatic tools with friendly-to-user interfaces are available in both offline software like TreeViewJ, and MEGA and integrative web-based like Galaxy (Peterson and Colosimo 2007; Sohpal et al. 2010; Jalili et al. 2020). Recently published tools also provide better specificity and sensitivity compared to before. However, the linkage between variants and changes in phenotype is still unclear. Although standardized phenotype ontologies such as the Human Phenotype Ontology (HPO) have been constructed to resolve these issues (Köhler et al. 2017), further research is needed to clarify the clinical interpretation of genetic testing.

4.3 Computational Tools in Gene and Genome Analysis

The use of genomic and genomic analysis is crucial in biotechnology. It includes a range of methods and equipment that use biological systems to modify and examine genetic data (Diniz and Canduri 2017). Here are some crucial features of gene and genome study and analysis in biotechnology: Application of genetic and biomedical engineering; Analysis of genomic data; Gene cloning and modification (Zhou et al. 1994; Apolinario et al. 1993); Diagnostic and forensic applications (Aly and Aldeyarbi 2020); Gene expression and protein production analysis; individualized healthcare and genomic medicine; A better knowledge of the genetic foundation of characteristics and illnesses is now possible thanks to the advancements in biotechnology tools and methods (Table 4.1), which are also advancing research in other areas including agriculture, medicines, and environmental sciences.

Genome assembly algorithms: De Bruijn graph-based assemblers, overlap-layout-consensus methods, and hybrid assembly techniques are just a few of the algorithms that have been instrumental in producing high-quality genome assemblies (Bankevich et al. 2012). A breakthrough in our knowledge of the evolutionary links between various species has been comparative genomics. Researchers can find conserved areas, analyze genetic variants, and learn more about the functional components and regulatory networks that underlie genes and genomes by comparing genomes from different species (Massey 2016). Function

Table 4.1 The overview information of currently available tools

No.	Name	Type	Features	Interface	Organism	References
1	Basic Local Alignment Search Tool (BLAST+)	Assemble tool	Used to find similarities between DNA or protein sequences and identify homologous regions	Web-based	Plants, viruses, bacteria, animals	Camacho et al. (2009)
2	Smith-Waterman Algorithm		Provides local sequence alignment for more sensitive searches	Web-based, command-line	Animals, bacteria, plants, viruses	Xia et al. (2022)
3	Ensemble	Genome browsers and annotation tools	Offers a comprehensive database of genomes with advanced annotation and visualization capabilities	Web-based	Human, Mouse, Fruit fly, Worm, Yeast Animals: Cow, Pig, Sheep, Horse Plant: Rice, Maize, Soybean	Birney et al. (2004)
4	UCSC Genome Browser		Provides an extensive collection of annotated genomes and allows for interactive exploration of genomic data	Web-based (main), command-line	Arabidopsis, Human, Mouse, Yeast, E. coli, Fruit fly, Rat, Dog	Zweig et al. (2008)
5	GATK (Genome Analysis Toolkit)	Variant Calling and Annotation Tools	A widely used tool for variant calling and genomic analysis	Command-line (main), web-based	Humans, crops, organisms, other species	McKenna et al. (2010)
6	Annovar		Enables functional annotation of genetic variants by integrating information from various databases	Command-line (main), web-based	Humans, microbes, organisms, animals, plants	Wang et al. (2010)
7	Maximum Likelihood (ML) and Bayesian Inference (BI) methods	Phylogenetic Analysis Tools	Used for inferring evolutionary relationships and constructing phylogenetic trees	Command-line tools or web-based	Bacteria, viruses, plants, animals, humans	Cole et al. (2014)
8	MEGA (Molecular Evolutionary Genetics Analysis)		Provides a comprehensive suite of tools for phylogenetic analysis	Command-line	Humans, others	Tamura et al. (2021)
9	Cytoscape					

		Network and pathway analysis	Enables visualization and analysis of biological networks, including protein-protein interactions and gene regulatory networks	Desktop-based	Bacteria, plants, humans, animals	Shannon et al. (2003)
10	String		Facilitates the exploration of protein-protein interactions and functional associations	Web-based (main)	Bacteria, plants, animals, humans	Zhao and Sahni (2019)

annotation: new developments in bioinformatics have spurred the creation of computer resources and methods for functional annotation, enabling scientists to anticipate the roles of genes, locate regulatory components, and annotate non-coding portions of genomes (Butkiewicz and Bush 2016). Improved statistical models, databases, and variant calling algorithms have improved our ability to recognize and decipher genomic variations, such as single nucleotide polymorphisms (SNPs), insertions, deletions, and structural alterations (Zverinova and Guryev 2022). Researchers can gain deeper knowledge of genes and genomes by creating integrative analytic tools, which can reveal intricate interactions and linkages within biological systems (Jeong et al. 2014). Insights into the ecological functions, interconnections, and possible industrial uses of complex microbial communities have been gained thanks to the development of bioinformatics tools and pipelines.

Due to distinguished features in the genome structure of prokaryotes and eukaryotes, in this chapter, we mentioned the workflow of both types and mostly focused on humans & animals, and microorganisms. Regarding humans and animals, the genetic variation is mainly regulated by the genome. Variants arise as a result of differences in DNA sequence, deletions, insertions, and structural variations. Species diversity is strongly regulated by genetic variation, which contributes to individual evolution, adaptation, and disease resistance (Rentzsch et al. 2019). In addition, the human and animal genomes are a very complex and highly organized structure, which can have billions of DNA base pairs arranged into chromosomes. The regions of DNA that encode the instructions for building proteins that regulate genes are located on the genome itself. Besides, the genome has non-coding regions that will regulate genes, stabilize the genome and other regulatory functions (Filippakopoulos et al. 2012). Evolutionary conservation is a special feature in humans and animals that species have both heredity but still exist evolution to adapt to different conditions. Many genes and regulatory elements are conserved across species, suggesting their functional importance and common ancestry. Cell division, metabolism, and development are regulated by evolutionary conserved factors. The workflow of genome analysis is below (Fig. 4.2):

- Data Acquisition and Quality Control: Obtain the raw sequencing data, which may be produced using NGS methods. Assess the quality of the raw data by performing basic quality control tests, such as deleting adapter sequences or low-quality reads, and evaluating read quality scores.
- Read Alignment and Mapping: Utilize tools for alignment such as BWA, Bowtie, or STAR to match the sequencing reads to a reference genome or assembly. Create alignment files in the SAM or BAM formats that provide the genomic coordinates of the mapped reads.
- Variant Calling: Compare the aligned reads to the reference genome to find genetic changes (such as single nucleotide polymorphisms, or SNPs, insertions, and deletions). For this, variant calling tools like GATK, SAMtools, or FreeBayes are frequently employed. Apply quality filters to eliminate false-positive variations and keep variants with a high degree of confidence.

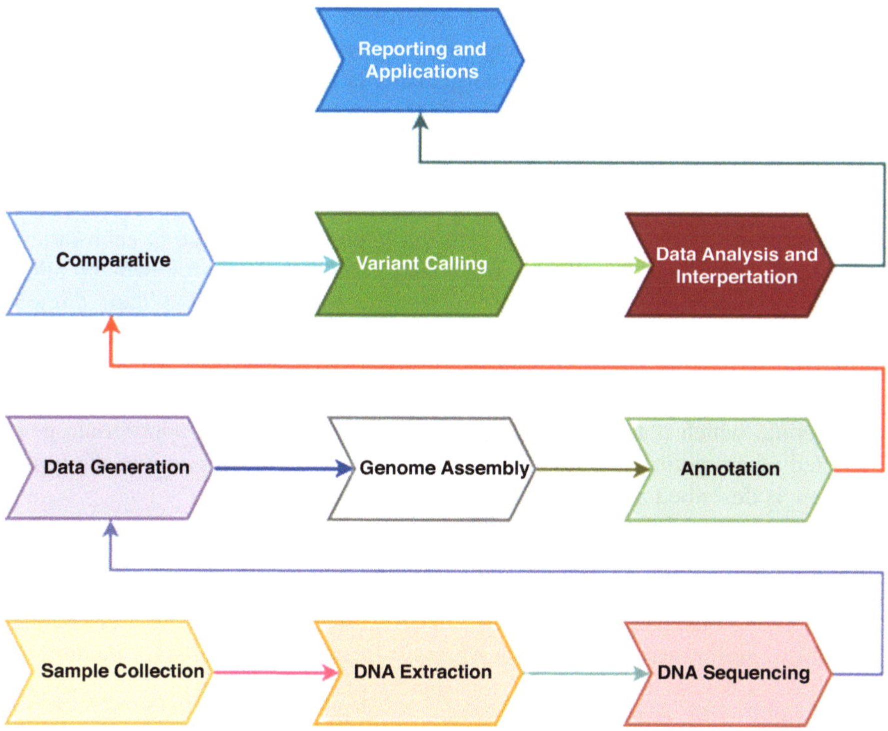

Fig. 4.2 Basic workflow of genome sequencing processing

- Annotation and Functional Analysis (Wang et al. 2010): Use tools like ANNOVAR, VEP, or SnpEff to annotate the discovered variations to ascertain their possible functional implications. Include data from databases like dbSNP, ClinVar, or COSMIC to evaluate the known relationships or clinical relevance of the variations. Utilize tools like DAVID, Enrichr, or GSEA to perform functional enrichment analysis to comprehend the biological pathways, gene ontology words, or protein domains enriched in the discovered variations.
- Structural Variation Analysis (optional): Use specialist tools like CNVkit, DELLY, or BreakDancer to find large-scale structural variants (such as copy number variations or translocations). Consider the structural alterations in the context of chromosomal rearrangements, gene disruptions, or disease-related pathways for analysis and interpretation.
- Population Genetics and Phylogenetic Analysis (if applicable): Use methods like PLINK, ADMIXTURE, or BEAST to analyze genetic diversity among populations or individuals to research population genetics, demographic history, and evolutionary connections. Using programs like RAxML, PhyML, or IQ-TREE, create phylogenetic trees or carry out phylogenomic studies to comprehend species connections or genetic divergence.

Regarding the basic features of microbial genomes: The genome on microorganisms is a very special thing which is represented by three basic contents: Compact size: The genome of microorganisms is relatively small, compact compared to the genome of eukaryotes. The lower number of non-coding regions, such as introns or interstitial regions, results in size compactness (Zhu et al. 2020). The proportion of bacterial genes is not properly arranged because it contains a higher proportion of protein-coding genes, which maximizes the potential for inheritance (Karlsson et al. 2015). The distance between genes that are close to each other is interspersed by non-coding DNA fragments. This efficient arrangement allows bacteria to pack significant amounts of genetic information into their compact genomes. The horizontal gene transfer tendency of the microbial genome increases its dynamics. Horizontal gene transfer regulates the transfer of genetic material between different organisms, through mechanisms such as conjugating and transducting, which is to induce new heritable traits or generate appropriate genes suspected of promoting species survival and evolution. An overview of the typical workflow is described as follows (Fig. 4.2):

- Data Acquisition and Quality Control (Wang et al. 2001): Obtain the raw sequencing data, which may be produced using NGS (such as Illumina sequencing) or long-read sequencing (such as PacBio, Oxford Nanopore) technologies. Perform preliminary quality control tests to evaluate the data's quality, cut adapter sequences, and eliminate low-quality reads or sequences with inaccurate base calls.
- Genome Assembly (Bankevich et al. 2012): Utilizing specialist assembly tools like SPAdes, Velvet, or Canu, combine the sequencing reads into contiguous sequences (contigs) or whole genomes. Metrics like N50 length, contig coverage, and the prevalence of mis-assemblies may be used to control the quality of assemblies.
- Genome Annotation (Sargent et al. 2020): To discover genes, regulatory components, and other genomic characteristics, annotate the assembled genomes. To find protein-coding genes, do gene prediction using programs like Prokka, Glimmer, or GeneMark. Genome annotation uses databases like UniProt, the NR database from the NCBI, or specific databases for microbial genes to annotate the predicted genes with functional information. s.
- Comparative Genomics: To investigate genomic differences, gene content, and evolutionary linkages, the genomes of various microorganisms should be compared using programs like Roary, OrthoFinder, or BLAST to locate and examine genomic areas that are unique to certain strains or species.
- Functional Analysis and Pathway Reconstruction (Catozzi et al. 2022): Identify the possible activities of the predicted genes by assigning functional annotations to them using databases like COG, KEGG, or Pfam. Examine the gene clusters that are engaged in particular biological processes or metabolic pathways. Utilize resources like Pathway Tools, KEGG, or MetaCyc to reconstruct metabolic pathways and examine the microbes' possible metabolic capabilities.

- Virulence Factors and Antibiotic Resistance Analysis (if applicable) (Wang et al. 2020): Examine the microbial genomes for possible virulence factors and antibiotic-resistance genes using specialist databases s of virulence factors and antibiotic-resistant genes, such as VFDB, Victors, or CARD. The results can examine how these genes are distributed and how they vary between strains or species.
- Phylogenetic Analysis (if applicable): Build phylogenetic trees based on conserved genes or whole genome sequences to infer connections between different species. Tools for phylogenetic tree generation and visualization are RAxML, PhyML, or IQ-TREE.

4.4 The Application of Bioinformatics in Gene and Genome Analysis

High-throughput sequencing technology is necessary to comprehend the complexity of genes and genomes (Diniz and Canduri 2017). The capacity for research has been changed by genetic analysis tools, and scientists now have access to a variety of computer tools that support the analysis of genetic data. Several applications of bioinformatics in gene and genome analysis are assembly, annotation, variant calling, evolutionary linkage, etc. as the following: Genome assembly is a bioinformatics technique that is used to joint together whole genomes from the short reads produced by high-throughput sequencing technology. These tools use computational methods and algorithms to align and combine overlapping reads, clear out ambiguities, and provide consensus sequences (Ghurye et al. 2016). Functional annotation is a method forecasting how variations may affect regulatory components, non-coding RNAs, splice sites, and protein-coding regions using databases, algorithms, and machine learning techniques (Brunet et al. 2022). Variant calling and analysis are bioinformatics techniques using to extract genetic variants from genomic data like indels, single nucleotide polymorphisms (SNPs), and structural changes. These tools analyze the findings to find genuine variations by comparing sequencing reads to a reference genome, using statistical models (Robinson et al. 2017). The study of evolutionary linkages, the identification of conserved areas, and the detection of genomic re-arrangements are all made possible by comparative genomics which compares the genomes of various species (Armstrong et al. 2019). Bioinformatics is also essential for assessing epigenetic alterations like DNA methylation and histone modifications in the field of epigenomics. Differential methylation patterns, chromatin states, and the study of the epigenetic control of gene expression are all studied using various tools (Wang and Chang 2018). With a variety of applications in gene and genome analysis, the integration and interpretation of intricate biological data from fields like genomics, transcriptomics, proteomics, and metabolomics using bioinformatics tools is potential. However, further development is still needed to clarify the connections between genes, proteins, and other molecules, as well as building biological network (Ma'ayan 2008; Sunyaev and Roth 2013).

4.5 Conclusion

In conclusion, it is witnessed a dramatic development in bioinformatics in gene and genome analysis during recent years. Thanks to the advances in computational methods and molecular biological techniques, bioinformatics tools have overcome many hurdles in infrastructure, the volume of storage, and quality control of sequencing and assembling. The accuracy and specificity of variant detection have been enhanced. The abundance of web-based tools and command-line tools with specific workflows for each organism as well as online databases of sequencing data, expression level, and molecular pathways supports scientists a lot in data analyzing. However, the interchange from genetic analyzing results to clinical decisions is still unclear. More and more research and clinical trials are needed to fill this gap. In addition, with the availably huge database about sequencing, it might be time to focus on secondary data processing to depict a bigger picture for bioinformatics and clinical implementation.

References

Abou Ziki MD, Mani A (2016) Metabolic syndrome: genetic insights into disease pathogenesis. Curr Opin Lipidol 27(2):162–171

Akalın PK (2006) Introduction to bioinformatics. Mol Nutr Food Res 50(7):610–619

Aly SM, Aldeyarbi H (2020) Applications of forensic entomology: overview and update. Arch Med Sadowej Kryminol 70(1):44–77

Apolinario E et al (1993) Cloning and manipulation of the Schizosaccharomyces pombe his7+ gene as a new selectable marker for molecular genetic studies. Curr Genet 24(6):491–495

Armstrong J et al (2019) Whole-genome alignment and comparative annotation. Annu Rev Anim Biosci 7:41–64

Bankevich A et al (2012) SPAdes: a new genome assembly algorithm and its applications to single-cell sequencing. J Comput Biol 19(5):455–477

Benson DA et al (2013) GenBank. Nucleic Acids Res 41(D1):D36–D42

Birney E et al (2004) An overview of Ensembl. Genome Res 14(5):925–928

Bluestone JA, Herold K, Eisenbarth G (2010) Genetics, pathogenesis and clinical interventions in type 1 diabetes. Nature 464(7293):1293–1300

Bowdin S et al (2016) Recommendations for the integration of genomics into clinical practice. Genet Med 18(11):1075–1084

Brunet MA, Leblanc S, Roucou X (2022) OpenVar: functional annotation of variants in non-canonical open reading frames. Cell Biosci 12(1):130

Butkiewicz M, Bush WS (2016) In silico functional annotation of genomic variation. Curr Protoc Hum Genet 88:6.15.1–6.15.17

Camacho C et al (2009) BLAST+: architecture and applications. BMC Bioinformatics 10:421

Can T (2014) Introduction to bioinformatics. In: Yousef M, Allmer J (eds) miRNomics: microRNA biology and computational analysis. Humana Press, Totowa, pp 51–71

Cantacessi C et al (2010) A practical, bioinformatic workflow system for large data sets generated by next-generation sequencing. Nucleic Acids Res 38(17):e171

Caspi R et al (2020) The MetaCyc database of metabolic pathways and enzymes—a 2019 update. Nucleic Acids Res 48(D1):D445–D453

Catozzi S et al (2022) Reconstruction and analysis of a large-scale binary Ras-effector signaling network. Cell Commun Signal 20(1):24

Cole SR, Chu H, Greenland S (2014) Maximum likelihood, profile likelihood, and penalized likelihood: a primer. Am J Epidemiol 179(2):252–260
Crews KR et al (2012) Pharmacogenomics and individualized medicine: translating science into practice. Clin Pharmacol Ther 92(4):467–475
Danecek P et al (2011) The variant call format and VCFtools. Bioinformatics 27(15):2156–2158
Daoud M, Mayo M (2019) A survey of neural network-based cancer prediction models from microarray data. Artif Intell Med 97:204–214
Dayhoff MO, Eck RV (1972) Atlas of protein sequence and structure. National Biomedical Research Foundation
Diniz WJ, Canduri F (2017) REVIEW-ARTICLE bioinformatics: an overview and its applications. Genet Mol Res 16(1)
Endrullat C et al (2016) Standardization and quality management in next-generation sequencing. Appl Transl Genomics 10:2–9
Fabregat A et al (2016) The reactome pathway knowledgebase. Nucleic Acids Res 44(D1):D481–D487
Felsenstein J (1981) Evolutionary trees from DNA sequences: a maximum likelihood approach. J Mol Evol 17(6):368–376
Filippakopoulos P et al (2012) Histone recognition and large-scale structural analysis of the human bromodomain family. Cell 149(1):214–231
Fleischmann RD et al (1995) Whole-genome random sequencing and assembly of Haemophilus influenzae Rd. Science 269(5223):496–512
Gauthier J et al (2019) A brief history of bioinformatics. Brief Bioinform 20(6):1981–1996
Ghurye JS, Cepeda-Espinoza V, Pop M (2016) Metagenomic assembly: overview, challenges and applications. Yale J Biol Med 89(3):353–362
Haeckel E (1866) Generelle morphologie der organismen, vol 2. Georg Reimer, Berlin
Hesper B, Hogeweg PDLLBC (1970) Bioinformatica: een werkconcept. Kameleon 1(6):28–29
Hu T et al (2021) Next-generation sequencing technologies: an overview. Hum Immunol 82(11): 801–811
Jalili V et al (2020) The galaxy platform for accessible, reproducible and collaborative biomedical analyses: 2020 update. Nucleic Acids Res 48(W1):W395–W402
Jeong Y, Choi J, Lee KH (2014) Technology advancement for integrative stem cell analyses. Tissue Eng Part B Rev 20(6):669–682
Kanehisa M (2002) The KEGG database. In: 'In silico' simulation of biological processes, pp 91–103
Kanz C et al (2005) The EMBL nucleotide sequence database. Nucleic Acids Res 33(suppl_1): D29–D33
Karlsson M et al (2015) Insights on the evolution of mycoparasitism from the genome of Clonostachys rosea. Genome Biol Evol 7(2):465–480
Köhler S et al (2017) The human phenotype ontology in 2017. Nucleic Acids Res 45(D1):D865–D876
Kumar R, Gupta M, Sarwat M (2022) Bioinformatics in drug design and delivery. In: Saharan VA (ed) Computer aided pharmaceutics and drug delivery: an application guide for students and researchers of pharmaceutical sciences. Springer Nature Singapore, Singapore, pp 641–664
Lazaridis KN et al (2014) Implementing individualized medicine into the medical practice. Am J Med Genet C Semin Med Genet 166(1):15–23
Lelieveld SH, Veltman JA, Gilissen C (2016) Novel bioinformatic developments for exome sequencing. Hum Genet 135(6):603–614
Li H et al (2009) The sequence alignment/map format and SAMtools. Bioinformatics 25(16): 2078–2079
Ma'ayan A (2008) Network integration and graph analysis in mammalian molecular systems biology. IET Syst Biol 2(5):206–221
Marco-Puche G et al (2019) RNA-Seq perspectives to improve clinical diagnosis. Front Genet 10: 1152

Massey SE (2016) Comparative microbial genomics and forensics. Microbiol Spectr 4(4)
McKenna A et al (2010) The genome analysis toolkit: a MapReduce framework for analyzing next-generation DNA sequencing data. Genome Res 20(9):1297–1303
Olson MV (1993) The human genome project. Proc Natl Acad Sci U S A 90(10):4338–4344
Pereira R, Oliveira J, Sousa M (2020) Bioinformatics and computational tools for next-generation sequencing analysis in clinical genetics. J Clin Med 9:132. https://doi.org/10.3390/jcm9010132
Peter JAC et al. (2015) SAM/BAM format v1.5 extensions for *de novo* assemblies. bioRxiv, pp 020024
Peterson MW, Colosimo ME (2007) TreeViewJ: an application for viewing and analyzing phylogenetic trees. Source Code Biol Med 2(1):7
Rentzsch P et al (2019) CADD: predicting the deleteriousness of variants throughout the human genome. Nucleic Acids Res 47(D1):D886–D894
Robinson JT et al (2017) Variant review with the integrative genomics viewer. Cancer Res 77(21): e31–e34
Sam B, Patrick ST (2013) What is next generation sequencing? Arch Dis Child Educ Pract Ed 98(6):236
Sargent L et al (2020) G-OnRamp: generating genome browsers to facilitate undergraduate-driven collaborative genome annotation. PLoS Comput Biol 16(6):e1007863
Schochetman G, Ou C-Y, Jones WK (1988) Polymerase chain reaction. J Infect Dis 158(6): 1154–1157
Shannon P et al (2003) Cytoscape: a software environment for integrated models of biomolecular interaction networks. Genome Res 13(11):2498–2504
Sohpal VK, Dey A, Singh A (2010) MEGA biocentric software for sequence and phylogenetic analysis: a review. Int J Bioinforma Res Appl 6(3):230–240
Staden R (1979) A strategy of DNA sequencing employing computer programs. Nucleic Acids Res 6(7):2601–2610
Stallman R (2003) Free software foundation (FSF). In: Encyclopedia of computer science. Wiley, Hoboken, pp 732–733
Sunyaev SR, Roth FP (2013) Systems biology and the analysis of genetic variation. Curr Opin Genet Dev 23(6):599–601
Tamura K, Stecher G, Kumar S (2021) MEGA11: molecular evolutionary genetics analysis version 11. Mol Biol Evol 38(7):3022–3027
Tateno Y et al (2002) DNA data Bank of Japan (DDBJ) for genome scale research in life science. Nucleic Acids Res 30(1):27–30
Wang KC, Chang HY (2018) Epigenomics: technologies and applications. Circ Res 122(9): 1191–1199
Wang X, Ghosh S, Guo S-W (2001) Quantitative quality control in microarray image processing and data acquisition. Nucleic Acids Res 29(15):e75
Wang K, Li M, Hakonarson H (2010) ANNOVAR: functional annotation of genetic variants from high-throughput sequencing data. Nucleic Acids Res 38(16):e164
Wang G et al (2020) The characteristic of virulence, biofilm and antibiotic resistance of Klebsiella pneumoniae. Int J Environ Res Public Health 17:6278. https://doi.org/10.3390/ijerph17176278
Womble DD (1999) GCG. In: Misener S, Krawetz SA (eds) Bioinformatics methods and protocols. Humana Press, Totowa, pp 3–22
Xia Z et al (2022) A review of parallel implementations for the Smith-Waterman algorithm. Interdiscip Sci 14(1):1–14
Zhao C, Sahni S (2019) String correction using the Damerau-Levenshtein distance. BMC Bioinformatics 20(Suppl 11):277
Zhou P et al (1994) A system for gene cloning and manipulation in the yeast Candida glabrata. Gene 142(1):135–140

Zhu F et al (2020) Metagenome-wide association of gut microbiome features for schizophrenia. Nat Commun 11(1):1612

Zverinova S, Guryev V (2022) Variant calling: considerations, practices, and developments. Hum Mutat 43(8):976–985

Zweig AS et al (2008) UCSC genome browser tutorial. Genomics 92(2):75–84

Role of Bioinformatics in Non-coding RNA Analysis

5

Anshu Mathuria, Mehak, and Indra Mani

Abstract

The transcriptome, comprising RNA molecules expressed in cells or tissues, is predominantly composed of non-coding RNAs (ncRNAs), which has most of the region in the genome of humans. The classification of ncRNAs includes housekeeping and regulatory ncRNAs, with the latter encompassing long-ncRNAs (lncRNAs), microRNAs (miRNAs), and small interfering RNAs (siRNAs). These ncRNAs, including lncRNAs, play a crucial role in various levels of gene regulation, like transcription, RNA processing, translation, and chromatin modification. By interacting with RNA, DNA, and proteins, lncRNAs influence chromatin structure and the localization and activity of various protein complexes and RNA processing. The study of lncRNAs presents both challenges and opportunities, as they exhibit complex sequence and structural characteristics. The application of bioinformatics in the study of ncRNAs highlights how computational methods have contributed to the prediction and identification of novel ncRNAs, target gene prediction, RNA structure prediction, evolutionary analysis, functional prediction, and the construction of regulatory networks. This chapter briefly discusses the databases and tools that aid in the analysis and interpretation of ncRNA data, including LncTarD, LnCeVar, MirGeneDB, miRTarBase, SEAweb, DIANA-LncBase, miRPathDB, RNAInter, oRNAment, miRDB, ENCORI, NPInter, etc. These resources provide valuable information on ncRNA interactions, targets, functions, and regulation, enabling researchers to explore the complex world of ncRNAs.

Anshu Mathuria and Mehak contributed equally the first authors.

A. Mathuria · Mehak · I. Mani (✉)
Department of Microbiology, Gargi College, University of Delhi, New Delhi, India
e-mail: indra.mani@gargi.du.ac.in

V. Singh, A. Kumar (eds.), *Advances in Bioinformatics*,
https://doi.org/10.1007/978-981-99-8401-5_5

Keywords

ncRNAs · lncRNAs · siRNAs · miRNAs · Tools · Databases

5.1 Introduction

The set of RNA molecules manifest in a cell or tissue is known as a transcriptome (Wang et al. 2019). It is truthful that over 90% of the human genome undergoes transcription, yet a minority of this activity is associated with genes coding for proteins, which make up less than 2% of the total transcription (Pertea 2012; Li and Liu 2019). As a result, the majority of transcribed genes give rise to non-coding RNAs (ncRNAs).

The terminology non-coding RNA (ncRNA) is commonly used to describe molecules of RNA that do not carry instructions for protein synthesis. However, this does not signify that such RNAs lack functions or information. Recent research has challenged the conventional belief that protein-coding genes are solely responsible for genetic information processing. Studies have shown that a substantial portion of mammalian and other complex organism genomes is transcribed into ncRNAs (Mattick and Makunin 2006). These ncRNAs can undergo alternative splicing and processing, resulting in the production of smaller RNA molecules. Moreover, there are numerous longer transcripts, characterized by intricate patterns of overlapping and interlacing sense and antisense strands, the majority of which have unknown functions.

Among these ncRNAs, including those originating from introns, there exists a concealed network of internal signals that exert influence over different aspects of gene expression during physiological processes and development. These ncRNAs play a crucial role in regulating chromatin structure, maintaining epigenetic information, controlling transcription, RNA splicing, RNA editing, RNA degradation, and translation. The intricate networks which are formed by these RNA molecules likely play a necessary role in determining complex characteristics, and they may also contribute to disease processes. Exploring the vast landscape of RNA regulatory networks is necessary for understanding the underlying mechanisms that shape various biological processes and diseases.

The ncRNAs can be broadly categorized based on their functions into two groups: regulatory and housekeeping ncRNAs (Pertea 2012; Li and Liu 2019). They play roles in various levels of gene regulation, including mRNA processing, transcription, translation, and chromatin modification (Chan and Tay 2018; Fernandes et al. 2019). Regulatory ncRNAs enclose diverse types of molecules, such as long non-coding RNAs (lncRNAs), microRNAs (miRNAs), and small interfering RNAs (siRNAs), among others. These ncRNAs can influence gene expression (Yamamura et al. 2018; Grillone et al. 2020).

LncRNAs are increasingly recognized as crucial regulators in gene expression networks and shows a diverse range of sizes and shapes. Eukaryotic genomes produce various classes of lncRNAs transcribed from different DNA regions, including enhancers, promoters, and intergenic regions. Some lncRNAs originate from lengthy primary transcripts through unconventional RNA processing pathways, leading to the generation of novel RNA species with unexpected structures (Wu et al. 2017). These lncRNAs can undergo distinct processing mechanisms, such as cleavage by ribonuclease P (RNase P), to produce mature 3′ ends and capping through small nucleolar RNA (snoRNA)–protein (snoRNP) complexes at their ends, or the formation of circular structures.

Through intricate mechanisms, lncRNAs play a crucial role in regulating gene expression, genomic imprinting, dosage compensation, nuclear organization, and nuclear-cytoplasmic trafficking (Zhang et al. 2014). The association of lncRNAs with diseases and their distinct expression patterns in various tissues suggest that they are fundamental components of the transcriptional regulatory circuitry. The specific structural and sequence characteristics of lncRNAs mediate their functions. They can interact with RNA, DNA, and proteins in both the cytoplasm and the nucleus, exerting their regulatory effects through these interactions.

In 2003, after the completion of the human genome project (HGP), further exploration of the non-coding regions and their significance in the traditional definition of genes became necessary. It was discovered through the ENCyclopedia Of DNA Elements (ENCODE) project, which began in 2003, that approximately 80% of the human genome exhibits biochemical functionality. Within this functional portion, it was revealed that 76% of the DNA is transcribed into RNA, with about 2% of this transcribed RNA being expressed as functional proteins (Denham et al. 2022). This substantial difference between the estimated 20,000 protein-coding genes and the over 100,000 distinct transcripts identified in mammalian transcriptomes indicates the possibility of uncovering a novel category of non-translated RNAs.

5.2 Classifications of Non-coding RNAs

Non-coding RNAs (ncRNAs) are primarily divided into two groups: housekeeping and regulatory ncRNAs (Hombach and Kretz 2016; Natsidis et al. 2019; Krahn et al. 2020; Winkle et al. 2021). Further, the classification of regulatory ncRNAs is done based on their size in which ncRNAs with less than 200 nucleotides are referred to as small ncRNAs, whereas those exceeding this threshold are called lncRNAs (Dahariya et al. 2019; Sikora et al. 2020). The small ncRNA category encompasses miRNAs, Piwi-interacting RNAs (piRNAs) and siRNAs. Illustrations of the structural classification of ncRNAs (rRNA, tRNA, miRNAs, siRNAs, piRNAs, circRNA, snoRNA, shRNA, and lncRNAs) are given in Fig. 5.1 (Zhao et al. 2022).

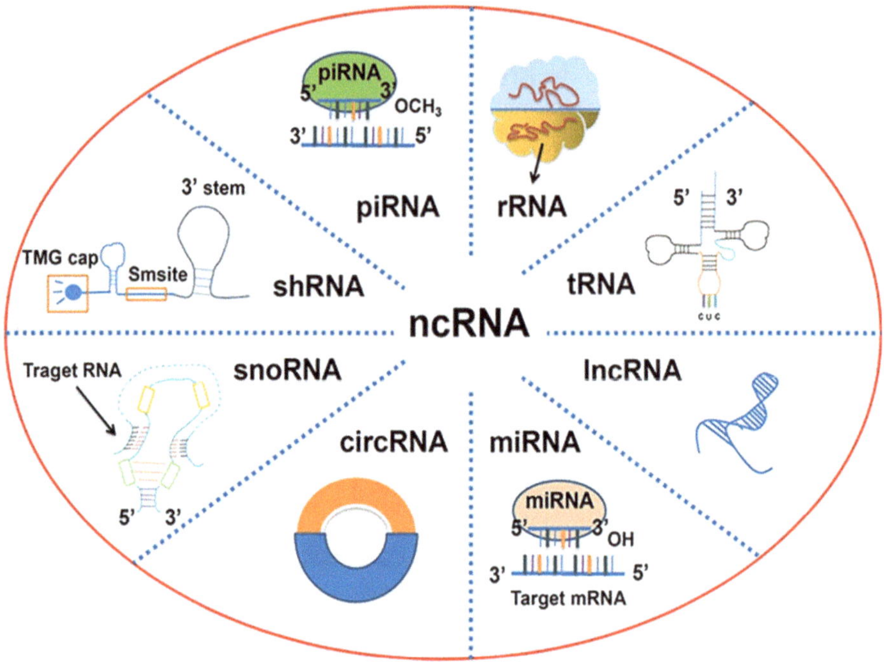

Fig. 5.1 Schematic diagram of different non-coding RNA (ncRNA) and their structure, including ribosomal RNA (rRNA), transfer RNA (tRNA), long non-coding RNA (lncRNA), microRNA (miRNA), circular RNA (circRNA), small nucleolar RNA (snoRNA), short hairpin RNA (shRNA), and piwi-interacting RNA (piRNA). (Adapted from Zhao et al. 2022)

5.3 Functions of Non-coding RNAs

The functions of various types of non-coding RNA include miRNAs, which have been extensively studied across different biological contexts. Moreover, the functions of many lncRNAs, including large intervening non-coding RNAs (lincRNAs), remain unclear. However, several evidence suggests the functionality of lncRNAs: (1) The functional significance of lncRNA promoters, exons, spliced junctions, predicted structures, and genomic locations is evident from their conservation. (2) The presence of specific chromatin signatures associated with active transcription in lncRNA promoters, exons, splice junctions, predicted structures, and genomic locations suggests their involvement in gene regulation. (3) Additionally, the regulation of lncRNAs by crucial transcription factors and molecular signals further strengthens their functional importance. (4) LncRNAs possess patterns of alternative splicing and dynamic expression during cellular differentiation, inferring their regulatory roles. (5) They manifest specific patterns of expression and subcellular localization in different tissues and cells, indicating their context-specific functions. (6) LncRNAs express splicing patterns or altered expression in diseases, including cancer, suggesting their potential involvement in disease processes (Mattick 2009).

Indeed, lncRNAs have been shown to exercise regulatory functions at the transcriptional, post-transcriptional and epigenetic levels through different mechanisms. Here are some examples:

5.3.1 Transcriptional Regulation

lncRNAs can interact with transcriptional activators or repressors, as well as different components of the transcriptional machinery such as the DNA duplex and RNA polymerase II at the transcriptional level (Goodrich and Kugel 2006). By doing so, lncRNAs can influence gene expression and transcription.

5.3.2 Post-transcriptional Regulation

LncRNAs engage in several post-transcriptional processes, such as pre-mRNA processing, alternative splicing transport, translation, and degradation. They can influence other RNA molecules or proteins involved in these processes, thereby impacting gene expression at the post-transcriptional level.

5.3.3 Epigenetic Regulation

LncRNAs plays an important role in various epigenetic processes. They are involved in gene imprinting, which is the differential expression of genes based on their parental origin. LncRNAs also contribute to X-chromosome inactivation, a process which equalizes gene expression between males and females by silencing one of the two X chromosomes in females. Additionally, lncRNAs are implicated in other epigenetic mechanisms like histone modifications and chromatin remodeling, which can influence gene expression patterns.

Indeed, multiple regulatory mechanisms of lncRNAs have been identified (Rinn and Chang 2012). It includes:

(a) **Decoy function**: Certain lncRNAs can proceed as decoys by resembling the DNA-binding sites of regulatory proteins. By binding to these proteins, lncRNAs prevent them from interacting with their target DNA sequences, leading to the inhibition of gene transcription. For instance, the lncRNA Gas5 possesses a secondary structure with a hairpin sequence motif. This motif bears a resemblance to the site of DNA binding of the glucocorticoid receptor (GR). As a result, Gas5 play a role as a decoy for the GR, effectively impeding the transcription process of the GR's target genes (Kino et al. 2010).
(b) **Adaptor function:** Some lncRNAs perform the function of adaptors to facilitate the formation of protein complexes. By interacting with multiple proteins, these

lncRNAs bring them together, enabling the assembly of specific protein complexes that perform regulatory functions.

(c) **Localization guides**: Certain lncRNAs play a role in proceeding with the proper localization of specific protein complexes within the cell. By interacting with these complexes or other cellular components, lncRNAs help in directing them to their correct subcellular locations, ensuring their proper functioning.

(d) **miRNA regulation**: Some lncRNAs can engage with miRNAs for binding sites or sequester miRNAs away from their mRNA targets which act as "miRNA sponges." Thus, this interaction between lncRNAs and miRNAs can regulate the availability and activity of miRNAs, thereby impacting post-transcriptional gene regulation (Hansen et al. 2013).

5.4 Regulatory Role of Non-coding RNAs

Gene expression is controlled by lncRNAs through various mechanisms. These molecules interact with proteins, DNA and RNA which influence the transcription of nearby and distant genes as well as structure and function of chromatin. They also impact stability, RNA splicing, and translation. Additionally, lncRNAs play a role in the regulation and formation of nuclear condensates and organelles.

5.4.1 Chromatin Regulation

LncRNAs play a complex role in regulating gene expression and chromatin structure, as demonstrated by the identification of RNA–chromatin interactions using genome-wide approaches (Bonetti et al. 2020) and chromatin conformation capture methods (Isoda et al. 2017). The regulatory potential of RNA itself has been revealed by the extensive study of these mechanisms. The de-compaction of chromatin is caused by the ability of RNA's negative charge to balance the positively charged histone tails (Dueva et al. 2019). Therefore, opening and closing of RNA-mediated chromatin can act as a quick switch to regulate gene expression. LncRNAs exert their regulatory effects through a combination of cis- and trans-acting mechanisms. In some circumstances, lncRNAs have direct interactions with DNA, changing the chromatin environment. In some cases, lncRNAs bind DNA in a way that is specific to a particular sequence, but in other cases, this can happen indirectly due to their affinity for proteins that can associate with both DNA and RNA.

5.4.2 Protein–lncRNA Localization and Function on Chromatin

It is known that a variety of lncRNAs confine on chromatin, where they interact with proteins and affect their binding and activity to particular DNA regions. The function of proteins may be facilitated or inhibited by this interaction between lncRNAs and proteins. The transcriptional effects of lncRNAs on target genes can also be mediated by long-distance chromatin interactions that are facilitated by proteins like CCCTC-binding factor (CTCF) (Saldana-Meyer et al. 2019). It is significant to note that careful consideration and rigorous methods should be used in these studies to assess lncRNA–chromatin factor interactions. The degree to which lncRNAs have an impact on the targeted chromatin can also be determined by the expression levels of a specific lncRNA in relation to the interacting factors (Schertzer et al. 2019). Unraveling the intricate network of interactions between lncRNAs, chromatin factors, and proteins holds great promise for advancing our understanding of gene regulation and has the potential to unveil novel therapeutic targets and strategies for various diseases.

5.4.3 Direct Interactions Between lncRNAs and DNA

This is the ability of lncRNAs to combine with DNA to form hybrid structures, which can affect chromatin accessibility, is a crucial property of these molecules. These interactions can hold the form of R-loops or triple helices (triplexes). Although it is still unclear how common these structures are in vivo, it is thought that their formation is common and essential for many lncRNAs regulatory functions. It has been suggested that RNA–DNA–DNA triplexes are examples of non-coding RNA–DNA interaction which mediates gene activation or silencing (Schmitz et al. 2010; Grote et al. 2013; O'Leary et al. 2015). Triplex formation is primarily influenced by the RNA sequence (Li et al. 2016; Blank-Giwojna et al. 2019).

In order to study sequences forming triplex in vivo, a method known as TrIP-seq (targeted RNA immunoprecipitation sequencing) has been created (Maldonado et al. 2019). The lncRNA KHPS1, which integrates into a triple helix upstream of the sphingosine kinase 1 (SPHK1) enhancer, is an illustration of triplex-mediated gene regulation. This triple helix aids in the recruitment of chromatin modifiers that activate transcription of the RNA which has been derived from the SPHK1 enhancer (eRNA-SPHK1), thereby enhancing SPHK1 expression. Notably, the specificity of triplex formation was shown by switching the region of KHPS1 that forms triplexes with the region of another lncRNA, MEG3, which led to a change in the target gene's specificity (Blank-Giwojna et al. 2019).

Another thoroughly investigated method of lncRNA interaction with chromatin is through R-loops. R-loops were once thought to be a threat to genome stability, but recent research (Tan-Wong et al. 2019; Niehrs and Luke 2020) suggests that they act as regulatory hubs and coordinators of DNA repair and gene expression. With the help of proteins that recognize and bind to R-loops, several lncRNAs control gene expression in conjunction with these structures. Numerous different outcomes can result from this interaction (Gibbons et al. 2018). For instance, the lncRNA TARID causes the transcription of the *TCF21* gene to proceed in the opposite direction by creating an R-loop at the gene's CpG-rich promoter. The DNA demethylating factor TET1 is attracted towards the R-loop by GADD45A, which activates the transcription of *TCF21* (Ariel et al. 2020). Additionally, R-loop-forming lncRNAs can control the expression of genes coding for proteins in either a cis or a trans manner. For instance, the lncRNA APOLO participates in a widespread regulation of genes which are auxin-responsive in *Arabidopsis thaliana*, which in turn forms trans-acting R-loops.

5.4.4 Transcription Regulation

A lncRNA's placement in relation to nearby genes is critical to how they regulate one another. The conservation of lncRNAs widespread antisense and bidirectional transcription hints that their non-random genomic distribution may be an evolutionary adaptation for genes to control their own expression in a context-specific manner (Seila et al. 2008). The genomic arrangement is crucial for the cis-regulation of gene expression in the case of divergent lncRNAs (Luo et al. 2016).

This regulation may be governed by two main mechanisms. The lncRNA transcript itself can first control nearby genomic loci. Second, specific chromatin states or steric hindrances generally affects the expression of nearby genes can be produced during the transcription or splicing of the lncRNA. Multiple independent gain-of-function and loss-of-function experiments must be carried out in order to fully understand the functionality of lncRNAs. These tests aid in separating the various potential mechanisms by which lncRNAs could exert their regulatory effects.

5.5 Application of Bioinformatics to Studies of Non-coding RNAs

The computational techniques have been employed to examine ncRNA through various avenues, including: (1) Computational methods and integration with the experimental data (e.g., RNA-seq data, tiling array) have made it easier to anticipate and identify new ncRNAs from genomic sequence analysis. (2) These methods have been utilized to forecast the target genes of miRNAs, providing insights into their regulatory roles. (3) It has been employed to predict the tertiary and secondary structures of RNA molecules, aiding in understanding their functional properties. (4) The utilization of computational tools has facilitated the exploration of the preservation and progression of ncRNA genes and miRNA target genes, shedding light on their evolutionary dynamics and functional significance. (5) Computational analysis techniques, such as the "guilt by association" approach, have been employed to predict the functions of ncRNAs by examining their associations with other genes or biological processes. (6) Computational approaches have facilitated the formation of regulatory networks that incorporate ncRNA regulatory layers, providing a comprehensive understanding of gene regulation. (7) Techniques have been instrumental in the creation of databases and web servers that serve as valuable resources for ncRNA research, facilitating data access and analysis (Cheng et al.

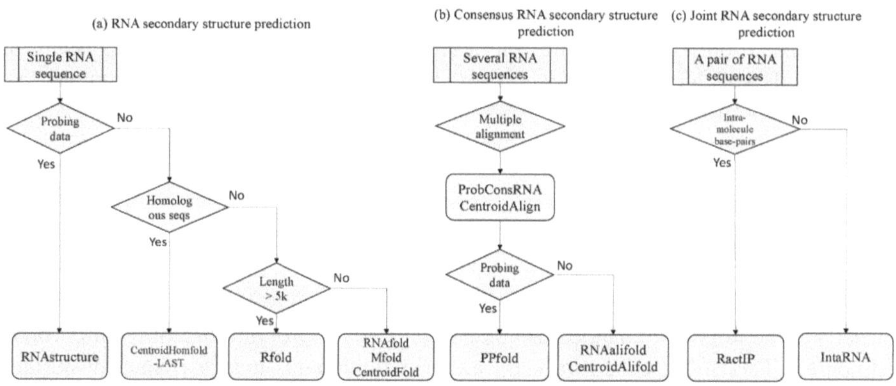

Fig. 5.2 Flowcharts of structure predictions for lncRNAs: (**a**) RNA secondary structure predictions for a single RNA sequence; (**b**) Consensus/common RNA secondary structure predictions for several RNA sequences; (**c**) Joint RNA secondary structure predictions for a pair of RNA sequences. (Adapted with permission Iwakiri et al. 2016)

2014). For the prediction of secondary structures of ncRNAs, different *in-silico* tools have been used (Iwakiri et al. 2016; Gupta et al. 2021; Mani 2021). Flowcharts of structure predictions for lncRNAs are given in Fig. 5.2 (Iwakiri et al. 2016).

Recent advancements in technology, such as microarray and NGS, have enabled the high-throughput discovery of genes implicated in the molecular pathology of Schizophrenia (Lanz et al. 2019). In 2011, Pier Paolo Pandolfi and his colleagues proposed an innovative regulatory mechanism called competing endogenous RNA (ceRNA) (Salmena et al. 2011). This hypothesis suggests that coding RNA and ncRNAs, including circRNAs, lncRNAs, and pseudogenes, can interface with each other through miRNA response elements (MREs), which are the sequences complementary to miRNAs. This interaction forms a complex regulatory network throughout the transcriptome (Ghafouri-Fard et al. 2021). The ceRNA hypothesis suggests that RNA transcripts can impact expression levels by competing for shared miRNAs, leading to a positive correlation in their expression levels. Disrupting the crosstalk between ceRNAs has been linked to a range of developmental processes and pathological conditions, including neurodegenerative diseases like Alzheimer's disease, tumorigenesis, and mental disorders such as depression and SCZ (Lang et al. 2019; Ala 2020). This highlights the importance of investigating the ceRNA regulatory network in understanding the underlying mechanisms of these conditions.

5.6 Databases of Non-coding RNAs

Recently, several databases have emerged that provide valuable information on the interactions between ncRNAs such as lncRNAs and miRNAs with proteins or genes (Rigden and Fernández 2021; Gangotia et al. 2021). These databases facilitate the acquisition of information related to bipartite interactions. Most of these databases rely on techniques such as manual or automatic text mining of documented interactions found in scientific literature. Some of these databases, such as MirGeneDB and NPInter v4.0, go beyond and include additional information like sequences.

The following databases are discussed in this section.

5.6.1 LncTarD

LncTar is a recently created repository that focuses on integrative disease–lncRNA–target interactions (http://biocc.hrbmu.edu.cn/LncTarD/or http://biobigdata.hrbmu.edu.cn/LncTarD). Based entirely on experimental data, it is a manually curated database with 2822 interactions related to 177 diseases and 475 lncRNAs. LncTar seeks to improve our comprehension of regulatory networks involved in the pathogenesis of human diseases (Zhao et al. 2020; Naipauer et al. 2021). By providing a comprehensive collection of experimentally validated disease–lncRNA–target interactions, LncTar emerges as a valuable resource for researchers and clinicians, offering new avenues for investigating the intricate molecular mechanisms underlying human diseases and potentially enabling the development of targeted therapeutic interventions.

5.6.2 LnCeVar

A database called LnCeVar (http://www.bio-bigdata.net/LnCeVar/) was created to provide genomic data on lncRNA variations that may have an impact on ceRNA interactions. This database contains a variety of variations that can affect ceRNA interactions, which includes single nucleotide polymorphisms (SNPs), copy number variations, and somatic mutations. LnCeVar makes use of a dataset derived from experimental studies as well as curated published data. LnCeVar gathers information from well-known sources like the Catalogue of Somatic Mutations in Cancer (COSMIC), the Cancer Genome Atlas (TCGA), and the 1000 Genomes Project in order to enhance its content. Users can access and explore the gathered data using the database's user-friendly interface. Users also have the choice to download the data for additional analysis. LnCeVar makes it easier to find and visualize dysregulated variation-ceRNA networks, facilitating the investigation of these intricate interactions (Xu et al. 2021). LnCeVar serves as a valuable resource for researchers and bioinformaticians, offering a comprehensive collection of lncRNA variations that have the potential to influence ceRNA interactions, thereby enabling the exploration of dysregulated variation-ceRNA networks and providing insights into the molecular mechanisms underlying complex diseases.

5.6.3 MirGeneDB

An open-source program called MirGeneDB (https://mirgenedb.org) focuses on miRNAs and offers the best nomenclature and annotation. The recent version, MirGeneDB 2.0, includes information on more than 45 different organisms, including *Mus musculus* and *Homo sapiens*. MirGeneDB forms a trustworthy source for research on miRNAs when combined with other databases like miRCarta and miRBase. A more user-friendly web interface that makes it easier to browse, search for, and download pertinent FASTA and annotation files for each organism is also part of the most recent update (Fromm et al. 2020). MirGeneDB serves as a comprehensive and reliable resource for miRNA research, providing standardized nomenclature, annotation, and access to miRNA sequences and associated information across multiple organisms, thus enhancing our understanding of the regulatory roles and potential applications of miRNAs in various biological processes and diseases.

5.6.4 miRTarBase

A database on miRNA–target interactions (MTIs) called miRTarBase (https://mirtarbase.cuhk.edu.cn/miRTarBase/miRTarBase_2022/php/index.php) has undergone experimental validation that focuses on miRNA–target interactions (MTIs). It gathers data from high-throughput technologies like CLIP-Seq. Additionally, miRTarBase incorporates information from databases like the TCGA atlas, SommamiR, miRBase, and miRSponge. MiRTarBase's main goal is to offer a vast array of validated MTIs for building networks and forecasting miRNA interactions (Huang et al. 2020, 2021). In conclusion, miRTarBase serves as a comprehensive and reliable repository of experimentally validated miRNA–target interactions, providing valuable insights into the complex regulatory networks involving miRNAs and their targets, thus facilitating the exploration of miRNA-mediated gene regulation and the development of potential therapeutic strategies for various diseases.

5.6.5 SEAweb

The small-RNA Expression Atlas, or SEAweb (https://bio.tools/SEAweb), is a web application that contains about 4200 small RNA (sRNA) sequence (miRNA, piRNA, snoRNA, snRNA, siRNA) datasets. It makes it possible to use the Oasis 2 metadata search tool to analyze published data. SEAweb distinguishes itself by integrating pathological data to identify potential associations with a wide range of datasets containing tissue-specific miRNAs across various conditions. Users can also contrast their own data with the information in the atlas which has the ability to download data on differential expression (Rahman et al. 2020; Meng et al. 2021). SEAweb serves as a valuable resource for researchers studying small RNA expression, offering a comprehensive collection of sRNA datasets and integrating pathological data, thereby enabling the identification of tissue-specific miRNAs and potential associations with different conditions, ultimately fostering a deeper understanding of small RNA-mediated regulatory processes in health and disease.

5.6.6 DIANA-LncBase

Within the DIANA tools initiative, DIANA-LncBase v3.0 (http://www.microrna.gr/LncBase) is a comprehensive repository. The information provided by this database, which focuses on miRNA–lncRNA interactions, is based on the experimental data collected from both mice and humans. The DIANA tools initiative provides applications for various other molecules, such as DNA, mRNA, and transcription factors, in addition to miRNA–lncRNA interactions. Approximately 300,000 throughput CLIP-seq (crosslinking and immunoprecipitation followed by high-throughput sequencing) datasets are used in an algorithmic approach to build the DIANA-LncBase database. In order to locate Argonaute (AGO) protein binding events, these datasets are examined. The database can now offer insightful information about miRNA–lncRNA interactions (Karagkouni et al. 2019; Perdikopanis et al. 2021). DIANA-LncBase v3.0, as part of the DIANA tools initiative, provides researchers with a comprehensive and reliable resource for exploring miRNA–lncRNA interactions, leveraging extensive experimental data and employing advanced algorithms to identify Argonaute protein binding events, thereby facilitating the investigation of the intricate regulatory roles played by miRNAs and lncRNAs in gene expression and cellular processes.

5.6.7 miRPathDB 2.0

In its brand-new release 2.0, miRPathDB (https://mpd.bioinf.uni-sb.de/) offers access to target genes and pathways linked to all miRNAs from miRBase and miRCarta for both mice and humans. The targetome suggested by this database is based on miRNAs and was developed using an integer linear program (ILP). In order to improve its functionality and integration, miRPathDB is also connected to other freely accessible resources like miRTarBase, TargetScan, and miRanda (Kehl et al. 2019; Strafella et al. 2021). In conclusion, miRPathDB 2.0 serves as a valuable resource for researchers studying miRNA-mediated gene regulation, providing comprehensive information on target genes and pathways associated with miRNAs, and integrating multiple databases and resources, thus enabling a deeper understanding of the functional implications of miRNAs in various biological processes and diseases.

5.7 In Silico Prediction of Non-coding RNAs

There are several databases have developed for the *in-silico* prediction of non-coding RNAs target. It is established based on information collected from scientific literature, diverse datasets, and nucleotide sequence information. The following databases are discussed in this section.

5.7.1 RNAInter

RNAInter (http://www.rnainter.org/) is a database that focuses on collecting and organizing the interactome data involving diverse biomolecules, with a specific emphasis on RNA–protein interactions. The database contains data that has been experimentally generated and vetted, which are used to generate predictions. Additionally, RNAInter integrates information from 35 other interaction resources through a unique pipeline. The most recent update of RNAInter was released in 2019, and it introduced several related tools and applications. One of these tools is RAID v2.0, which is linked to RNAInter and provides an integrated platform for exploring RNA-associated interactions. The update also included the integration of RIscoper (Zhang et al. 2019), IntaRNA (Mann et al. 2017), PRIdictor (Tuvshinjargal et al. 2016), and DeepBind (Alipanahi et al. 2015), which are computational tools used for analyzing and predicting RNA interactions.

RNAInter offers a comprehensive collection of nearly 40 million RNA interactions spanning 154 different species (Lin et al. 2020). These interactions include a wide spectrum of biomolecules (Sabaie et al. 2021) and give important insights into the intricate regulatory networks involving RNA molecules and their interactions with proteins and other molecules. By consolidating experimental and predicted data from various sources and applying computational tools, RNAInter serves as a valuable source for researchers who are interested in studying and understanding the interactions between RNA molecules and other biomolecules. It facilitates the exploration and analysis of RNA interactomes across different species, contributing to the advancement of knowledge in RNA biology and its functional implications.

5.7.2 oRNAment

Transcription and translation modulation are crucial processes for maintaining cellular homeostasis. These processes involve complex machinery that continuously interacts with RNA and proteins. RBPs play significant roles in regulating the metabolism of RNA and facilitating communication with other molecules (Gerstberger et al. 2014). To better understand the dynamics of RBPs, researchers from the McGill University and University of Montreal in Canada have developed the oRNAment (oRNA motifs enrichment in transcriptomes) database (http://rnabiology.ircm.qc.ca/oRNAment/). This database contains experimentally validated motifs of 223 RBPs, obtained through techniques such as RNA compete and RBNS (RNA Bind-n-Seq) platforms.

Specifically, oRNAment covers humans, *Mus musculus*, *Caenorhabditis elegans*, *Drosophila melanogaster*, and *Danio rerio* (Bouvrette et al. 2020; Zhou et al. 2020). oRNAment offers researchers with useful insights into the possible involvement of RBPs in a wide spectrum of RNA molecules, including both protein-coding and non-coding RNAs, by encompassing both coding and non-coding transcriptomes. This broader coverage helps researchers explore the intricate regulatory mechanisms involving RBPs and their interactions with various types of RNAs across different species. Overall, the oRNAment database offers a valuable resource for investigating RBP-mediated regulation of communication and RNA metabolism. Its inclusion of putative motifs across non-coding and coding transcriptomes in multiple species contributes to a deeper understanding of RBP dynamics and their functional implications in diverse biological contexts.

The oRNAment database addresses the challenges posed by the computationally intensive analysis and large data output by employing specific methodologies and thresholds. To address these concerns, the database generated all potential instances in advance utilizing high-performance computing capabilities. Furthermore, the data is kept in a cutting-edge column-oriented database management system (DBMS), allowing for the efficient retrieval and processing of massive datasets. When compared to typical data management approaches, this methodology has exhibited processing rates up to 1000 times quicker. On computers and tablets, the oRNAment database has a user-friendly design with fully interactive and responsive capabilities. Users can perform searches and explore the results through interactive figures, facilitating data interpretation and hypothesis generation. One significant aspect of oRNAment is that it is the first database to provide detailed information about the transcriptome-wide distribution features of putative RBP target motifs across multiple species. This comprehensive coverage across species enhances its utility for researchers interested in studying post-transcriptional gene regulation and designing experiments to investigate RBP-mediated processes (Bouvrette et al. 2020).

As the database evolves, future versions of oRNAment will expand its coverage to include the complete transcriptomes of additional species. Moreover, the database will continue to incorporate motifs of other RBPs as they are experimentally defined, further enriching the available resources for studying RNA–protein interactions and post-transcriptional gene regulation. Overall, oRNAment offers a powerful tool for researchers to address hypotheses, design experiments, and delve into the transcriptome-wide distribution of putative RBP target motifs. Its continuous development and expansion will provide an even more comprehensive resource for studying post-transcriptional gene regulation across various species.

5.7.3 miRDB

It is true that the free online database miRDB (https://mirdb.org) provides an improved computational model for predicting miRNA targets and providing annotations for five different species. A support vector machine (SVM) model-based prediction tool called miRTarget is part of the database. Numerous RNA studies and publicly available CLIP-seq data were used to train this model. The ability of miRTarget to calculate a probability score for each prediction is one of its standout features. The modelling tool determines this score, which is used to show the degree of statistical support for the prediction (Chen and Wang 2020; Tokumaru et al. 2021). It offers important details regarding the dependability and degree of confidence of the predicted miRNA targets. Researchers have used miRDB and its miRTarget tool extensively in the field of miRNA research to help them find potential targets and comprehend the regulatory functions of miRNAs in various organisms. Significant updates to miRDB have recently been made in order to improve its functionality. Implementing a better algorithm for miRNA target prediction is one notable improvement.

Updated transcriptome-wide target prediction data from the miRDB as a result of this update includes 3.5 million predicted targets that are regulated by 7000 miRNAs in five different species. Additionally, miRDB now comes with a web server that uses the updated prediction algorithm. This increases the database's flexibility and usability by allowing users to perform customized target prediction using their own provided sequences (Chen and Wang 2020).

The miRDB's prediction of cell-specific miRNA targets is another new feature. In order to provide tailored target prediction for particular cellular models, the database now contains expression profiles from more than 1000 different cell lines. This knowledge offers useful insights into how miRNAs work in particular cellular contexts. A new web query interface for predicting miRNA functions has also been released by miRDB.

Through the integration of target prediction data and Gene Ontology knowledge, this interface enables researchers to gain thorough understandings of the functional roles of miRNAs in particular biological processes. MiRDB's miRNA target prediction capabilities, including transcriptome-wide target prediction, customized target prediction, and prediction of cell-specific miRNA targets, have all been significantly improved by recent updates. The database's usefulness for researching miRNA functions and their regulatory networks has been increased by the addition of a web query interface for miRNA function prediction (Kozomara et al. 2019). In conclusion, miRDB, with its improved computational model, comprehensive target prediction data, and added features such as cell-specific target prediction and functional analysis, serves as a valuable tool for researchers in the field of miRNA research, providing reliable predictions and insights into the regulatory functions of miRNAs in different species and cellular contexts.

5.7.4 ENCORI: The Encyclopedia of RNA Interactomes

ENCORI, formerly known as StarBase, is a well-known database (https://ngdc.cncb.ac.cn/databasecommons/database/id/169). It functions as a comprehensive resource for integrating various RNA species-related data, primarily from high-throughput sequencing (HTS) studies. A variety of immunoprecipitated RNAs, including PAR-CLIP (Photoactivatable Ribonucleoside-Enhanced Crosslinking and Immunoprecipitation), CLIP-Seq, HITS-CLIP (HTS of RNA Isolated by Cross-Linking Immunoprecipitation), iCLIP (Individual-nucleotide Resolution Crosslinking and Immunoprecipitation), CLASH (Cross-linking, Ligation, and Sequencing of Hybrids) and furthermore, ENCORI uses gene expression data from more than 30 different cancer types, enabling pan-cancer analyses.

ENCORI primarily concentrates on the interactome data involving various RNA molecules, such as interactions involving miRNA–ncRNA, RNA binding protein (RBP)–ncRNA, miRNA–mRNA, and RBP–mRNA. Visualizing these interactions can help shed light on the regulatory networks involving these RNA species. Additional studies that are part of ENCORI include ones that analyze survival data and differentially expressed genes (Li et al. 2014; Yang et al. 2021). In the aforementioned study, the researchers developed starBase v2.0 (http://starbase.sysu.edu.cn/), an updated database, to systematically examine the regulatory interaction networks among various classes of RNAs.

The database made use of information from 108 CLIP-Seq experiments (PAR-CLIP, iCLIP, HITS-CLIP, and CLASH) derived from 37 separate studies. HTS methods have recently advanced which leads to the development of methods like CLIP-Seq, HITS-CLIP, PAR-CLIP, CLASH, and iCLIP. In these methods, RNA molecules are cross-linked to the molecules that interact with them, and then the bound RNAs are identified through immunoprecipitation and sequencing.

The precise RNA molecules that interact with miRNAs or RBPs can be found by mapping the sequencing reads to the reference genome. The identification of thousands of binding sites between miRNAs and their target RNAs, as well as between RBPs and their RNA targets, is made possible by the use of HTS in CLIP-based methods. This makes it possible to systematically and genome-wide explore the interactions between miRNAs and their target mRNAs, as well as between RBPs and different kinds of RNAs, such as ncRNAs and other functional RNA molecules, that are relevant to biology.

The researchers discovered various networks of RNA–RNA and protein–RNA interactions by examining the binding sites of millions of RBPs. They discovered 285,000 protein–RNA regulatory relationships, 9000 miRNA–circRNA regulatory relationships, and 16,000 miRNA–pseudogene regulatory relationships. The addition of extensive CLIP-Seq experimentally supported miRNA–mRNA and miRNA–lncRNA interaction networks to starBase v2.0 further increased its capabilities. These networks were created using the RNA-binding protein binding sites that had been identified (Yang et al. 2011). Around 10,000 ceRNA pairs were also discovered by the researchers from the miRNA target sites supported by CLIP-Seq data. CeRNA pairs are interactions between different RNA molecules in which one RNA molecule serves as a sponge for miRNAs, modifying the availability of miRNAs for other target RNAs (Yang et al. 2021).

Overall, starBase v2.0 offers an invaluable resource for researching the complex regulatory networks involving various classes of RNAs, including protein–RNA interactions, ncRNAs, and miRNAs. Researchers can explore and examine these interactions in relation to various biological processes and diseases.

5.7.5 NPInter v4.0

NPInter (http://bigdata.ibp.ac.cn/npinter) is indeed a well-known database that has gained recognition in the field. In 2019, NPInter launched a major update, NPInter v4.0, which expanded its curated interactions to include over 600,000 interactions. The database collects information through various approaches, including text mining and processing of experimental data from techniques like CLIP-seq, CLASH, PARIS (Psoralen Analysis of RNA Interactions and Structures), and CHIRP-seq (Chromatin Isolation by RNA Purification followed by sequencing) (Xu et al. 2021).

NPInter v4.0 operates in two primary ways. First, it retrieves data from public repositories such as GEO (Gene Expression Omnibus) and ENCODE to gather relevant information. Second, it utilizes data from RISE database (http://rise.zhanglab.net) and conducts mining of literature to obtain additional information. The main objective of NPInter is to identify and provide comprehensive annotations for interactions that occurs among ncRNAs and other biomolecules, particularly in context of diseases. The database aims to offer detailed annotation and prediction scores for these interactions, enabling researchers to gain insights into the functional roles and regulatory mechanisms of ncRNAs in various disease contexts (Teng et al. 2019). The release of NPInter v4.0 marks a significant expansion in the database's data size, achieved by incorporating recently identified ncRNA interactions reported in the data collections and literature. The interaction entries have been carefully arranged and annotated in detail, with prediction scores included. Each and every molecule involved in the interactions has been annotated with relevant identifiers, and nucleotide sequence-based searches can be performed using the basic local alignment search tool (BLAST).

In addition, ncRNA–DNA interactions and circRNA interactions obtained by ChIRP-seq data have been included into NPInter v4.0. The inclusion of ncRNA binding areas on the genome, as given by the newly added BioCircos.js module, broadens the scope of the NPInter ncRNA regulation network. Disease associations have been annotated for the molecules involved to emphasize the links between ncRNA interactions and disorders.

The updated website interface offers more convenient services to users. In comparison to similar databases like starBase and regression analysis-based inductive DNA microarray (RAID) (Yi et al. 2016), NPInter emphasizes providing detailed annotations for interactions rather than solely focusing on molecules. Visualization of modules and predictive scores have been integrated to enhance confidence in the interactions. Given the continuous advancements in high-throughput methods, which contribute to an increasing number of interactions being discovered across various organisms and cell types, NPInter is committed to regular updates and maintenance of the database. In conjunction with their online ncRNA research platform that includes NONCODE (Fang et al. 2017), Coding-Non-

Coding Index (CNCI) (Sun et al. 2013), and ncFANs (Liao et al. 2011), the goal is to provide a comprehensive and useful data source on the ncRNA interaction network. In addition, a set of web resources for RNA research, ranging from identification to function, is accessible to help researchers explore ncRNA-related topics.

5.8 Tools for Analyzing Interactions of Non-coding RNAs

There are various tools available to study non-coding RNA data. An overview of such tools, especially RNA interactions, is given in Table 5.1.

5.9 Conclusion and Future Perspectives

The detailed study on ncRNAs has revealed a complex landscape of RNA transcribed from the genome without coding for proteins. The human genome predominantly transcribes ncRNAs, which play vital roles in gene regulation at various levels. These ncRNAs, including lncRNAs, act as decoys, adaptors, guides, or regulators, influencing gene expression and protein complex assembly. They also regulate chromatin, impacting structure of chromatin and gene expression by interactions with proteins and DNA. Although the precise functions and mechanisms of many lncRNAs are still being investigated, their importance in gene regulation and disease processes is increasingly evident. NGS techniques have significantly contributed to expanding our knowledge of gene expression, regulation, and biomolecular organization in health and disease. Computational tools that facilitate experimental validation of molecular interactions and reduce operational expenses are essential. It is crucial to carefully choose appropriate computational tools depending on the input data, biological inquiries, and the information available to investigate biomolecular interactions. Various databases and tools, such as LncTarD, LnCeVar, miRTarBase, MirGeneDB, miRPathDB, SEAweb, DIANA-LncBase, RNAInter, oRNAment, miRDB, ENCORI, Pinter, etc., provide valuable resources for researchers to explore ncRNA interactions, targets, functions, and regulation. These resources contribute to the analysis and interpretation of ncRNA data, enhancing our understanding of the intricate regulatory networks involving ncRNAs. Continued research and the development of bioinformatics resources will further advance our knowledge of ncRNAs and their roles in cellular processes and disease.

Conflict of Interest None.

Acknowledgment: Authors are thankful to the Principal, Gargi College for providing the infrastructural support.

Table 5.1 Compilation of deep learning methodologies commonly employed in the field of RNomics

S. no.	Tool	Approach	Target	References
1	GCLMI (Graph Convolution for novel lncRNA–miRNA Interactions)	Convolution of an encoder and the graph	miRNA–lncRNA interactions	Huang et al. (2019)
2	RPI-SAN	Neural networks for automatic coding	Protein–ncRNA interaction pairs	Yi et al. (2018)
3	lncRScan	Examine the complex assemblies for lncRNA	Distinguish lncRNA from mRNAs	Sun et al. (2012)
4	LncRNA2 Function	On the basis of idea that genes with similar expression patterns under various conditions may have related biological pathways and functions, annotate lncRNA	Biological pathway enrichment	Jiang et al. (2015)
5	DeePathology	Deep neural networks	Prediction of the mRNA–miRNA interactions origin	Azarkhalili et al. (2019)
6	DeepTarget	In order to train a deep recurrent neural network for auto encoding and sequence-sequence interaction learning, expression data is used	Identifying mRNA–miRNA interactions	Lee et al. (2016)
7	RPITER	Stacked auto-encoder (SAE) and Convolution neural network (CNN)	The origin of mRNA–miRNA interactions can be predicted	Peng et al. (2019)
8	deepMirGene	Training recurrent neural networks (RNNs) with expression data, in particular long short-term memory (LSTM) networks	A systematic process to identify precursor miRNAs	Park et al. (2016)

References

Ala U (2020) Competing endogenous RNAs, non-coding RNAs and diseases: an intertwined story. Cells 9:1574. https://doi.org/10.3390/cells9071574

Alipanahi B, Delong A, Weirauch MT, Frey BJ (2015) Predicting the sequence specificities of DNA- and RNA-binding proteins by deep learning. Nat Biotechnol 33:831–838. https://doi.org/10.1038/nbt.3300

Ariel F et al (2020) R-loop mediated transaction of the APOLO long noncoding RNA. Mol Cell 77: 1055–1065.e4

Azarkhalili B, Saberi A, Chitsaz H, Sharifi-Zarchi A (2019) DeePathology: deep multi-task learning for inferring molecular pathology from cancer transcriptome. Sci Rep 9:16526. https://doi.org/10.1038/s41598-019-52937-5

Blank-Giwojna A, Postepska-Igielska A, Grummt I (2019) lncRNA KHPS1 activates a poised enhancer by triplex-dependent recruitment of Epigenomic regulators. Cell Rep 26(11): 2904–2915.e4. https://doi.org/10.1016/j.celrep.2019.02.059

Bonetti A, Agostini F, Suzuki AM, Hashimoto K, Pascarella G et al (2020) RADICL-seq identifies general and cell type-specific principles of genome-wide RNA-chromatin interactions. Nat Commun 11(1):1018. https://doi.org/10.1038/s41467-020-14337-6. Erratum in: Nat Commun. 2021 May 19;12(1):3128

Bouvrette LPB, Bovaird S, Blanchette M, Lécuyer E (2020) oRNAment: a database of putative RNA binding protein target sites in the transcriptomes of model species. Nucleic Acids Res 48 (D1):D166–D173. https://doi.org/10.1093/nar/gkz986

Chan JJ, Tay Y (2018) Noncoding RNA:RNA regulatory networks in cancer. Int J Mol Sci 19:1310. https://doi.org/10.3390/ijms19051310

Chen Y, Wang X (2020) miRDB: an online database for prediction of functional microRNA targets. Nucleic Acids Res 48(D1):D127–D131. https://doi.org/10.1093/nar/gkz757

Cheng C, Moore J, Greene C (2014) Applications of bioinformatics to non-coding RNAs in the era of next-generation sequencing. Pac Symp Biocomput 412–6

Dahariya S, Paddibhatla I, Kumar S, Raghuwanshi S, Pallepati A, Gutti RK (2019) Long non-coding RNA: classification, biogenesis and functions in blood cells. Mol Immunol 112: 82–92. https://doi.org/10.1016/j.molimm.2019.04.011

Denham AN, Drake J, Gavrilov M, Taylor ZN, Bacanu S-A, Vladimirov VI (2022) Long non-coding RNAs: the new frontier into understanding the etiology of alcohol use disorder. Noncoding RNA 8:59. https://doi.org/10.3390/ncrna8040059

Dueva R et al (2019) Neutralization of the positive charges on histone tails by RNA promotes an open chromatin structure. Cell Chem Biol 26:1436–1449.e5

Fang S, Zhang L, Guo J, Niu Y, Wu Y, Li H, Zhao L, Li X, Teng X, Sun X et al (2017) NONCODEV5: a comprehensive annotation database for long non-coding RNAs. Nucleic Acids Res 46:D308–D314

Fernandes JCR, Acuña SM, Aoki JI, Floeter-Winter LM, Muxel SM (2019) Long non-coding RNAs in the regulation of gene expression: physiology and disease. Noncoding RNA 5:17. https://doi.org/10.3390/ncrna5010017

Fromm B, Domanska D, Høye E, Ovchinnikov V, Kang W, Aparicio-Puerta E, Johansen M, Flatmark K, Mathelier A, Hovig E et al (2020) MirGeneDB 2.0: the metazoan microRNA complement. Nucleic Acids Res 48:D132–D141. https://doi.org/10.1093/nar/gkz885

Gangotia D, Gupta A, Mani I (2021) Role of bioinformatics in biological sciences. In: Singh V, Kumar A (eds) Advances in bioinformatics. Springer, Singapore. https://doi.org/10.1007/978-981-33-6191-1_3

Gerstberger S, Hafner M, Tuschl T (2014) A census of human RNA-binding proteins. Nat Rev Genet 15:829–845. https://doi.org/10.1038/nrg3813

Ghafouri-Fard S et al (2021) A review on the expression pattern of non-coding RNAs in patients with schizophrenia: with a special focus on peripheral blood as a source of expression analysis. Front Psychiatry 12:640463. https://doi.org/10.3389/fpsyt.2021.640463

Gibbons HR, Shaginurova G, Kim LC, Chapman N, Spurlock CF 3rd, Aune TM (2018) Divergent lncRNA *GATA3-AS1* regulates *GATA3* transcription in T-helper 2 cells. Front Immunol 9:2512. https://doi.org/10.3389/fimmu.2018.02512

Goodrich JA, Kugel JF (2006) Non-coding-RNA regulators of RNA polymerase II transcription. Nat Rev Mol Cell Biol 7:612–616

Grillone K, Riillo C, Scionti F, Rocca R, Tradigo G, Guzzi PH, Alcaro S, Di Martino MT, Tagliaferri P, Tassone P (2020) Non-coding RNAs in cancer: platforms and strategies for investigating the genomic "dark matter". J Exp Clin Cancer Res 39:1–19. https://doi.org/10.1186/s13046-020-01622-x

Grote P et al (2013) The tissue-specific lncRNA Fendrr is an essential regulator of heart and body wall development in the mouse. Dev Cell 24:206–214

Gupta A, Gangotia D, Mani I (2021) Bioinformatics tools and software. In: Singh V, Kumar A (eds) Advances in bioinformatics. Springer, Singapore. https://doi.org/10.1007/978-981-33-6191-1_2

Hansen TB, Jensen TI, Clausen BH, Bramsen JB, Finsen B, Damgaard CK, Kjems J (2013) Natural RNA circles function as efficient microRNA sponges. Nature 495:384–388. PubMed: 23446346

Hombach S, Kretz M (2016) Non-coding RNAs: classification, biology and functioning. Adv Exp Med Biol 937:3–17. https://doi.org/10.1007/978-3-319-42059-2_1

Huang Y-A, Huang Z-A, You Z-H, Zhu Z, Huang W-Z, Guo J-X, Yu C-Q (2019) Predicting lncRNA-miRNA interaction via graph convolution auto-encoder. Front Genet 10:758. https://doi.org/10.3389/fgene.2019.00758

Huang H-Y, Lin Y-C-D, Li J, Huang K-Y, Shrestha S, Hong H-C, Tang Y, Chen Y-G, Jin C-N, Yu Y et al (2020) miRTarBase 2020: updates to the experimentally validated microRNA–target interaction database. Nucleic Acids Res 48:D148–D154. https://doi.org/10.1093/nar/gkz896

Huang J, Song N, Xia L, Tian L, Tan J, Chen Q, Zhu J, Wu Q (2021) Construction of lncRNA-related competing endogenous RNA network and identification of hub genes in recurrent implantation failure. Reprod Biol Endocrinol 19:108. https://doi.org/10.1186/s12958-021-00778-1

Isoda T, Moore AJ, He Z, Chandra V, Aida M, Denholtz M, Piet van Hamburg J, Fisch KM, Chang AN, Fahl SP, Wiest DL, Murre C (2017) Non-coding transcription instructs chromatin folding and compartmentalization to dictate enhancer-promoter communication and T cell fate. Cell 171(1):103–119.e18. https://doi.org/10.1016/j.cell.2017.09.001

Iwakiri J, Hamada M, Asai K (2016) Bioinformatics tools for lncRNA research. Biochim Biophys Acta 1859(1):23–30. https://doi.org/10.1016/j.bbagrm.2015.07.014

Jiang Q, Ma R, Wang J et al (2015) LncRNA2Function: a comprehensive resource for functional investigation of human lncRNAs based on RNA-seq data. BMC Genomics 16(3):S2

Karagkouni D, Paraskevopoulou MD, Tastsoglou S, Skoufos G, Karavangeli A, Pierros V, Zacharopoulou E, Hatzigeorgiou AG (2019) DIANA-LncBase v3: indexing experimentally supported miRNA targets on non-coding transcripts. Nucleic Acids Res 48:D101–D110. https://doi.org/10.1093/nar/gkz1036

Kehl T, Kern F, Backes C, Fehlmann T, Stöckel D, Meese E, Lenhof H-P, Keller A (2019) miRPathDB 2.0: a novel release of the miRNA pathway dictionary database. Nucleic Acids Res 48:D142–D147. https://doi.org/10.1093/nar/gkz1022

Kino T, Hurt DE, Ichijo T, Nader N, Chrousos GP (2010) Noncoding RNA gas5 is a growth arrest- and starvation-associated repressor of the glucocorticoid receptor. Sci Signal 3:ra8

Kozomara A, Birgaoanu M, Griffiths-Jones S (2019) miRBase: from microRNA sequences to function. Nucleic Acids Res 47:D155–D162

Krahn N, Fischer JT, Söll D (2020) Naturally occurring tRNAs with non-canonical structures. Front Microbiol 11:596914. https://doi.org/10.3389/fmicb.2020.596914

Lang Y, Zhang J, Yuan Z (2019) Construction and dissection of the ceRNA-ceRNA network reveals critical modules in depression. Mol Med Rep 19:3411–3420. https://doi.org/10.3892/mmr.2019.10009

Lanz TA et al (2019) Postmortem transcriptional profiling reveals widespread increase in inflammation in schizophrenia: a comparison of prefrontal cortex, striatum, and hippocampus among matched tetrads of controls with subjects diagnosed with schizophrenia, bipolar or major depressive disorder. Transl Psychiatry 9:151. https://doi.org/10.1038/s41398-019-0492-8

Lee B, Baek J, Park S, Yoon S deepTarget: end-to-end learning framework for microrna target prediction using deep recurrent neural networks. In: Proceedings of the 7th ACM international conference on bioinformatics, computational biology, and health informatics; Seattle, WA, USA. 2 October–5 October 2016, pp 434–442

Li J, Liu C (2019) Coding or noncoding, the converging concepts of RNAs. Front Genet 10:496. https://doi.org/10.3389/fgene.2019.00496

Li J-H, Liu S, Zhou H, Qu L-H, Yang J-H (2014) starBase v2.0: decoding miRNA-ceRNA, miRNA-ncRNA and protein–RNA interaction networks from large-scale CLIP-Seq data. Nucleic Acids Res 42:D92–D97. https://doi.org/10.1093/nar/gkt1248

Li Y, Syed J, Sugiyama H (2016) RNA–DNA triplex formation by long noncoding RNAs. Cell Chem Biol 23:1325–1333

Liao Q, Xiao H, Bu D, Xie C, Miao R, Luo H, Zhao G, Yu K, Zhao H, Skogerbø G et al (2011) ncFANs: a web server for functional annotation of long non-coding RNAs. Nucleic Acids Res 39:W118–W124

Lin Y, Liu T, Cui T, Wang Z, Zhang Y, Tan P, Huang Y, Yu J, Wang D (2020) RNAInter in 2020: RNA interactome repository with increased coverage and annotation. Nucleic Acids Res 48: D189–D197. https://doi.org/10.1093/nar/gkz804

Luo S et al (2016) Divergent lncRNAs regulate gene expression and lineage differentiation in pluripotent cells. Cell Stem Cell 18:637–652

Maldonado R, Schwartz U, Silberhorn E, Langst G (2019) Nucleosomes stabilize ssRNA–dsDNA triple helices in human cells. Mol Cell 73:1243–1254.e6

Mani I (2021) Role of bioinformatics in MicroRNA analysis. In: Singh V, Kumar A (eds) Advances in bioinformatics. Springer, Singapore. https://doi.org/10.1007/978-981-33-6191-1_19

Mann M, Wright PR, Backofen R (2017) IntaRNA 2.0: enhanced and customizable prediction of RNA-RNA interactions. Nucleic Acids Res 45(W1):W435–W439. https://doi.org/10.1093/nar/gkx279

Mattick JS (2009) The genetic signatures of noncoding RNAs. PLoS Genet 5:e1000459

Mattick JS, Makunin IV (2006) Non-coding RNA. Hum Mol Genet 15 Spec No 1:R17–R29. https://doi.org/10.1093/hmg/ddl046

Meng Q, Chu Y, Shao C, Chen J, Wang J, Gao Z, Yu J, Kang Y (2021) Roles of host small RNAs in the evolution and host tropism of coronaviruses. Brief Bioinform 22:1096–1105. https://doi.org/10.1093/bib/bbab027

Naipauer J, Solá MEG, Salyakina D, Rosario S, Williams S, Coso O, Abba MC, Mesri EA, Lacunza E (2021) A non-coding RNA network involved in KSHV tumorigenesis. Front Oncol 11: 687629. https://doi.org/10.3389/fonc.2021.687629

Natsidis P, Schiffer PH, Salvador-Martínez I, Telford MJ (2019) Computational discovery of hidden breaks in 28S ribosomal RNAs across eukaryotes and consequences for RNA integrity numbers. Sci Rep 9:1944. https://doi.org/10.1038/s41598-019-55573-1

Niehrs C, Luke B (2020) Regulatory R-loops as facilitators of gene expression and genome stability. Nat Rev Mol Cell Biol 21(3):167–178. https://doi.org/10.1038/s41580-019-0206-3

O'Leary VB et al (2015) PARTICLE, a triplex-forming long ncRNA, regulates locus-specific methylation in response to low-dose irradiation. Cell Rep 11:474–485

Park S, Min S, Choi H, Yoon S (2016) deepMiRGene: deep neural network based precursor MicroRNA prediction. arXiv:1605.00017. https://doi.org/10.48550/arXiv.1605.00017

Peng C, Han S, Zhang H, Li Y (2019) RPITER: a hierarchical deep learning framework for ncRNA–protein interaction pre-diction. Int J Mol Sci 20:1070. https://doi.org/10.3390/ijms20051070

Perdikopanis N, Georgakilas GK, Grigoriadis D, Pierros V, Kavakiotis I, Alexiou P, Hatzigeorgiou A (2021) DIANA-miRGen v4: indexing promoters and regulators for more than 1500 microRNAs. Nucleic Acids Res 49:D151–D159. https://doi.org/10.1093/nar/gkaa1060

Pertea M (2012) The human transcriptome: an unfinished story. Genes (Basel) 3(3):344–360. https://doi.org/10.3390/genes3030344

Rahman R-U, Liebhoff A-M, Bansal V, Fiosins M, Rajput A, Sattar A, Magruder DS, Madan S, Sun T, Gautam A et al (2020) SEAweb: the small RNA expression atlas web application. Nucleic Acids Res 48:D204–D219. https://doi.org/10.1093/nar/gkz869

Rigden DJ, Fernández XM (2021) The 2021 nucleic acids research database issue and the online molecular biology database collection. Nucleic Acids Res 49:D1–D9. https://doi.org/10.1093/nar/gkaa1216

Rinn JL, Chang HY (2012) Genome regulation by long noncoding RNAs. Annu Rev Biochem 81: 145–166
Sabaie H, Moghaddam MM, Moghaddam MM, Ahangar NK, Asadi MR, Hussen BM, Taheri M, Rezazadeh M (2021) Bioinformatics analysis of long non-coding RNA-associated competing endogenous RNA network in schizophrenia. Sci Rep 11(1):24413. https://doi.org/10.1038/s41598-021-03993-3
Saldana-Meyer R et al (2019) RNA interactions are essential for CTCF-mediated genome organization. Mol Cell 76:412–422.e5
Salmena L, Poliseno L, Tay Y, Kats L, Pandolfi PP (2011) A ceRNA hypothesis: the Rosetta stone of a hidden RNA language? Cell 146:353–358. https://doi.org/10.1016/j.cell.2011.07.014
Schertzer MD et al (2019) lncRNA-induced spread of polycomb controlled by genome architecture, RNA abundance, and CpG Island DNA. Mol Cell 75:523–537.e10
Schmitz KM, Mayer C, Postepska A, Grummt I (2010) Interaction of noncoding RNA with the rDNA promoter mediates recruitment of DNMT3b and silencing of rRNA genes. Genes Dev 24: 2264–2269
Seila AC, Calabrese JM, Levine SS, Yeo GW, Rahl PB, Flynn RA, Young RA, Sharp PA (2008) Divergent transcription from active promoters. Science 322(5909):1849–1851. https://doi.org/10.1126/science.1162253
Sikora M, Marycz K, Smieszek A (2020) Small and long non-coding RNAs as functional regulators of bone homeostasis, acting alone or cooperatively. Mol Ther Nucleic Acids 21:792–803. https://doi.org/10.1016/j.omtn.2020.07.017
Strafella C, Caputo V, Termine A, Fabrizio C, Ruffo P, Potenza S, Cusumano A, Ricci F, Caltagirone C, Giardina E et al (2021) Genetic determinants highlight the existence of shared etiopathogenetic mechanisms characterizing age-related macular degeneration and neurodegenerative disorders. Front Neurol 12:626066. https://doi.org/10.3389/fneur.2021.626066
Sun L, Zhang Z, Bailey TL et al (2012) Prediction of novel long non-coding RNAs based on RNA-Seq data of mouse Klf1 knockout study. BMC Bioinformatics 13(1):331
Sun L, Luo H, Bu D, Zhao G, Yu K, Zhang C, Liu Y, Chen R, Zhao Y (2013) Utilizing sequence intrinsic composition to classify protein-coding and long non-coding transcripts. Nucleic Acids Res 41:e166
Tan-Wong SM, Dhir S, Proudfoot NJ (2019) R-loops promote antisense transcription across the mammalian genome. Mol Cell 76:600–616.e6
Teng X, Chen X, Xue H, Tang Y, Zhang P, Kang Q, Hao Y, Chen R, Zhao Y, He S (2019) NPInter v4.0: an integrated database of ncRNA interactions. Nucleic Acids Res 48:D160–D165. https://doi.org/10.1093/nar/gkz969
Tokumaru Y, Oshi M, Patel A, Katsuta E, Yan L, Angarita FA, Dasgupta S, Nagahashi M, Matsuhashi N, Futamura M et al (2021) Low expression of miR-195 is associated with cell proliferation, glycolysis and poor survival in estrogen receptor (ER)-positive but not in triple negative breast cancer. Am J Cancer Res 11:3320–3334
Tuvshinjargal N, Lee W, Park B, Han K (2016) PRIdictor: protein-RNA interaction predictor. Biosystems 139:17–22. https://doi.org/10.1016/j.biosystems.2015.10.004
Wang B, Kumar V, Olson A, Ware D (2019) Reviving the transcriptome studies: an insight into the emergence of single-molecule transcriptome sequencing. Front Genet 10:384. https://doi.org/10.3389/fgene.2019.00384
Winkle M, El-Daly SM, Fabbri M, Calin GA (2021) Noncoding RNA therapeutics—challenges and potential solutions. Nat Rev Drug Discov 20:629–651. https://doi.org/10.1038/s41573-021-00219-z
Wu H, Yang L, Chen LL (2017) The diversity of long noncoding RNAs and their generation. Trends Genet 33(8):540–552. https://doi.org/10.1016/j.tig.2017.05.004
Xu D, Wang L, Pang S, Cao M, Wang W, Yu X, Xu Z, Xu J, Wang H, Lu J et al (2021) The functional characterization of epigenetically related lncRNAs involved in dysregulated CeRNA–CeRNA networks across eight cancer types. Front Cell Dev Biol 9:649755. https://doi.org/10.3389/fcell.2021.649755

Yamamura S, Imai-Sumida M, Tanaka Y, Dahiya R (2018) Interaction and cross-talk between non-coding RNAs. Cell Mol Life Sci 75:467–484. https://doi.org/10.1007/s00018-017-2626-6
Yang JH, Li JH, Shao P, Zhou H, Chen YQ, Qu LH (2011) starBase: a database for exploring microRNA-mRNA interaction maps from Argonaute CLIP-Seq and Degradome-Seq data. Nucleic Acids Res 39(Database issue):D202–D209. https://doi.org/10.1093/nar/gkq1056
Yang X, Tian M, Zhang W, Chai T, Shen Z, Kang M, Lin J (2021) Identification of potential core genes in esophageal carcinoma using bioinformatics analysis. Medicine 100:e26428. https://doi.org/10.1097/MD.0000000000026428
Yi Y, Zhao Y, Li C, Zhang L, Huang H, Li Y, Liu L, Hou P, Cui T, Tan P et al (2016) RAID v2.0: an updated resource of RNA-associated interactions across organisms. Nucleic Acids Res 45: D115–D118
Yi H-C, You Z, Huang D-S, Li X, Jiang T-H, Li L-P (2018) A deep learning framework for robust and accurate prediction of ncRNA-protein interactions using evolutionary information. Mol Ther Nucleic Acids 11:337–344. https://doi.org/10.1016/j.omtn.2018.03.001
Zhang K, Shi ZM, Chang YN, Hu ZM, Qi HX, Hong W (2014) The ways of action of long non-coding RNAs in cytoplasm and nucleus. Gene 547(1):1–9. https://doi.org/10.1016/j.gene.2014.06.043
Zhang Y, Liu T, Chen L, Yang J, Yin J, Zhang Y, Yun Z, Xu H, Ning L, Guo F et al (2019) RIscoper: a tool for RNA–RNA interaction extraction from the literature. Bioinformatics 35: 3199–3202. https://doi.org/10.1093/bioinformatics/btz044
Zhao H, Shi J, Zhang Y, Xie A, Yu L, Zhang C, Lei J, Xu H, Leng Z, Li T et al (2020) LncTarD: a manually-curated database of experimentally-supported functional lncRNA–target regulations in human diseases. Nucleic Acids Res 48:D118–D126. https://doi.org/10.1093/nar/gkz985
Zhao C, Xie W, Zhu H, Zhao M, Liu W, Wu Z, Wang L, Zhu B, Li S, Zhou Y, Jiang X, Xu Q, Ren C (2022) LncRNAs and their RBPs: how to influence the fate of stem cells? Stem Cell Res Ther 13(1):175. https://doi.org/10.1186/s13287-022-02851-x
Zhou Y-K, Shen Z-A, Yu H, Luo T, Gao Y, Du P-F (2020) Predicting lncRNA–protein interactions with miRNAs as mediators in a heterogeneous network model. Front Genet 10:1341. https://doi.org/10.3389/fgene.2019.01341

Online Resources

http://www.microrna.gr/LncBase
http://www.bio-bigdata.net/LnCeVar/
http://biocc.hrbmu.edu.cn/LncTarD/
http://bio-bigdata.hrbmu.edu.cn/LncTarD
https://mirgenedb.org/
https://mpd.bioinf.uni-sb.de/
https://mirtarbase.cuhk.edu.cn/~miRTarBase/miRTarBase_2022/php/index.php
https://bio.tools/SEAweb
https://mirdb.org/
https://ngdc.cncb.ac.cn/databasecommons/database/id/169
http://starbase.sysu.edu.cn/
http://bigdata.ibp.ac.cn/npinter
http://www.rnainter.org/
http://rnabiology.ircm.qc.ca/oRNAment/
http://rise.zhanglab.net

Next Generation Sequencing in Healthcare 6

Duy Ha Nguyen, Yen Vy Nguyen Thi, and Dinh-Toi Chu

Abstract

With the fast development and broad application of next-generation sequencing (NGS) technology, data on genomic sequences is now reaching the aims of solving the mystery of life, producing better crops, detecting infections, and improving quality of life. NGS approaches have greatly sped human genome decipherment and extended our understanding of genetic variants, disease causes, and evolutionary linkages. NGS is based on the accordance sequencing of millions of DNA units, which produces vast amounts of sequence data. This technology has accelerated the transition away from the Sanger sequencing method, providing several benefits like greater speed, throughput, and lower costs. Current sequencing methods, including short-read and long-read analysis, focus on the clinical use of NGS in genetic disorders, cancers, infectious diseases, and pharmacokinetic domains. With a wide range of applications, NGS has been increasingly employed as the gold standard in diagnosis and therapeutic treatments, as well as in prognosis, particularly in uncommon diseases. Despite its obvious benefits, NGS has met certain obstacles. Data processing, storage volume, and clinical interpretation continue to be major obstacles, necessitating

D. H. Nguyen
Vietnam Military Medical University, Hanoi, Vietnam

Y. V. N. Thi
Center for Biomedicine and Community Health, International School, Vietnam National University, Hanoi, Vietnam

D.-T. Chu (✉)
Center for Biomedicine and Community Health, International School, Vietnam National University, Hanoi, Vietnam

Faculty of Applied Sciences, International School, Vietnam National University, Hanoi, Vietnam
e-mail: toicd@vnu.edu.vn

V. Singh, A. Kumar (eds.), *Advances in Bioinformatics*,
https://doi.org/10.1007/978-981-99-8401-5_6

the use of strong computing techniques and infrastructure. Furthermore, standardization and quality control methods are required to guarantee that results are reproducible and comparable across laboratories and sequencing platforms.

Keywords

Next-generation sequencing (NGS) · Genomic sequence · Whole-genome sequencing (WGS)

Abbreviations

CRC Colorectal cancer
NGS Next-generation sequencing
WGS Whole genome sequencing

6.1 Introduction

Next-generation sequencing (NGS) is a low-cost and rapid technology for sequencing DNA or RNA. NGS is also known as high-throughput sequencing. The NGS technology permits the concurrent sequencing of millions of DNA or RNA fragments (Matthijs et al. 2016). Instead of using size separation to organize fluorescent molecules, NGS employs positional segregation, in which millions of unique template DNA sequences attach to various regions on the slide and remain fixed there during the sequencing process (Muzzey et al. 2015). The NGS method includes sample preparation, library construction, shredding DNA/RNA into small pieces, concatenating these fragments into clusters, and sequencing them utilizing procedures (Gu et al. 2019). NGS is a vital tool in molecular biology and biomedicine. It has resulted in significant advances in DNA and RNA sequencing, offered insights into genetic variations and genetic patterns, and helped to a greater understanding of mass structure and function. This technology has largely superseded the conventional Sanger sequencing method and is now widely used in biological research, genetic medicine, and a variety of other sectors. Since the development of NGS technology, the number of newly discovered disease-related genes has increased considerably across all fields of medicine. Over time, the quantity of data in the Online Human Mendelian Genetics (OMIM) collection for which the genetic basis of a certain phenotype is known has grown. This information explosion has happened as an effect of NGS's capacity to fast and sensitively sequence any region of a person's genome, from a few genes to the complete genome (Fernandez-Marmiesse et al. 2018). The ability of NGS to elucidate the full spectrum of variants in a given individual will also encourage the discovery of genetic or polygenic causes of disease because, once the pathogenic mutation has been identified, additional analyses can be performed. After all, the data is already available. NGS is crucial in understanding the genomic underpinnings of cancer (Tuna and Amos 2013; Stephens et al. 2009). It allows for full analysis of the tumor genome, as well as the discovery of somatic mutations, copy number changes, and structural variations. NGS allows for the quick and precise identification of pathogens such as bacteria, viruses, fungi, and parasites (Murase et al. 1995; Revez et al. 2017). It can detect both known and unknown diseases, discover antibiotic resistance genes, and shed light on transmission dynamics. NGS enables thorough profiling of the microbiome, or microbial populations that dwell in the human body. It aids in understanding the function of the microbiome in health and disease, such as obesity, mental health issues and inflammatory bowel disease. Metagenomic sequencing using NGS gives information on microbial diversity, functional potential, and host interactions (Wensel et al. 2022). NGS can detect genetic variations that influence drug response and metabolism (Cousin et al. 2017). This data facilitates precision medicine by directing drug selection, adjusting doses, and reducing adverse responses. NGS-based pharmacokinetic studies may enhance treatment results and decrease adverse effects.

6.2 Advances in Next Generation Sequencing

NGS has been flourishing in recent years. With the outstanding superiority of reducing the cost of DNA sequencing, producing large numbers of DNA readings, and lengths ranging from 25 to more than 750 bp (Barba et al. 2014). NGS originated from the work of two major American and British scientists, James Watson and Francis Crick, in 1953 who found the structure of the DNA double helix (Watson and Crick 1953). In 1964 and 1965, Robert Holley, an American scientist, developed a method for sequencing tRNA, which was the first to sequence nucleic acids (Holley et al. 1964, 1965). A method of sequencing long DNA was also developed independently in 1977 by two British and American scientists, Frederick Sanger and Walter Gilbert (Sanger et al. 1977; Maxam and Gilbert 1977).

Robert Holley solved the sequence of the first RNA molecule in 1964. Specifically, in this research he identified the complete sequence and structural morphology of 77 ribonucleotides, from which he opened up many opportunities for other scientists to discover and determine the sequence of DNA and RNA molecules. An analysis of the nucleotide sequence of 24 bp within the 27 bp of the lac operon in human DNA was published in 1973 (Gilbert and Maxam 1973). Frederick Sanger sequenced the complete genome of phages by 1977 (Sanger et al. 1977).

Over the past 3 decades, DNA sequencing has had different milestones (Table 6.1), especially in important milestones such as:

In 1990 A project considered as the beginning of a breakthrough and development of NGS is the project on the human genome. The project has already been launched and scientists are expected to last for 15 years. The genome project involves many different countries such as the United States, France, the United Kingdom, Germany, India, Japan and China (Barba et al. 2014).

Table 6.1 Milestones in NGS over the past three decades

Time	Author	Contents
1984	Medical Research Council	Successfully sequenced the DNA of the *Epstein-Barr virus* (EBV), the cause of mononucleosis, with a length of 172.282 bp The method used is sequencing dideoxynucleotide/M13 EBV (Baer et al. 1984)
1986	California Institute of Technology	Announcing the invention of the world's first semi-automatic DNA sequencing machine Automate the enzyme sequence termination process for the purpose of analyzing Sanger DNA sequences (Barba et al. 2014)
1987	Applied Biosystems	Bringing to market an improved machine, faster sequencing than the original ABI 370 model (Barba et al. 2014)
1990	The United States, France, the United Kingdom, Germany, India, Japan and China	The project on the human genome (Barba et al. 2014)
1990	U.S. National Institutes of Health	Larger-scale sequencing of subjects *Mycoplasma capricolum, Saccharomyces cerevisiae, Escherichia coli*
1995	Craig Venter, Hamilton Smith et al.	Successfully solved the complete genome completion of the bacterium *Haemophilus influenza* This species has a round chromosome containing 1.830.137 bp (Fleischmann et al. 1995)
1996	Pal Nyren và Mostafa Ronaghi	Accelerate the development of methods for burning DNA without electrophoresis (Ronaghi et al. 1996)
1998	Eric Kawashima, Laurent Farinelli và Pascal Mayer	"Method of nucleic acid amplification"
1998	Joint British and American projects	Sequencing the whole genome of the nematode *Caenorhabditis elegans*. Specifically, the 97-megabase pair revealed more than 19,000 genes (Barba et al. 2014)
2000	Many countries	Complete human genome project
2001–2003	Many countries	Completion of the human genome project with 3.3 billion base pairs, about 23,000 genes

In 1998 A group of 3 scientists including Eric Kawashima, Laurent Farinelli, and Pascal Mayer developed the "Nucleic Acid Amplification Method". This method provides information describing the colonial sequence of DNA and was an important basement in the development of later technologies such as parallel sequencing. This method has been patented WO 98/44151 (Kawashima et al. 1998).

In 2000 Expanding international cooperation and promoting the application of the latest sequencing analysis technologies, the latest computer technology Human genome project was completed (Barba et al. 2014).

From 2001 to 2003 New findings in the draft and the complete human genome were published with 3.3 billion base pairs and about 23,000 genes (Barba et al. 2014).

6.3 Next Generation Sequencing Technologies

NGS has evolved rapidly over the past 15 years and new methods have constantly been developed and commercialized in Fig. 6.1. The Sanger dideoxy synthesis and the Maxam–Gilbert chemical separation method were considered the founding methods in DNA sequencing (Sanger et al. 1977; Sanger and Coulson 1975; Maxam and Gilbert 1980). Following was the second-generation sequencing method. The use of the second and third terms to describe these methods was the next generation of the Sanger method (Maxam and Gilbert 1980). Developing a method for sequencing DNA fragments by hybridization with DNA oligonucleotides arranged in a known sequence on filters in 1980. It is used to diagnose genetic diseases or chromosomal abnormalities (Drmanac et al. 2002; Hanna et al. 2000). The method does not use dideoxy terminals and relies on allowing nucleotide synthesis reactions to proceed normally and in combination with imaging of nucleotides and then removing synthetic blocking radicals on labeled labels, that method is called Sequencing by Synthesis (SBS) (Fuller et al.

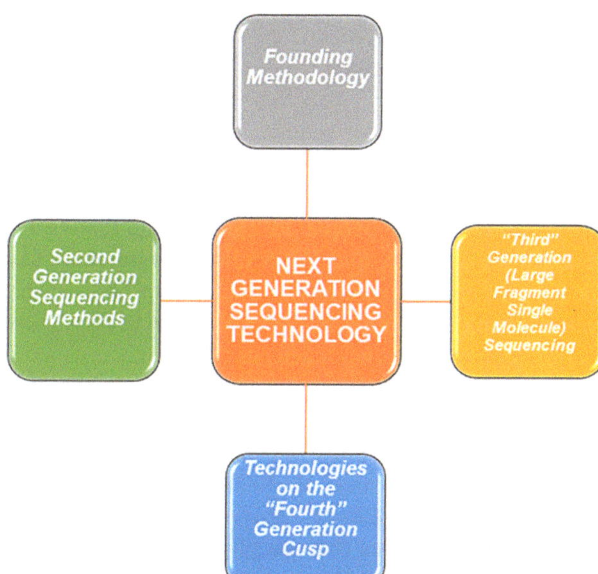

Fig. 6.1 Next generation sequencing technologies

2009; Mardis 2008). Building on the limitations of second-generation sequencing, the third-generation (large single-piece molecule) method was born with the most important purpose of sequencing long DNA (and RNA) molecules (Slatko et al. 2018). And the most modern is the "fourth" Cusp technology, with which the user can pass long DNA molecules through special "holes" of small diameter and then measure different currents as each nucleotide passes through the use of a bond detector (Slatko et al. 2018).

With the development and improvement of NGS, it is possible to solve the entire human genome within a short time and at an appropriate price. Therefore, higher requirements to manage, analyze and interpret big data sources NGS requires computational skills and bioinformatics (Pereira et al. 2020). The use of bioinformatics supports important steps such as: processing raw data to help analyze detailed data to explain clinical variations using computations (hardware), algorithms and software applications during operation (Pereira et al. 2020). Bioinformatics in NGS is divided into several levels: primary, secondary, and tertiary analysis. The analysis results of each level are basically the same on NGS platforms. However, each platform will also have its own characteristics (Pereira et al. 2020).

In addition, NGS still has some limitations such as: the sequencing process still has an error rate and many complicated operations that lead to a lot of waiting time and the accuracy rate is not yet reached the absolute value (Barba et al. 2014). There are also some requirements in NGS such as: use smaller platforms, use less energy or maybe if the platform can run on batteries, use less reagents and can be used in healthcare, ecology, agriculture (Barba et al. 2014). Strengths in robotics, liquid handling technology and samples can be applied to make NGS more advanced and superior (Barba et al. 2014).

6.4 The Application of Next Generation Sequencing in Healthcare

Cancer Whole-genome sequencing (WGS) enables researchers to identify all point alterations and structure rearrangements, which were previously both expensive and ineffectual due to their ability to target just specific traits. Some germline genome changes predispose people to cancer, but the vast majority of cancer genome changes are somatic, and WGS allows researchers to identify all point mutations, indels, and reorganizations of structures in both germ and somatic cells that cause cancer (Tuna and Amos 2013). WGS aids in the detection of other disruptive gene rearrangement patterns, such as tandem duplication, inversion, and deletion. Rearrangements have been discovered in numerous well-known cancer genes. The discovery of rearrangements in the RB, APC, FBXW7, and other recessive cancer genes may have led to gene inactivation, which may have resulted in cancer development (Stephens et al. 2009). Other gene rearrangements in prostate cancer include *CADM2, PTEN*, and *MAGI2*. Furthermore, many rearrangements can occur preferentially in genes that are physically adjacent to transcriptional or chromatin compartments, most likely as a result of DNA strand breakage and degradation. The

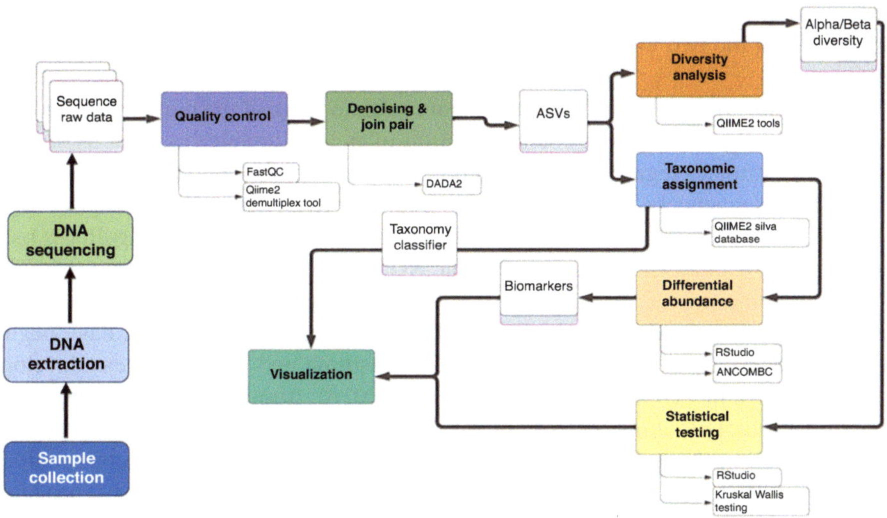

Fig. 6.2 Metagenomic analyses

intricacy of "closed chains" and other rearrangements implies that whole-genome sequencing—rather than exon-focused approaches or gene fusion—may be necessary to build a Spectrum of processes guiding the genesis and progression of prostate cancer (Berger et al. 2011). WGS is also useful for detecting complicated rearrangements. For example, PTEN, a tumor suppressor gene, has been identified to interact with and influence the action of other genes. *PREX2* is a frequently changed gene in melanoma samples (Berger et al. 2012).

Diagnosis and Surveillance of Infectious Diseases Current NGS methods may be employed in three main clinical microbiology laboratory applications: WGS, metagenomics, and shotgun metagenomics (Fig. 6.2) (Hilt and Ferrieri 2022). Previously, public health laboratories used pulsed-field gel electrophoresis (PFGE) to determine the serotype of *Salmonella* (Murase et al. 1995). However, comparative studies on WGS have demonstrated that its results was similar to PFGE method, thus WGS became the gold standard for reference technology (Harbottle et al. 2006; Leekitcharoenphon et al. 2014). WGS is additionally employed to track illness prevented by vaccines including *N. meningitides*, *S. pneumoniae*, and antibiotic-resistant infections like multidrug-resistant *M. tuberculosis* (Revez et al. 2017). Deep amplicon sequencing is a PCR technology expansion that allows for deeper coverage of the desired gene(s). Deep amplification sequencing is widely recognized for bacterial identification by amplification of 16S ribosomal RNA (16S rRNA) genes (Janda and Abbott 2007) and 18S rRNA or ribosome ITS genes for fungal identification (Wagner et al. 2018). Furthermore, 16S deep amplification sequencing makes it easier to find difficult-to-grow species, including tick-borne bacteria that are commonly missed by normal bacterial cultures (Thoendel 2020). Deep amplicon sequencing is used in clinical virological diagnosis, particularly to detect antiviral resistance in the two viruses cytomegalovirus and HIV (Rhee et al. 2022).

Microbiome Analysis NGS has aided our understanding of the human microbiome by enabling the finding and characterization of non-culturable bacteria, as well as informed estimates regarding their function. The most prevalent NGS techniques are 16S rRNA sequencers, metagenomic shotgun, and RNA sequencing (Wensel et al. 2022). The NGS microbiome may have a therapeutic utility in predicting illness risk, similar to how established human genomic NGS is used to estimate disease risk. The NGS microbiome for disease risk prediction has yet to be validated, although progress is being made. Longitudinal studies in children, for example, have begun to correlate bacteria with an increased risk of asthma and allergic illness development (Lynch and Vercelli 2021; Fujimura et al. 2016). Another example is the use of colon microbiota (colonic mucosa or stool samples) to predict the risk of colorectal cancer (CRC). Although metagenomic analyses have found microbial communities reflecting CRC, identification of precancerous lesions (e.g., colonic polyps) has been restricted (Wirbel et al. 2019; Thomas et al. 2019). Similarly, blood-based transcriptomes are the most effective in detecting advanced cancer (Poore et al. 2020). This means that NGS methods must be improved in order to detect sickness at an early stage when action can improve the prognosis for patients.

Pharmacogenomics Much study has been conducted to investigate the accuracy of NGS methods in pharmacogenomics, along with the usage of NGS sequenced or the reuse of diagnostic NGS data in pharmacogenomics (Cousin et al. 2017; Yang et al. 2016; Londin et al. 2014). Duong et al. used DMET, WES, and WGS in a three-way research to investigate the concordance of pharmacogenomics genotype findings based on these different technologies. They found a 94% agreement between DMET and WES and a 96% agreement between DMET and WGS (Yang et al. 2016).

NGS has revolutionized genomic research by enabling massive amounts sequencing operations including the Human Genome Project. It has sped up the finding of illness-related genetic variants, functional regions in the genome, and disease processes. NGS also makes population-scale investigations, evolutionary genomics, and the discovery of novel medicinal targets possible.

6.5 Conclusion

Rapid developments in DNA sequencing technology have resulted in huge cost savings as well as significant gains in throughput and accuracy. Every day, a deluge of genetic data is being released to the public as more creatures are sequenced. Genomic advances have been progressively increasing as a result of a revolution in sequencing technology. Furthermore, large-scale investigations in economics, metagenomics, epigenomics, and transcription come to completion. These investigations give not just knowledge for fundamental research but also directly applicable advantages. Scientists in various industries are using this data to improve agricultural, crop, and animal production, as well as to improve the identification, prognosis, and management of cancer and other ailments. NGS is a game-changing technology that offers up new avenues for molecular diagnostics. Many clinical laboratories have used NGS technology to find causal variations for physical ailments, precise genomic profiling for cancer, and pathogen identification for infectious diseases. To address the expanding needs of precision medicine, NGS technology, and bioinformatics tools will continue to advance and become the major method of diagnosis and standard of care for genetic analysis. NGS has both virtues and weaknesses, as well as restrictions and obstacles. To begin with, NGS delivers horizontal coverage and 100% accuracy, resulting in a lack of variants and false positives. Another difficulty is selecting and understanding the data because there is frequently more than one possible variety. Indeed, the more kilobases that are sequenced, the more likely it is that several candidates will be discovered. There are other ethical issues to consider, such as how to send impacted testers to direct customer testing. As a result, defining standards for running quality, interpreting variance, and controlling quality are critical. The conduct of NGS testing in clinical diagnostic laboratories requires a lot of resources. Experiment validation, bioinformatics support, and directed data archiving are required prior to implementing NGS testing, and these are too expensive for many small labs.

References

Baer R et al (1984) DNA sequence and expression of the B95-8 Epstein-Barr virus genome. Nature 310(5974):207–211

Barba M, Czosnek H, Hadidi A (2014) Historical perspective, development and applications of next-generation sequencing in plant virology. Viruses 6(1):106–136

Berger MF et al (2011) The genomic complexity of primary human prostate cancer. Nature 470(7333):214–220

Berger MF et al (2012) Melanoma genome sequencing reveals frequent PREX2 mutations. Nature 485(7399):502–506

Cousin MA et al (2017) Pharmacogenomic findings from clinical whole exome sequencing of diagnostic odyssey patients. Mol Genet Genom Med 5(3):269–279

Drmanac R et al (2002) Sequencing by hybridization (SBH): advantages, achievements, and opportunities. Chip Technol 77:75–101

Fernandez-Marmiesse A, Gouveia S, Couce ML (2018) NGS technologies as a turning point in rare disease research, diagnosis and treatment. Curr Med Chem 25(3):404–432

Fleischmann RD et al (1995) Whole-genome random sequencing and assembly of *Haemophilus influenzae* Rd. Science 269(5223):496–512

Fujimura KE et al (2016) Neonatal gut microbiota associates with childhood multisensitized atopy and T cell differentiation. Nat Med 22(10):1187–1191

Fuller CW et al (2009) The challenges of sequencing by synthesis. Nat Biotechnol 27(11): 1013–1023

Gilbert W, Maxam A (1973) The nucleotide sequence of the lac operator. Proc Natl Acad Sci 70(12):3581–3584

Gu W, Miller S, Chiu CY (2019) Clinical metagenomic next-generation sequencing for pathogen detection. Annu Rev Pathol 14:319–338

Hanna GJ et al (2000) Comparison of sequencing by hybridization and cycle sequencing for genotyping of human immunodeficiency virus type 1 reverse transcriptase. J Clin Microbiol 38(7):2715–2721

Harbottle H et al (2006) Comparison of multilocus sequence typing, pulsed-field gel electrophoresis, and antimicrobial susceptibility typing for characterization of *Salmonella enterica* serotype Newport isolates. J Clin Microbiol 44(7):2449–2457

Hilt EE, Ferrieri P (2022) Next generation and other sequencing technologies in diagnostic microbiology and infectious diseases. Genes (Basel) 13(9):1566

Holley RW, Madison JT, Zamir A (1964) A new method for sequence determination of large oligonucleotides. Biochem Biophys Res Commun 17(4):389–394

Holley RW et al (1965) Structure of a ribonucleic acid. Science 147(3664):1462–1465

Janda JM, Abbott SL (2007) 16S rRNA gene sequencing for bacterial identification in the diagnostic laboratory: pluses, perils, and pitfalls. J Clin Microbiol 45(9):2761–2764

Kawashima E, Farinelli L, Mayer P (1998) Method of nucleic acid amplification. PCT/GB98/00961

Leekitcharoenphon P et al (2014) Evaluation of whole genome sequencing for outbreak detection of *Salmonella enterica*. PLoS One 9(2):e87991

Londin ER et al (2014) Performance of exome sequencing for pharmacogenomics. Pers Med 12(2): 109–115

Lynch SV, Vercelli D (2021) Microbiota, epigenetics, and trained immunity. Convergent drivers and mediators of the asthma trajectory from pregnancy to childhood. Am J Respir Crit Care Med 203(7):802–808

Mardis ER (2008) Next-generation DNA sequencing methods. Annu Rev Genomics Hum Genet 9: 387–402

Matthijs G et al (2016) Guidelines for diagnostic next-generation sequencing. Eur J Hum Genet 24(1):2–5

Maxam AM, Gilbert W (1977) A new method for sequencing DNA. Proc Natl Acad Sci 74(2): 560–564

Maxam AM, Gilbert W (1980) [57] Sequencing end-labeled DNA with base-specific chemical cleavages. In: Methods in enzymology. Elsevier, pp 499–560

Murase T, Matsushima A, Okitsu T, Suzuki R (1995) Evaluation of DNA fingerprinting by PFGE as an epidemiologic tool for *Salmonella* infections. Microbiol Immunol 39:673–676

Muzzey D, Evans EA, Lieber C (2015) Understanding the basics of NGS: from mechanism to variant calling. Curr Genet Med Rep 3(4):158–165

Pereira RA-O, Oliveira JA-O, Sousa M (2020) Bioinformatics and computational tools for next-generation sequencing analysis in clinical genetics. J Clin Med 9(1):132. https://doi.org/10.3390/jcm9010132

Poore GD et al (2020) Microbiome analyses of blood and tissues suggest cancer diagnostic approach. Nature 579(7800):567–574

Revez J et al (2017) Survey on the use of whole-genome sequencing for infectious diseases surveillance: rapid expansion of European National Capacities, 2015–2016. Front Public Health 5:347

Rhee SY et al (2022) Public availability of HIV-1 drug resistance sequence and treatment data: a systematic review. Lancet Microbe 3(5):e392–e398

Ronaghi M et al (1996) Real-time DNA sequencing using detection of pyrophosphate release. Anal Biochem 242(1):84–89

Sanger F, Coulson AR (1975) A rapid method for determining sequences in DNA by primed synthesis with DNA polymerase. J Mol Biol 94(3):441–448

Sanger F, Nicklen S, Coulson AR (1977) DNA sequencing with chain-terminating inhibitors. Proc Natl Acad Sci 74(12):5463–5467

Slatko BE, Gardner AF, Ausubel FM (2018) Overview of next-generation sequencing technologies. Curr Protoc Mol Biol 122(1):e59

Stephens PJ et al (2009) Complex landscapes of somatic rearrangement in human breast cancer genomes. Nature 462(7276):1005–1010

Thoendel M (2020) Targeted metagenomics offers insights into potential tick-borne pathogens. J Clin Microbiol 58(11):10–128

Thomas AM et al (2019) Metagenomic analysis of colorectal cancer datasets identifies cross-cohort microbial diagnostic signatures and a link with choline degradation. Nat Med 25(4):667–678

Tuna M, Amos CI (2013) Genomic sequencing in cancer. Cancer Lett 340(2):161–170

Wagner K et al (2018) Molecular detection of fungal pathogens in clinical specimens by 18S rDNA high-throughput screening in comparison to ITS PCR and culture. Sci Rep 8(1):6964

Watson JD, Crick FH (1953) Molecular structure of nucleic acids: a structure for deoxyribose nucleic acid. Nature 171(4356):737–738

Wensel CR et al (2022) Next-generation sequencing: insights to advance clinical investigations of the microbiome. J Clin Invest 132(7):e154944

Wirbel J et al (2019) Meta-analysis of fecal metagenomes reveals global microbial signatures that are specific for colorectal cancer. Nat Med 25(4):679–689

Yang W et al (2016) Comparison of genome sequencing and clinical genotyping for pharmacogenes. Clin Pharmacol Ther 100(4):380–388

Genome Scale Modeling for Novel Drug Targets

7

Hara Prasad Mishra, Indrajeet Singh, and Ajay Kumar

Abstract

System-level metabolic research can make use of computational simulations of genome-scale modeling (GSM), this depict the relationships between genes for proteins and actions for all functional genes in an organism. Possible therapeutic targets can be found using genome-scale modeling (GSM). These computational frameworks can replace time-consuming and expensive gene knockdown investigations or morphological assessment of medicines in cancer cell lines. One of the cornerstones of systems biology, metabolic models has just recently begun adding clarity to the molecular link between genotype and phenotype.

Current reconstructed GSM are provided, and their uses are discussed. Some of these applications include drug discovery and targeting, enzyme function prediction, modeling interactions between numerous cells or animals, and human illness comprehension.

Keywords

Genome-scale metabolic models · Bioinformatics · In silico · Knockdown · KEGG

H. P. Mishra
Department of Pharmacology, University College of Medical Sciences, University of Delhi, Delhi, India

I. Singh · A. Kumar (✉)
Department of Biotechnology, Rama University, Kanpur, Uttar Pradesh, India
e-mail: drajay.fet@ramauniversity.ac.in

V. Singh, A. Kumar (eds.), *Advances in Bioinformatics*,
https://doi.org/10.1007/978-981-99-8401-5_7

7.1 Introduction

Finding and validating novel therapeutic targets is a difficult and time-consuming step in the development of new medications. We now have access to enormous volumes of data on genes, proteins, and metabolites because of the development of high-throughput technology, which can be used to find new therapeutic targets. It is challenging to comprehend the underlying mechanisms of disease and to pinpoint certain targets that can be affected by medications due to the extreme complexity of biological systems (Paul et al. 2021).

Genome-scale modeling (GSM) is a potent technology that can be utilized to get beyond these obstacles and hasten the identification of new drugs. Building computational simulations to replicate the behavior of complete organism, from the molecular to the cellular level, is a component of GSM. These models take into account information on how genes, proteins, and metabolites interact as well as external elements like pH and temperature. GSM can assist us in understanding the fundamental mechanisms of disease and in identifying possible therapeutic targets by replicating the behavior of complex systems (Wahi and Holst 2019). Reconstructing GSM has become one of the main modeling tools for systems-level metabolic investigations. Since the first GEM of *Haemophilus influenzae* RD was described in 1999 (Edwards and Palsson 1999). Due to its capacity to combine huge datasets and offer insights into the intricate relationships that underpin biological systems, GSM has grown in significance as a tool for the identification of new drugs. In a variety of conditions, including diabetes, cancer, osteoporosis, infectious diseases, and metabolic disorders, GSM has been utilized to pinpoint new therapeutic targets.

The ability to replicate the behavior of biological systems under a variety of situations, which can be challenging or impossible to achieve experimentally, is one of the main benefits of GSM. As a result, it is now possible to examine biological system's behavior in great detail and find new therapeutic targets that conventional experimental techniques could miss. Before investing in costly and time-consuming clinical studies, we can use GSM to test the efficacy and toxicity of medications in silico. This can speed up medication development, save costs, and lead to better patient results. Gene-protein-reaction (GPR) relationships are created on the basis of genome annotation data and experimentally collected information, and a GSM uses these to provide a computational description of an organism's full set of stoichiometry-based, mass-balanced metabolic reactions (Thiele and Palsson 2010).

GSM can be utilized to estimate the future outlook of breast and lung cancer and to uncover therapeutic targets that block the proliferation of certain cell lines. In order to find new targets for antimicrobial drugs, GSM have been employed. They have been useful in identifying novel pharmacological targets for use in individualized healthcare (Pacheco et al. 2019). A sequence-based drug target prediction approach has been devised for the rapid identification of novel drug targets by comparing their chemical properties to those of established drug targets (Li and Lai 2007). Regarding the rapid discovery of new drug targets, a sequence-based drug target prediction technique has been devised by comparing the chemical properties of existing drug targets. Using GSM to find new therapeutic targets, however, is not without its share of difficulties. For instance, the quality of a context-aware model depends on the quality of the input model, and most input models ought to be characterized as reconstructions rather than models. Despite these obstacles, GSM have been employed successfully to identify new pharmacological targets, and model quality has increased significantly in recent years (Pacheco et al. 2016).

In the topic that follows, we'll examine GSM fundamental ideas and how it might be used to find new drugs. We will also go over a number of case studies that show how effective GSM is at locating new therapeutic targets.

7.2 Bioinformatics in Genome Scale Modeling for Novel Drug Targets

Employing computational approaches, biologists simulate organisms' metabolic pathways from their annotated genome sequences to identify potential new drug targets (Fig. 7.1). When studying organisms, these models can shed light on, clarify, analyze, optimize, and even uncover previously unknown cellular functions (Passi

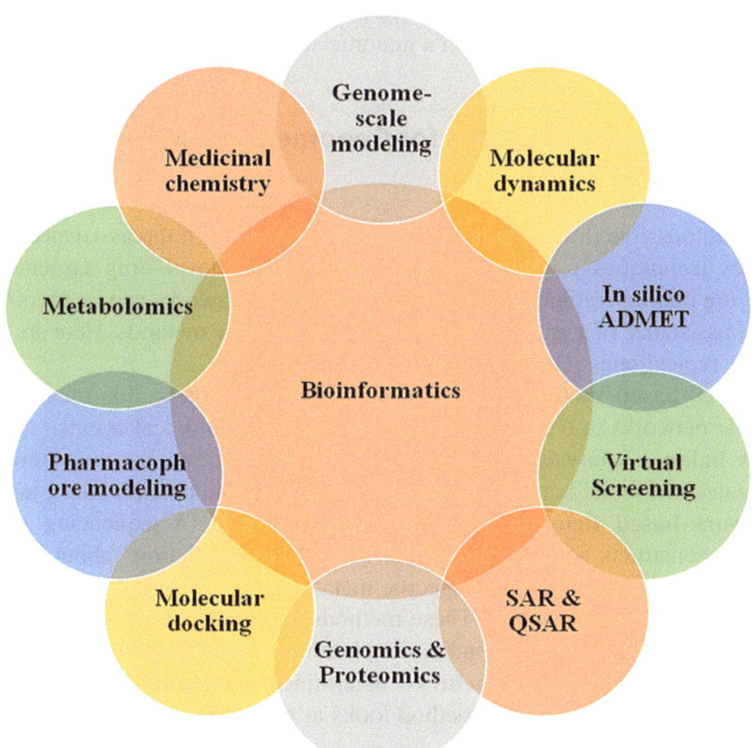

Fig. 7.1 Bioinformatics in genome scale modeling for novel drug targets

et al. 2021). Some applications of bioinformatics and metabolic models on a genome-wide scale in the field of drug development include the following:

1. By simulating the impact of knockout genes on the proliferation rate of cancerous cells, genomic metabolic models can be used to discover new curative targets (Guimerà et al. 2007).
2. Models of metabolism that are scaled to the genome can be used to locate metabolic pathways that could be the focus of pharmacological re-engineering efforts (Li et al. 2011).
3. Under some circumstances, metabolic models that are scaled to the genome can make predictions about possible biological targets of an organism (Passi et al. 2021).
4. Reconstructing metabolic systems at the individual, species, and community scales in a quick and flexible way can be accomplished through the use of automated tools such as CarveMe (Chavali et al. 2012).
5. Mathematical programming provides a potent set of tools for the analysis of "multi-modality" models; nevertheless, one must pay particular attention to the links that exist between the various modeling kinds (Machado et al. 2018).
6. The use of statistical methods to quantify the confidence of model predictions is one way to deal with the uncertainty that arises during the reconstruction and analysis of metabolic models on a genome-scale (Fell et al. 2010).

7.3 Bioinformatics Tools for Genome-Scale Modeling for Novel Drug Targets

Identifying possible drug targets can be accomplished with the assistance of bioinformatics techniques for genome-scale modeling of novel drug targets (GSM). Identifying pharmacological targets inside metabolic networks can be accomplished with the assistance of a number of different bioinformatic methods. Here are some of the more typical ones (Guimerà et al. 2007).

Network-based method: to find new therapeutic targets, this method utilizes metabolic networks to chart the connections between biological agents.

Flux balance analysis: Using a topological perspective, this two-step method investigates metabolic network features in order to locate potential drug targets.

Network-based drug target prediction: Using RNA-sequencing data, this method reconstructs a metabolic model and makes predictions about therapeutic targets and medicines for cancer-specific metabolism.

Bioinformatic approaches: These methods expand our understanding of effective target families in order to find novel pharmacological targets.

Metabolic network approach: To determine which antimicrobial medication targets are most important, this method looks at metabolic networks of pathogens.

7.4 Applications of Genome Scale Modeling in Drug Discovery

Genome-scale modeling (GSM) has many uses in the drug development process, such as finding new drug targets, perfecting medication combinations, and predicting drug toxicity shown in Table 7.1. We will look at some of the ways that GSM is being utilized to speed up drug discovery in this chapter.

Table 7.1 Examples of genome scale modeling approaches for novel drug targets

Methodology	Description	Advantages	Limitations
Flux Balance Analysis (FBA)	Constraint-based modeling approach that predicts cellular metabolism by optimizing metabolic fluxes	– Captures metabolic phenotypes accurately – Allows prediction of metabolic changes upon drug perturbations	Assumes steady-state conditions Limited incorporation of regulatory information
Constraint-based modeling (CBM)	Integrates metabolic, regulatory, and signaling networks to predict cellular behavior	Enables analysis of metabolic and regulatory interactions – Predicts drug–target interactions Can be coupled with omics data	– Requires comprehensive knowledge of network topology Computational challenges in large-scale networks
Genome-scale Metabolic Models (GEMs)	Mathematical representations of the entirety of metabolic reactions in an organism	Provides a comprehensive view of cellular metabolism Enables prediction of metabolic alterations in diseases Useful for target identification and drug repurposing	Requires accurate genome annotations Difficulties in integrating multi-omics data
Systems Biology Models	Integrative models that combine multiple biological networks to study complex cellular processes	Captures interactions between genes, proteins, and signaling pathways Predicts emergent properties of cellular systems	Requires extensive experimental data for parameterization Computationally intensive for large-scale models
Network-based approaches	Analyzes molecular interaction networks to identify key targets and pathways	Identifies potential drug targets based on network properties Enables exploration of disease modules and drug repurposing	Limited by the quality and coverage of interaction data Challenging to account for dynamic changes in the network

7.4.1 Identifying New Drug Targets

Finding new drug targets is one of GSM's main uses in the drug discovery process. GSM can uncover previously unidentified pathways and mechanisms to facilitate may be used for remedial purposes by modeling the behavior of biological systems at genome size.

Finding new drug targets is one of GSM's main uses in the drug discovery process. GSM can uncover previously unidentified pathways and mechanisms that may be used for therapeutic purposes by modeling the behavior of biological systems at genome size.

GSM has been used, for instance, to find potential therapeutic targets in cancer cells. Researchers have discovered distinct metabolic characteristics that set cancer cells apart from healthy ones by modeling the metabolic pathways in cancer cells. Drugs that precisely inhibit cancer cells while preserving healthy cells can target these metabolic properties.

The identification of possible therapeutic targets in infectious diseases has also been done using GSM. Researchers have identified weaknesses that can be addressed with antimicrobial medications by studying the metabolic pathways of pathogens including bacteria and viruses (Chung et al. 2013).

7.4.2 Optimizing Drug Combinations

The optimization of medication combinations is another way that GSM is used in drug discovery. Multiple routes or mechanisms are targeted medication combinations that are used to treat a variety of disorders, including cancer. Finding the best medicine combination, though, can be difficult.

The behavior of biological systems in response to medication combinations can be modeled using GSM. Researchers can determine the best drug combinations for a particular condition by simulating the impact of several drug combinations on the body.

For treating cancer, GSM has been used to determine the best medication combinations. Researchers have found drug combinations that work better together than they do alone by simulating the impact of several drug combinations on cancer cells.

7.4.3 Predicting Drug Toxicity

Predicting the toxicity of medications is one of the main difficulties in drug discovery. Due to unforeseen toxicities, many medications with promise in preclinical investigations fail in clinical trials.

Through the simulation of the impact of the medications on biological systems, GSM can be used to forecast the toxicity of drugs. Researchers can detect possible toxicities and create safer medications by predicting how biological systems behave in response to various treatments (Aguayo-Orozco et al. 2017).

GSM has been used, for instance, to forecast the cardio toxicity of anticancer medications. Researchers have been able to spot possible cardiac toxicities and develop safer medications by predicting how these compounds affect the metabolic pathways in heart cells.

7.5 Genome-Scale Modeling Help Identify Personalized Drug Targets

The creation of tailored medicine is another potential use for GSM in drug discovery. GSM can identify individual differences in metabolic pathways and mechanisms that may affect drug response by mimicking the behavior of biological systems at the genome size. GSM has been used, for instance, to pinpoint individual variations in medication metabolism. Researchers have been able to pinpoint genetic differences that affect drug metabolism and response by modeling the metabolic pathways in various individuals. These variants can be utilized to create custom drug regimens that are genetically customized to the patient, leading to more efficient and secure therapies (Raškevičius et al. 2018).

The following are some of the ways in which genome-scale modeling can assist in the identification of personalized therapeutic targets:

Systems biology approach: A technique in systems biology that uses genome-scale modeling of human metabolism to find new therapeutic agents and pharmacological targets is called "systems genomics." Personalized medication targets can be determined for an individual by conducting an analysis of the individual's metabolic pathways (Mardinoglu et al. 2013).

Computational framework: Genome-scale metabolic models, often known as GSM, are a type of computational framework that can be utilized to locate prospective drug targets. GSMMs are able to anticipate important genes (or processes) and critical metabolites for a pathogen because they simulate the metabolic network of the pathogen. Each of these predictions can lead to a distinct medication discovery (Viana et al. 2020).

Tools for drug design: The invention of novel medicine and personalized medicine both make use of GSMMs as useful tools. GSM are able to predict prospective therapeutic targets by matching the chemical structures of existing medications to the metabolic pathways of a particular disease (Duarte et al. 2007).

Promising platform: It is possible to use genome-scale models as a potentially useful framework for drug target identification. For instance, in order to find prospective therapeutic targets, a genome-scale model of the common fungus *Candida albicans* was utilized (Viana et al. 2020).

7.6 Genome-Scale Modeling is Used to Predict Drug Efficacy for Specific Patients

The following are some of the different ways that genome-scale modeling can be used to predict therapeutic efficacy for specific patients:

Constraint-based models: A statistical instruction is provided by constraint-based models like genome-scale modeling (GSM), which can be used to acquire an improved considerate of the metabolic capabilities of a cell. This makes it possible to conduct a study of genetic alterations across the entire system, as well as research metabolic illnesses and discover the critical enzyme responses and therapeutic targets (Collin et al. 2022).

Systems biology approach: A technique in systems biology that uses genome-scale modeling of human metabolism to find new therapeutic agents and pharmacological targets is called “systems genomics.” The conclusions that may be drawn from the application of this mechanistic modeling strategy can be put to use in the search for novel medication agents in addition to drug targets, as well as in the development of more effective drug–diagnostic combinations (Mardinoglu et al. 2013).

Tools for drug design: GSM are useful tools that can be applied to the design of drugs and personalized medicine. GSM are able to predict prospective therapeutic targets by matching the chemical structures of existing medications to the metabolic pathways of a particular disease (Folger et al. 2011).

Predictive capabilities: Personalized medicine has the potential to be equipped with predictive skills to research clinically relevant topics in silico, which might be achieved through the use of computational models. This can make it possible for personalized medicine to make predictions about the efficacy and safety of medications based on the micro biome and metabolism of a specific individual (Kanehisa et al. 2016).

In general, modeling on a genome-scale can be utilized as a constraint-based model, a systems biology method, a tool for drug design, and a predictive capability to forecast medication efficacy for particular patients. All of these applications are discussed more below.

7.7 Genome-Scale Modeling Models Based Drugs that Were Developed

There are numerous examples of pharmaceuticals that were created with the assistance of genome-scale modeling (GSM). Here are several examples:

Identification of metabolic targets for glioblastoma: The kind of brain cancer known as glioblastoma was studied using GSMMs in order to locate potential metabolic targets for the disease. Researchers were able to zero in on a metabolic target known as pyruvate kinase M2 (PKM2) and create a medication that specifically targets this enzyme (Heinken et al. 2023).

Drug design for human metabolites: In the process of designing medications targeting human metabolism, GSMMs were utilized. Scientists discovered substances that are most probable to bind the enzymes which are metabolizing the analyzed metabolite through a comparison of the chemical compositions of human by products to those contained in the DrugBank database. This was accomplished by analyzing the molecular structures of human metabolites to those included in DrugBank (Folger et al. 2011).

Identification of potential drug targets for cancer: In the search for possible targets of therapy for cancer treatment, GSM were utilized. Researchers looked into whether or not gene knockdown procedures could be useful in the process of finding therapeutic targets by employing GSM. Within the models, they found a total of 202 metabolic targets that have the potential to be employed as pharmacological targets (Gu et al. 2019).

Identification of potential drug targets against hepatocellular carcinoma: In order to discover new possible prevention strategies toward hepatic cancer therapy, GSM were put to use. Researchers identified metabolic targets that are unique to hepatocellular carcinoma by using GSM to do so. They then designed medicines that target these metabolic pathways (Thafar et al. 2021).

7.8 Challenges in Predicting Drug-Target Interactions Using Network-Based Approaches

Applying network-based methodologies for predicting how a medicine will interact with a certain target can be difficult for a number of reasons. The following is a list of some of the difficulties:

Data sparsity: In many cases, there is a lack of drug–target association information, and the amount of recognized relationships is significantly less than the entire number of possible connections. Because of this, it is challenging to construct reliable models that are able to anticipate future interactions (Jung et al. 2022).

Network heterogeneity: The network which depicts the relationship of medicines and their intended effects is typically heterogeneous, with various kinds of nodes and edges expressing various kinds of drug-related information. The process of incorporating all of this data into a single model might be complex (Bagherian et al. 2021).

Model complexity: The procession that depend on networks typically call for complicated models, which are not only difficult to interpret but also expensive to compute. Because of this, applying these strategies to datasets that are more extensive can be difficult (Cheng et al. 2012).

Lack of negative examples: The vast majority of drug–target interaction information sets only include positive cases, making it impossible to differentiate among true and false positive (El-Behery et al. 2022).

Limited validation: Because network-based systems typically don't collect their own experimental data, validating their predictions can be a difficult task. Because of this, determining the degree to which these procedures are accurate can be challenging (Moumbock et al. 2019).

In general, employing network-based methodologies to predict drug–target interactions might be difficult due to the limited availability of data, the heterogeneity of the network, the complexity of the model, the absence of negative instances, and the restricted validation. In order to address these obstacles, it is necessary to develop new approaches that are capable of dealing with these concerns and to integrate numerous sources of data in order to increase the accuracy of forecasts.

7.9 Conclusion

In conclusion, genome-scale modeling is proving as a potent tool in the search for of novel drug targets. This approach integrates diverse omics data, computational algorithms, and network-based analyses to gain a comprehensive understanding of cellular metabolism, regulatory networks, and disease pathways. Through the systematic exploration of these models, researchers can uncover potential therapeutic targets and repurpose existing drugs for various diseases.

The application of genome-scale modeling has provided valuable insights into personalized medicine, allowing for the development of tailored treatment strategies based on individual patient characteristics. By integrating genomic data and clinical information, researchers can predict drug responses, optimize treatment regimens, and improve patient outcomes. Additionally, the integration of multi-omics data and advanced computational techniques has enhanced the predictive power of these models, enabling the identification of synergistic drug combinations and therapeutic biomarkers.

Furthermore, the integration of genome-scale modeling with other fields, such as network pharmacology, artificial intelligence, and structural biology, holds great promise for future advancements. These interdisciplinary approaches have the potential to accelerate drug discovery and development processes, reduce costs, and improve the success rates of clinical trials. Moreover, the advent of in silico clinical trials and virtual patient models presents exciting opportunities for predicting drug efficacy and optimizing trial design, ultimately leading to more efficient and targeted therapies. However, it is important to acknowledge the limitations and challenges associated with genome-scale modeling. These include the need for accurate genome annotations, the integration of complex regulatory information, the scalability of models to larger systems, and the requirement for comprehensive experimental validation. Addressing these challenges will require ongoing collaboration between computational biologists, experimental researchers, and clinicians to refine and validate these models.

In conclusion, genome-scale modeling for novel drug targets offers immense potential to revolutionize drug discovery and development. As our understanding of biological systems and computational methods continues to advance, genome-scale modeling will play a pivotal role in identifying effective and personalized therapies for a wide range of diseases. By harnessing the power of this approach, we can pave the way for more precise, targeted, and efficacious treatments, ultimately improving patient outcomes and transforming healthcare practices.

References

Aguayo-Orozco A et al (2017) In silico systems pharmacology to assess drug's therapeutic and toxic effects. Curr Pharm Des 22(46):6895–6902

Bagherian M et al (2021) Machine learning approaches and databases for prediction of drug–target interaction: a survey paper. Brief Bioinform 22(1):247–269

Chavali AK, D'Auria KM, Hewlett EL, Pearson RD, Papin JA (2012) A metabolic network approach for the identification and prioritization of antimicrobial drug targets. Trends Microbiol 20(3):113–123

Cheng F, Liu C, Jiang J, Lu W, Li W, Liu G et al (2012) Prediction of drug–target interactions and drug repositioning via network-based inference. PLoS Comput Biol 8(5):e1002503

Chung BK-S, Dick T, Lee D-Y (2013) In silico analyses for the discovery of tuberculosis drug targets. J Antimicrob Chemother 68(12):2701–2709

Collin CB, Gebhardt T, Golebiewski M, Karaderi T, Hillemanns M, Khan FM, Salehzadeh-Yazdi-A, Kirschner M, Krobitsch S, EU-STANDS4PM Consortium, Kuepfer L (2022) Computational models for clinical applications in personalized medicine-guidelines and recommendations for data integration and model validation. J Pers Med 12(2):166

Duarte NC, Becker SA, Jamshidi N, Thiele I, Mo ML, Vo TD et al (2007) Global reconstruction of the human metabolic network based on genomic and bibliomic data. Proc Natl Acad Sci U S A 104(6):1777–1782

Edwards JS, Palsson BO (1999) Systems properties of the *Haemophilus influenzae* Rd metabolic genotype. J Biol Chem 274:17410–17416

El-Behery H, Attia AF, El-Fishawy N et al (2022) An ensemble-based drug–target interaction prediction approach using multiple feature information with data balancing. J Biol Eng 16:21

Fell DA, Poolman MG, Gevorgyan A (2010) Building and analysing genome-scale metabolic models. Biochem Soc Trans 38(5):1197–1201

Folger O, Jerby L, Frezza C, Gottlieb E, Ruppin E, Shlomi T (2011) Predicting selective drug targets in cancer through metabolic networks. Mol Syst Biol 7(1):501

Gu C, Kim GB, Kim WJ et al (2019) Current status and applications of genome-scale metabolic models. Genome Biol 20:121

Guimerà R et al (2007) A network-based method for target selection in metabolic networks. Bioinformatics 23(13):1616–1622

Heinken A, Hertel J, Acharya G et al (2023) Genome-scale metabolic reconstruction of 7302 human microorganisms for personalized medicine. Nat Biotechnol 41:1320

Jung Y-S, Kim Y, Cho Y-R (2022) Comparative analysis of network-based approaches and machine learning algorithms for predicting drug–target interactions. Methods 198:19–31

Kanehisa M, Sato Y, Kawashima M, Furumichi M, Tanabe M (2016) KEGG as a reference resource for gene and protein annotation. Nucleic Acids Res 44(D1):D457–D462

Li Q, Lai L (2007) Prediction of potential drug targets based on simple sequence properties. BMC Bioinform 8:353

Li Z, Wang RS, Zhang XS (2011) Two-stage flux balance analysis of metabolic networks for drug target identification. BMC Syst Biol 5(Suppl 1):S11

Machado D, Andrejev S, Tramontano M, Patil KR (2018) Fast automated reconstruction of genome-scale metabolic models for microbial species and communities. Nucleic Acids Res 46(15):7542–7553

Mardinoglu A et al (2013) Genome-scale modeling of human metabolism—a systems biology approach. Biotechnol J 8(9):985–996

Moumbock AFA et al (2019) Current computational methods for predicting protein interactions of natural products. Comput Struct Biotechnol J 17:1367–1376
Pacheco MP, Pfau T, Sauter T (2016) Benchmarking procedures for high-throughput context specific reconstruction algorithms. Front Physiol 6(6):1–19
Pacheco MP, Bintener T, Sauter T (2019) Towards the network-based prediction of repurposed drugs using patient-specific metabolic models. EBioMedicine 43:26–27
Passi A, Tibocha-Bonilla JD, Kumar M, Tec-Campos D, Zengler K, Zuniga C (2021) Genome-scale metabolic modeling enables in-depth understanding of big data. Metabolites 12(1):14
Paul A, Anand R, Karmakar SP et al (2021) Exploring gene knockout strategies to identify potential drug targets using genome-scale metabolic models. Sci Rep 11:213
Raškevičius V, Mikalayeva V, Antanavičiūtė I, Ceslevičienė I, Skeberdis VA, Kairys V, Bordel S (2018) Genome scale metabolic models as tools for drug design and personalized medicine. PLoS One 13(1):e0190636
Thafar MA, Olayan RS, Albaradei S et al (2021) DTi2Vec: Drug–target interaction prediction using network embedding and ensemble learning. J Cheminform 13:71
Thiele I, Palsson BO (2010) A protocol for generating a high-quality genome-scale metabolic reconstruction. Nat Protoc 5:93–121
Viana R, Dias O, Lagoa D, Galocha M, Rocha I, Teixeira MC (2020) Genome-scale metabolic model of the human pathogen *Candida albicans*: a promising platform for drug target prediction. J Fungi 6:171
Wahi K, Holst J (2019) Asct2: a potential cancer drug target. Expert Opin Ther Targets 23(7): 555–558

Role of Bioinformatics in Genome Editing

8

Amit Joshi, Ajay Kumar, Vikas Kaushik, Prashant Kumar, and Sushma Dubey

Abstract

Bioinformatics plays a crucial role in advancing genome editing techniques by utilizing computational tools and algorithms to analyze biological data. This chapter provides an overview of the role of bioinformatics in genome editing, focusing on key areas where it is applied. These areas include computational analysis of target sequences, prediction, and evaluation of off-target effects, designing and optimizing CRISPR systems, functional annotation of genomic variants, comparative genomics, homology analysis, and integration of multi-omics data. By leveraging bioinformatics, researchers can identify target sites for gene editing, predict and minimize off-target effects, enhance the efficiency of CRISPR systems, understand the functional consequences of genetic alterations, explore evolutionary relationships, and gain comprehensive insights into biological systems. Moreover, comparative genomics and homology analysis are discussed as vital approaches that leverage bioinformatics to understand evolutionary relationships and identify conserved elements across species. Integration of multi-omics data, such as genomics, transcriptomics, and proteomics,

A. Joshi (✉)
Department of Biochemistry, Kalinga University, Naya Raipur, Chhattisgarh, India
e-mail: amit.joshi@kalingauniversity.ac.in

A. Kumar
Department of Biotechnology, Rama University, Kanpur, Uttar Pradesh, India

V. Kaushik
Science Habitat, Markham, ON, Canada

P. Kumar
Department of Bioinformatics, Kalinga University, Naya Raipur, Chhattisgarh, India

S. Dubey
Department of Biotechnology, Kalinga University, Naya Raipur, Chhattisgarh, India

V. Singh, A. Kumar (eds.), *Advances in Bioinformatics*,
https://doi.org/10.1007/978-981-99-8401-5_8

is highlighted as a powerful strategy for gaining comprehensive insights into biological systems. While challenges exist, such as accurate off-target prediction and data management, future directions involving machine learning and user-friendly tools hold promise. Bioinformatics continues to revolutionize genome editing, advancing precision medicine, agriculture, and biological research.

Keywords

Bioinformatics · Genome editing · Computational analysis · Off-target effects · Multi-omics integration

8.1 Introduction to Genome Editing

Genome editing is a revolutionary field in genetics that has opened up unprecedented opportunities for manipulating the DNA sequences of living organisms. It allows scientists to make precise modifications to the genetic material of cells, organisms, and even entire populations. This breakthrough technology has the potential to revolutionize various fields, including medicine, agriculture, and biotechnology (Khalil 2020). At its core, genome editing involves the deliberate alteration of DNA sequences within the genome. It enables scientists to add, delete, or replace specific genetic information, thereby modifying the characteristics of an organism (Zhang and Zhou 2014). This capability has far-reaching implications, as it can lead to the development of new treatments for genetic disorders, enhance crop productivity and nutritional value, and contribute to the understanding of fundamental biological processes. One of the most widely used and powerful genome editing techniques is CRISPR-Cas9, which stands for Clustered Regularly Interspaced Short Palindromic Repeats and CRISPR-associated protein 9. CRISPR-Cas9 utilizes a guide RNA molecule to target a specific DNA sequence and Cas9, a DNA-cutting enzyme, to introduce the desired modifications. The simplicity, versatility, and efficiency of CRISPR-Cas9 have revolutionized genome editing research and applications. However, genome editing is a complex process that requires careful planning and execution. This is where bioinformatics plays a crucial role. Bioinformatics is an interdisciplinary field that combines biology, computer science, and statistics to analyze and interpret biological data. In the context of genome editing, bioinformatics provides the necessary tools and computational resources to facilitate the design, analysis, and optimization of genome editing experiments.

Bioinformatics aids in the identification and selection of target sequences within the genome that can be modified using CRISPR-Cas9 or other genome editing tools. By analyzing the genomic context, bioinformatics algorithms can predict the potential off-target effects of genome editing and help researchers mitigate any unintended consequences (Akram et al. 2022). This computational analysis enables scientists to make informed decisions regarding target selection and minimize the risks associated with genome editing. Furthermore, bioinformatics assists in the annotation and functional characterization of genomic variants. It helps researchers understand the impact of specific genetic changes on gene function, protein structure, and biological pathways. By integrating multiple omics data, such as genomics, transcriptomics, and proteomics, bioinformatics enables comprehensive analysis of the effects of genome editing on various levels of biological regulation. Moreover, bioinformatics plays a crucial role in managing and integrating the vast amount of data generated during genome editing experiments. It provides the necessary tools and databases for storing, organizing, and analyzing genomic information. This data management infrastructure ensures that researchers can access and share their findings, promoting collaboration and accelerating scientific progress in the field of genome editing. Genome editing has emerged as a powerful tool for manipulating genetic information, with far-reaching implications across various domains. Bioinformatics serves as an essential component in the field of genome editing, providing computational tools and resources to facilitate target selection, optimize experimental design, predict off-target effects, annotate genomic variants, integrate multi-omics data, and manage the vast amount of generated information. By harnessing the power of bioinformatics, scientists can unlock the full potential of genome editing and contribute to advancements in medicine, agriculture, and biotechnology.

8.2 Overview of Bioinformatics

Bioinformatics is an interdisciplinary field that combines biology, computer science, and statistics to analyze and interpret biological data. It encompasses a wide range of techniques, tools, and methodologies aimed at extracting meaningful information from vast amounts of biological data. With the advancements in high-throughput technologies, such as next-generation sequencing, bioinformatics has become an indispensable discipline in modern biological research. At its core, bioinformatics focuses on developing computational algorithms and models to study biological systems. It involves the collection, storage, retrieval, and analysis of biological data, including genomic sequences, protein structures, gene expression profiles, and metabolic pathways. By leveraging computational techniques, bioinformatics enables researchers to gain valuable insights into the complex and intricate workings of living organisms. One of the primary areas of bioinformatics is genomics, which involves the study of an organism's entire set of genes, known as its genome (Ratan et al. 2018). Bioinformatics tools and algorithms are extensively used to analyze and interpret genomic data. This includes tasks such as DNA sequence assembly, annotation of genes and regulatory elements, identification of genetic variations, and comparative genomics to understand the evolutionary relationships between species.

Another vital aspect of bioinformatics is proteomics, which focuses on the study of proteins, their structures, and functions. Bioinformatics plays a critical role in protein sequence analysis, predicting protein structures, and identifying protein–protein interactions (Fernandez-Recio et al. 2005). These insights are invaluable in understanding the complex mechanisms underlying various biological processes and diseases. Bioinformatics also contributes to the field of transcriptomics, which involves the study of gene expression patterns. By analyzing RNA sequencing data, bioinformatics can provide valuable information about which genes are active under specific conditions, identify alternative splicing events, and uncover regulatory networks governing gene expression. Furthermore, bioinformatics is essential in the field of metabolomics, which involves the study of small molecules called metabolites in biological systems. Bioinformatics tools assist in the identification and quantification of metabolites, as well as the integration of metabolomic data with other omics data to gain a comprehensive understanding of cellular processes. In addition to these specific areas, bioinformatics also encompasses other important applications such as systems biology, drug discovery, and precision medicine. It facilitates the integration of multiple data types, including genomics, proteomics, transcriptomics, and metabolomics, to gain a holistic view of biological systems and unravel their complexities (Kibar and Vingron 2023).

The field of bioinformatics continues to evolve rapidly, driven by advancements in computational technologies, machine learning, and artificial intelligence. These advancements enable the development of more sophisticated algorithms and models, allowing researchers to extract deeper insights from biological data. Moreover, bioinformatics promotes collaboration and data sharing through the development of databases, software tools, and public repositories, facilitating the dissemination of knowledge and fostering scientific advancements. Bioinformatics plays a vital role in modern biological research by leveraging computational techniques to analyze and interpret biological data. It encompasses various sub-disciplines, including genomics, proteomics, transcriptomics, and metabolomics, enabling researchers to gain valuable insights into the structure, function, and regulation of biological systems. With the continuous advancements in technology and computational methods, bioinformatics will continue to drive innovation and discovery in the field of life sciences.

8.3 Importance of Bioinformatics in Genome Editing

Bioinformatics plays a crucial role in the field of genome editing, providing valuable tools and resources that aid in the design, analysis, and optimization of genome editing experiments. The integration of bioinformatics with genome editing techniques enhances the precision, efficiency, and safety of genetic modifications. Let's explore the importance of bioinformatics in genome editing. Firstly, bioinformatics enables the computational analysis of target sequences within the genome. By utilizing bioinformatics algorithms, researchers can identify and select optimal target sites for genome editing (Joshi et al. 2023). These algorithms take into account factors such as specificity, efficiency, and potential off-target effects, helping researchers make informed decisions in choosing suitable target sequences. This bioinformatics-guided target selection ensures precise and effective genome editing, minimizing unintended genetic alterations. Secondly, bioinformatics assists in the design and optimization of CRISPR systems. CRISPR-Cas9, the most widely used genome editing tool, requires the design of guide RNA molecules that can accurately target specific DNA sequences. Bioinformatics algorithms predict the efficiency and specificity of guide RNA sequences, facilitating the selection of optimal guides for successful genome editing. Additionally, bioinformatics aids in optimizing the delivery of CRISPR components into target cells, improving the overall efficiency of genome editing experiments.

Another vital role of bioinformatics in genome editing is the prediction and evaluation of off-target effects. Despite the remarkable specificity of CRISPR-Cas9, there is a possibility of unintended genetic modifications at off-target sites. Bioinformatics tools analyze the genomic context and sequence homology to predict potential off-target sites, allowing researchers to mitigate these risks. By identifying and evaluating off-target effects computationally, researchers can optimize their experimental designs and minimize unintended genetic alterations. Furthermore, bioinformatics provides essential resources for the functional annotation of genomic variants resulting from genome editing. Bioinformatics databases and tools assist in the identification and characterization of genetic alterations, enabling researchers to understand the impact of these modifications on gene function, regulatory elements, and protein structure. This functional annotation helps researchers assess the potential consequences of genome editing and aids in the interpretation of experimental results. Bioinformatics also facilitates comparative genomics and homology analysis, which are critical for studying the evolutionary relationships between species and identifying conserved regions within genomes. By comparing genomic sequences across different species, researchers can identify functional elements and target sites that are conserved, enhancing the effectiveness of genome editing strategies (Hatanaka et al. 2023).

Furthermore, bioinformatics enables the integration of multi-omics data generated during genome editing experiments. By integrating genomics, transcriptomics, proteomics, and metabolomics data, researchers can gain a comprehensive understanding of the effects of genome editing on various biological levels. Bioinformatics tools and algorithms assist in the analysis and interpretation of these multi-dimensional datasets, providing a holistic view of the genetic modifications and their impact on cellular processes. Lastly, bioinformatics plays a pivotal role in data management and integration. The vast amount of data generated during genome editing experiments requires efficient storage, organization, and retrieval. Bioinformatics provides the necessary infrastructure, including databases, software tools, and pipelines, to manage and integrate genomic information. This ensures that researchers can access and share their data, promoting collaboration and accelerating scientific progress in the field of genome editing. Bioinformatics is of paramount importance in the field of genome editing (Navaridas et al. 2023). It provides computational tools, resources, and algorithms that aid in target selection, guide RNA design, off-target prediction, functional annotation, comparative genomics, multi-omics integration, and data management. By leveraging bioinformatics, researchers can enhance the precision, efficiency, and safety of genome editing techniques, paving the way for advancements in medicine, agriculture, and biotechnology.

8.4 Computational Analysis of Target Sequences

Computational analysis of target sequences is a fundamental aspect of genome editing that relies on bioinformatics tools and algorithms to identify optimal sites for genetic modification. By utilizing computational techniques, researchers can select target sequences with high specificity and efficiency, maximizing the success of genome editing experiments. Let's delve into the significance of computational analysis in identifying target sequences. The first step in computational analysis is the identification of potential target sites within the genome. This involves searching for specific DNA sequences that are amenable to modification using genome editing tools such as CRISPR-Cas9. Bioinformatics algorithms analyze the genomic sequence to identify regions that meet specific criteria, such as the presence of suitable protospacer adjacent motifs (PAMs) for CRISPR-Cas9 recognition (Cancellieri et al. 2023). These algorithms ensure the selection of target sequences that are compatible with the chosen genome editing tool. Furthermore, computational analysis assists in evaluating the uniqueness of target sequences. It is crucial to select target sites that are specific to the desired genomic region to minimize off-target effects. Bioinformatics algorithms compare the target sequence against the entire genome to assess its uniqueness and potential for off-target binding. This analysis helps researchers identify target sequences with minimal homology to non-intended regions, reducing the likelihood of unintended genetic modifications. Another aspect of computational analysis is the prediction of target site efficiency. Not all target sequences are equally efficient for genome editing. Bioinformatics tools predict the efficiency of target sites based on various factors, such as the accessibility of the DNA sequence, secondary structure formation, and nucleotide composition (Sharma et al. 2023). By evaluating these parameters, researchers can prioritize target sequences with high editing efficiency, increasing the success rate of genome editing experiments. Additionally, computational analysis aids in assessing potential limitations and challenges associated with target sequences. For instance, certain genomic regions may have high levels of repetitive elements or structural complexities, making them less amenable to genome editing. Bioinformatics algorithms can identify such regions and provide insights into potential challenges that may arise during the editing process. This information allows researchers to make informed decisions and adjust their experimental designs accordingly.

Moreover, computational analysis contributes to the identification of functional elements within target sequences. Bioinformatics algorithms scan the target sequence for important genomic features, such as coding regions, regulatory elements, and non-coding RNAs. This analysis helps researchers identify target sites that have the desired functional impact, such as modifying a specific gene or disrupting a regulatory element. By incorporating functional annotations, researchers can select target sequences that align with their experimental objectives. Furthermore, computational analysis facilitates the design of guide RNA molecules for CRISPR-Cas9-mediated genome editing (Table 8.1).

Guide RNAs guide the Cas9 enzyme to the target sequence for precise DNA cleavage. Bioinformatics algorithms predict and optimize guide RNA sequences to

Table 8.1 List of tools and webservers required for various steps of computational analysis of target sequences

Step	Tools/webservers	Links
Define the search criteria	User-defined criteria	–
Retrieve the genomic sequence	Ensembl	https://www.ensembl.org/
	NCBI Entrez	https://www.ncbi.nlm.nih.gov/gquery/
	UCSC Genome Browser	https://genome.ucsc.edu/
Preprocess the genomic sequence	FASTX Toolkit	http://hannonlab.cshl.edu/fastx_toolkit/
	Trimmomatic	http://www.usadellab.org/cms/?page=trimmomatic
	BWA	http://bio-bwa.sourceforge.net/
Identify potential target sites	CRISPRseek	http://www.bioconductor.org/packages/release/bioc/html/CRISPRseek.html
	Cas-OFFinder	http://www.rgenome.net/cas-offinder/portable
	E-CRISP	http://www.e-crisp.org/E-CRISP/
Assess target site uniqueness	BLAST	https://blast.ncbi.nlm.nih.gov/Blast.cgi
	Bowtie	http://bowtie-bio.sourceforge.net/
Predict target site efficiency	E-CRISP	http://www.e-crisp.org/E-CRISP/
	CRISPRscan	https://www.crisprscan.org/
	sgRNAcas9	http://www.sgrnacas9.org/
Evaluate potential limitations	RepeatMasker	http://www.repeatmasker.org/
	Tandem Repeats Finder	https://tandem.bu.edu/trf/trf.html
	G4Hunter	http://bioinformatics.ibp.cz/data/g4hunter/
Consider functional elements	Ensembl Variant Effect Predictor (VEP)	https://www.ensembl.org/info/docs/tools/vep/index.html
	ANNOVAR	https://annovar.openbioinformatics.org/
	GenomicRanges	https://bioconductor.org/packages/release/bioc/html/GenomicRanges.html
Optimize guide RNA design	CRISPRdirect	https://crispr.dbcls.jp/
	CHOPCHOP	https://chopchop.cbu.uib.no/
	sgRNAcas9	http://www.sgrnacas9.org/
Prioritize target sequences	Custom ranking based on defined criteria	–

ensure their specificity and efficiency (Naeem and Alkhnbashi 2023). By analyzing factors such as off-target potential, secondary structure formation, and binding affinity, computational analysis helps design guide RNAs that can precisely target the desired genomic region. Computational analysis of target sequences is a critical step in genome editing, enabled by bioinformatics tools and algorithms. By leveraging computational techniques, researchers can identify optimal target sites with high specificity, efficiency, and functional impact. Computational analysis aids in target sequence identification, uniqueness assessment, efficiency prediction, identification of potential challenges, and guide RNA design. Through this computational approach, researchers can enhance the precision and success of genome editing experiments, driving advancements in various fields such as medicine, agriculture, and biotechnology.

Computational analysis of target sequences involves a series of steps to identify optimal sites for genetic modification. By leveraging bioinformatics tools and algorithms, researchers can perform an in-depth analysis of the genome to select target sequences with high specificity and efficiency (see Table 8.1). Here are the steps involved in computational analysis of target sequences:

- Define the search criteria: Determine the specific requirements for the target sequence, such as the desired genomic region, sequence length, and any specific motifs or features to consider.
- Retrieve the genomic sequence: Obtain the relevant genomic sequence from databases or sequencing experiments, ensuring it covers the region of interest.
- Preprocess the genomic sequence: Perform necessary preprocessing steps, such as removing ambiguous characters, correcting sequencing errors, or handling variations in genome assembly.
- Identify potential target sites: Utilize bioinformatics algorithms to scan the genomic sequence and identify potential target sites based on specific criteria. This may involve searching for suitable protospacer adjacent motifs (PAMs) for CRISPR-Cas9 or other recognition sequences for alternative genome editing tools.
- Assess target site uniqueness: Compare the potential target sites against the entire genome to evaluate their uniqueness. Bioinformatics tools can help identify regions with homology to other non-intended genomic locations, minimizing the risk of off-target effects.
- Predict target site efficiency: Utilize computational algorithms to predict the efficiency of target sites. Factors such as DNA accessibility, secondary structure formation, and nucleotide composition can be assessed to estimate the likelihood of successful genome editing.
- Evaluate potential limitations: Analyze the target sequences for any limitations or challenges that may impact the editing process. This could include the presence of repetitive elements, structural complexities, or other known constraints. Identifying such limitations helps researchers anticipate potential difficulties and adjust their experimental design accordingly.
- Consider functional elements: Scan the target sequences for important genomic features, such as coding regions, regulatory elements, or non-coding RNAs. This step allows researchers to select target sites that align with their specific experimental objectives and have the desired functional impact.
- Optimize guide RNA design (if applicable): If using CRISPR-Cas9, design and optimize guide RNA molecules to guide the Cas9 enzyme to the target sequence. Computational analysis can predict off-target potential, assess secondary structure formation, and optimize binding affinity to ensure guide RNAs are specific and efficient.

- Prioritize target sequences: Based on the results of the computational analysis, prioritize the identified target sequences according to their uniqueness, predicted efficiency, functional impact, and any other relevant criteria.

By following these steps, researchers can utilize computational analysis to identify and prioritize optimal target sequences for genome editing. This approach enhances the precision and success of genetic modifications, contributing to advancements in fields such as medicine, agriculture, and biotechnology.

8.5 Designing and Optimizing CRISPR Systems

Designing and optimizing CRISPR systems is a crucial step in genome editing, as it directly impacts the efficiency and precision of the gene editing process. CRISPR, or Clustered Regularly Interspaced Short Palindromic Repeats, is a revolutionary technology that allows researchers to precisely modify the DNA of organisms. The design and optimization of CRISPR systems involve several key considerations to ensure successful and accurate gene editing outcomes. The first step in designing a CRISPR system is the selection of the Cas9 protein or other nucleases that will be used to target the specific genomic region of interest. Cas9 is the most commonly used nuclease, but other nucleases such as Cpf1 are also employed. Factors such as the efficiency, specificity, and off-target effects of the nuclease need to be taken into account during the selection process. Once the nuclease is chosen, the next crucial step is designing the guide RNA (gRNA) that will guide the nuclease to the target DNA sequence (Tian et al. 2023). The gRNA is a short RNA molecule that binds to the target DNA and directs the nuclease to create a double-stranded break at the desired genomic location. Designing an effective gRNA involves identifying the protospacer adjacent motif (PAM) sequence, which is necessary for Cas9 binding, as well as optimizing the gRNA sequence to enhance its specificity and minimize off-target effects. Bioinformatics tools and algorithms play a vital role in the design and optimization of CRISPR systems. These tools help in identifying suitable target sites within the genome, predicting potential off-target effects, and optimizing the gRNA sequence for improved efficiency and specificity. Tools like CRISPR Design, E-CRISP, and CRISPRscan assist researchers in selecting optimal target sites and designing high-quality gRNAs.

Another important aspect of designing and optimizing CRISPR systems is the delivery method of the CRISPR components into the target cells or organisms. Different delivery methods, such as viral vectors, electroporation, or nanoparticle-mediated delivery, have varying efficiencies and capabilities to reach specific cell types or tissues. The choice of delivery method depends on factors such as the target organism, cell type, and intended application. Optimizing CRISPR systems also involves assessing and fine-tuning experimental parameters, such as the concentration of the CRISPR components, incubation time, and temperature. These parameters can significantly influence the editing efficiency and minimize potential off-target effects. Iterative optimization experiments are often performed to achieve the desired editing outcomes. Designing and optimizing CRISPR systems require careful consideration of various factors such as nuclease selection, gRNA design, delivery method, and experimental parameters. The use of bioinformatics tools and algorithms aids in efficient target site selection, gRNA design, and prediction of off-target effects. By optimizing these factors, researchers can enhance the efficiency, specificity, and accuracy of CRISPR-based genome editing, opening up new avenues for genetic research and potential therapeutic applications. The steps involved in designing and optimizing CRISPR systems:

- Identify the target genomic region: Determine the specific region of the genome that needs to be edited or modified. This can be a gene, regulatory element, or other genomic features.
- Select the appropriate nuclease: Choose the suitable nuclease for the intended application. Cas9 is commonly used, but other nucleases like Cpf1 or Cas12a can also be considered based on their specific properties.
- Design the guide RNA (gRNA): Design a gRNA sequence that targets the desired genomic region. The gRNA should be complementary to the target DNA sequence and contain the necessary protospacer adjacent motif (PAM) sequence required for nuclease binding.
- Evaluate potential off-target effects: Utilize bioinformatics tools to predict potential off-target sites where the gRNA may bind. Assess the specificity of the designed gRNA to minimize the risk of unintended modifications in other genomic regions.
- Optimize the gRNA sequence: Fine-tune the gRNA sequence to enhance its efficiency and specificity. Consider parameters such as length, secondary structure, and GC content to improve the binding affinity and minimize off-target effects.
- Determine the delivery method: Choose an appropriate delivery method for introducing the CRISPR components into the target cells or organisms. This can include viral vectors, electroporation, lipid-based transfection, or other specialized delivery systems.
- Validate and optimize experimental parameters: Conduct preliminary experiments to optimize key parameters such as the concentration of CRISPR components, incubation time, temperature, and cell density. These parameters can significantly influence the efficiency and specificity of the editing process.
- Assess editing efficiency: Evaluate the efficiency of the CRISPR system by analyzing the frequency and accuracy of desired edits in the target genomic region. Techniques like PCR, DNA sequencing, or reporter assays can be used for this purpose.
- Iterate and refine: Based on the results obtained, refine the design and experimental parameters if necessary. Iterative optimization may involve modifying the gRNA sequence, adjusting nuclease concentrations, or exploring alternative delivery methods.
- Validate the desired edits: Confirm the desired genomic modifications through thorough analysis, such as targeted sequencing or functional assays. Validate the

edited phenotype or functional outcome, depending on the specific objectives of the experiment.

8.6 Prediction and Evaluation of Off-Target Effects

Prediction and evaluation of off-target effects is a critical aspect of genome editing using CRISPR technology. While CRISPR systems offer remarkable precision, there is still a possibility of unintended modifications at genomic sites similar to the target sequence. Therefore, it is crucial to employ computational tools and experimental methods to predict and evaluate potential off-target effects. The first step in predicting off-target effects is the identification of potential off-target sites. This involves analyzing the genomic sequence for regions that share high similarity with the target sequence and the corresponding guide RNA (gRNA) (Spade 2023). Bioinformatics tools and algorithms have been developed to identify potential off-target sites based on sequence alignment and mismatch analysis. These tools search for sequences that possess similar protospacer adjacent motifs (PAMs) and exhibit only a few nucleotide mismatches with the target sequence. Once potential off-target sites are identified, the next step is to prioritize and evaluate their likelihood of being edited. Several factors come into play during this evaluation. One important consideration is the number and position of mismatches between the gRNA and the off-target site. Off-target sites with fewer mismatches and located near the PAM sequence are generally more prone to editing. Experimental validation is essential to confirm the presence and extent of off-target effects. Various techniques can be employed for this purpose, such as targeted sequencing, high-throughput sequencing, or genome-wide analyses. These approaches involve deep sequencing of the genomic regions surrounding the predicted off-target sites to detect any modifications or alterations.

Additionally, researchers can use control experiments to distinguish true off-target effects from potential artifacts. Control experiments involve comparing edited samples with appropriate negative controls, such as samples treated with an inactive nuclease or samples without any CRISPR components. This helps differentiate specific editing events from background noise or unintended modifications unrelated to CRISPR activity. Furthermore, advancements in CRISPR technology have led to the development of modified or engineered Cas proteins that exhibit improved specificity and reduced off-target effects. These modified nucleases, such as high-fidelity Cas9 variants or Cas9 fusion proteins offer enhanced targeting precision while minimizing unintended editing at off-target sites. Prediction and evaluation of off-target effects are crucial steps in CRISPR-based genome editing. Through the use of bioinformatics tools, computational analysis, and experimental validation, researchers can assess the likelihood and extent of off-target modifications. This knowledge enables the refinement of CRISPR designs and the development of strategies to minimize off-target effects, ultimately enhancing the specificity and accuracy of genome editing applications.

8.7 Functional Annotation of Genomic Variants

Functional annotation of genomic variants is a vital step in understanding the potential impact of genetic variations on gene function and disease susceptibility. With the advent of high-throughput sequencing technologies, numerous genomic variants can be identified in individuals, necessitating comprehensive annotation to interpret their functional significance. The process of functional annotation involves associating genomic variants with various functional elements in the genome. This includes identifying whether the variant falls within protein-coding regions, regulatory regions, non-coding RNA genes, or other important genomic features. Additionally, the annotation aims to determine the potential consequences of the variants, such as their impact on protein structure, gene expression, splicing, or regulatory interactions (Zhou et al. 2023).

Bioinformatics tools and databases play a crucial role in functional annotation. These resources provide comprehensive genomic annotations and integrate information from diverse data sources, including public databases, functional genomics experiments, evolutionary conservation analyses, and computational predictions. They assist in prioritizing variants for further investigation and provide insights into their potential functional consequences. Variant annotation typically involves the use of annotation tools that utilize reference genome sequences and incorporate variant calling data. These tools assign functional annotations based on known features, such as protein domains, DNA-binding motifs, and transcription factor binding sites. They can also predict the impact of variants on protein structure, function, and stability using algorithms and structural modeling approaches. Additionally, functional annotation often includes the analysis of allele frequencies in population databases. This information helps determine the prevalence of variants in different populations and it can be informative for studying genetic diversity, disease associations, or population-specific effects (Fig. 8.1).

Moreover, functional annotation extends beyond individual variants to consider their potential interactions within biological pathways or networks. Integration of variant data with functional pathway analysis allows for the identification of affected biological processes, enrichment of gene sets, and prioritization of pathways that may be dysregulated due to the presence of specific variants. Experimental validation is essential to confirm the functional impact of variants identified through annotation. Techniques such as functional assays, reporter assays, gene expression studies, or genome editing experiments can provide direct evidence of the effects of variants on gene function and cellular processes (Nagral et al. 2023). Functional annotation of genomic variants is a crucial step in interpreting their potential biological significance. By leveraging bioinformatics tools, databases, and experimental validations, researchers can gain insights into the functional consequences of genetic variations. This knowledge facilitates the understanding of disease mechanisms, identification of therapeutic targets, and personalized medicine approaches based on an individual's genetic makeup.

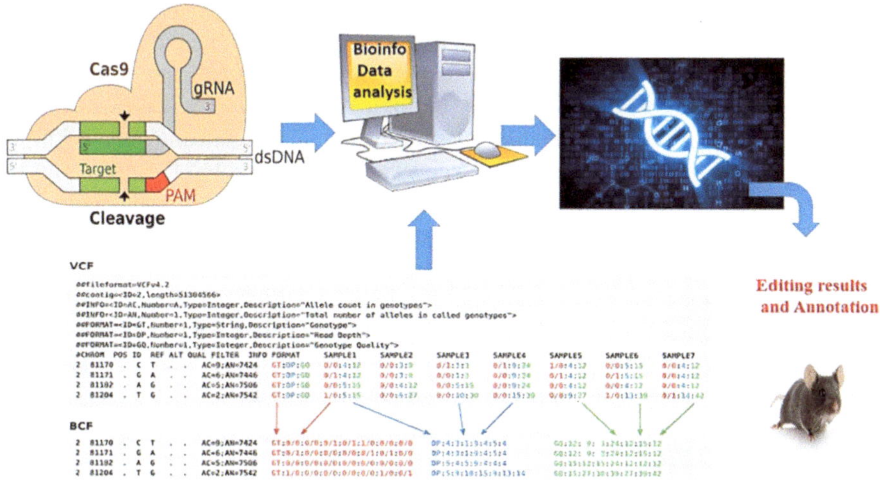

Fig. 8.1 Genomic data collection and its editing using bioinformatics tools to generate annotation of novel genetic functionality

8.8 Comparative Genomics and Homology Analysis

Comparative genomics and homology analysis are powerful approaches used to study the similarities and differences in genomic sequences among different organisms. These methods provide valuable insights into evolutionary relationships, functional conservation, and identification of important genomic elements across species. Comparative genomics involves the systematic comparison of genomic sequences from different organisms. By aligning and comparing DNA or protein sequences, researchers can identify regions of similarity and divergence. This analysis helps in understanding the evolutionary relationships between species and provides clues about the conservation of functional elements, such as protein-coding genes, regulatory sequences, or non-coding RNAs. Homology analysis is a fundamental aspect of comparative genomics. It aims to identify and characterize genes or genomic elements that have descended from a common ancestor. Through homology analysis, researchers can infer the presence of orthologous genes (genes in different species that have a common ancestor) or paralogous genes (genes that have arisen through gene duplication events within a species). Bioinformatics tools and algorithms are essential for conducting comparative genomics and homology analysis (Tao et al. 2023; Kaushik et al. 2022). These tools utilize sequence alignment algorithms, such as BLAST (Basic Local Alignment Search Tool), to compare sequences and identify regions of similarity. Multiple sequence alignment methods, such as ClustalW, MUSCLE, or MAFFT, are employed to align sequences from multiple organisms, allowing for the identification of conserved regions and the detection of evolutionary changes.

Comparative genomics and homology analysis have wide-ranging applications. They provide insights into gene function and regulation, identification of conserved non-coding elements, inference of gene regulatory networks, and discovery of candidate genes involved in specific biological processes or diseases. These approaches are particularly valuable for studying model organisms, as the knowledge gained from well-characterized species can be extrapolated to understand the biology of related organisms. Furthermore, comparative genomics and homology analysis have implications in fields such as evolutionary biology, phylogenetics, and drug discovery. By comparing genomes across species, researchers can trace the evolutionary history of genes and identify genetic variations that contribute to phenotypic differences or disease susceptibility. Comparative genomics and homology analysis provide valuable insights into the evolution and functional conservation of genomic sequences. Through the use of bioinformatics tools and algorithms, researchers can compare sequences, identify homologous genes, and unravel the relationships between different organisms. These approaches have broad applications in understanding gene function, unraveling evolutionary relationships, and advancing our knowledge of the biological processes underlying life.

8.9 Integration of Multi-Omics Data

Integration of multi-omics data has emerged as a powerful approach to unravel the complexities of biological systems by combining information from multiple molecular levels. Omics technologies, such as genomics, transcriptomics, proteomics, metabolomics, and epigenomics, generate vast amounts of data, providing a comprehensive view of cellular processes and their interconnections. Integration of these multi-omics datasets enables a deeper understanding of biological mechanisms, identification of biomarkers, and discovery of novel therapeutic targets. The integration of multi-omics data involves several key steps. First, data from different omics platforms need to be collected and preprocessed to ensure compatibility and quality. This includes data normalization, filtering, and transformation to account for technical variations and biases introduced during data generation. Next, bioinformatics methods and statistical algorithms are applied to integrate the multi-omics datasets (Cai et al. 2022). These methods aim to identify relationships, patterns, and associations between the different molecular layers. They can involve data fusion techniques, network analysis, machine learning algorithms, or statistical modeling approaches. The goal is to extract meaningful information and uncover molecular interactions, regulatory networks, and biological pathways that drive complex biological phenomena.

One of the main challenges in integrating multi-omics data is dealing with the high dimensionality and heterogeneity of the datasets. Various computational approaches have been developed to address these challenges, including dimensionality reduction techniques, feature selection methods, and data integration algorithms. These approaches help reduce noise, identify key features, and capture the underlying biological signals present in the data. Integration of multi-omics data has numerous applications across different fields of biology and medicine. In cancer research, for example, the integration of genomics, transcriptomics, and proteomics data can provide a comprehensive view of molecular alterations, identify driver mutations, and reveal potential therapeutic targets. In personalized medicine, integration of multi-omics data can aid in predicting treatment responses, stratifying patient populations, and guiding therapeutic decisions. Furthermore, the integration of multi-omics data has implications in systems biology and precision medicine. It enables the identification of biomarkers for early disease detection, understanding disease mechanisms, and discovering new drug targets (Zhang et al. 2022). By combining information from multiple molecular layers, researchers can gain a more comprehensive understanding of the complexity of biological systems and uncover novel insights that would be difficult to obtain from single-omics analyses. The integration of multi-omics data is a powerful approach that leverages the wealth of information provided by different omics technologies. By combining and analyzing data from genomics, transcriptomics, proteomics, metabolomics, and epigenomics, researchers can gain a deeper understanding of biological processes, identify molecular interactions, and discover new biomarkers and therapeutic targets. This integrative approach has the potential to revolutionize our understanding of complex diseases, drive precision medicine efforts, and advance our knowledge of the intricacies of living systems.

8.10 Data Management and Integration

Data management and integration play a crucial role in modern scientific research, especially in fields such as bioinformatics, genomics, and systems biology. As the volume and complexity of data generated from various sources continue to increase, effective data management and integration strategies are essential for organizing, storing, and analyzing large datasets and extracting meaningful insights. Data management involves the systematic organization and storage of data to ensure its accessibility, accuracy, and integrity. It encompasses various activities, including data acquisition, data cleaning, data storage, data documentation, and data sharing. Proper data management practices help researchers maintain data quality, enable reproducibility, and facilitate collaboration and data sharing within the scientific community (Yeo and Selvarajoo 2022). One of the key aspects of data management is data integration, which involves combining data from multiple sources or experiments to create a unified and comprehensive dataset. Integration enables researchers to merge diverse datasets, such as genomic data, clinical data, or environmental data, to gain a more holistic understanding of complex biological systems. It allows for the identification of patterns, correlations, and relationships that may not be apparent when analyzing individual datasets in isolation.

Bioinformatics tools and databases play a vital role in data management and integration. These resources provide platforms for data storage, retrieval, and analysis, as well as standardized formats and protocols for data exchange. Researchers can leverage these tools and databases to manage and integrate various types of biological data, including DNA sequences, gene expression profiles, protein structures, and functional annotations. Furthermore, data management and integration often involve the use of data integration frameworks and computational algorithms. These methods facilitate the seamless integration of diverse datasets by addressing issues such as data heterogeneity, data format conversion, and data mapping. Data integration frameworks enable researchers to merge datasets with different structures, ontologies, or data models, ensuring compatibility and consistency across the integrated dataset. Effective data management and integration have numerous benefits in scientific research. They enable researchers to uncover hidden insights, generate new hypotheses, and make data-driven decisions. Integration of diverse datasets enhances the power and robustness of analyses, allowing for a more comprehensive understanding of complex biological phenomena. Furthermore, proper data management practices ensure the long-term preservation and availability of valuable research data, promoting transparency and reproducibility. Data management and integration are essential components of scientific research in the era of big data. By implementing effective data management strategies, researchers can ensure data quality, accessibility, and reproducibility. Integration of diverse datasets enables researchers to extract meaningful insights and gain a deeper understanding of complex biological systems. Embracing proper data management and integration practices is crucial for advancing scientific knowledge, facilitating collaboration, and driving discoveries across various disciplines.

8.11 Challenges and Future Directions in Bioinformatics for Genome Editing

Bioinformatics has played a crucial role in advancing genome editing technologies, such as CRISPR-Cas9, and has greatly facilitated the design, analysis, and optimization of gene-editing experiments. However, several challenges remain, and future directions in bioinformatics are poised to address these challenges and further enhance the efficiency and precision of genome editing techniques. One of the primary challenges in bioinformatics for genome editing is the accurate prediction of off-target effects. While considerable progress has been made in developing computational tools and algorithms to predict potential off-target sites, there is still room for improvement. Enhancing the specificity and accuracy of off-target prediction algorithms will be crucial in minimizing unintended modifications and ensuring the safety of genome editing applications. Another challenge lies in the prediction of on-target editing efficiency. While bioinformatics tools can identify potential target sites for genome editing, accurately estimating the editing efficiency at these sites remains a challenge (Han et al. 2022). Factors such as chromatin accessibility, DNA structure, and epigenetic modifications can influence the editing outcomes. Integrating these factors into computational models will improve the prediction of on-target editing efficiency and aid in selecting optimal target sites. Furthermore, the analysis and interpretation of large-scale genomics and multi-omics datasets pose significant challenges. The integration of genomics, transcriptomics, proteomics, and other omics data requires sophisticated algorithms and computational methods. Developing comprehensive and scalable bioinformatics pipelines that can handle the vast amounts of data generated by high-throughput sequencing technologies will be crucial for leveraging these datasets in genome editing research.

Additionally, the interpretation of functional consequences resulting from genomic modifications is an ongoing challenge. While bioinformatics tools can predict the impact of genetic variants and editing events on protein function and gene regulation, accurately understanding the functional implications in complex biological systems remains complex. Integrating experimental validation, functional assays, and advanced computational approaches will be instrumental in unraveling the intricate relationship between genomic alterations and phenotypic outcomes. As for future directions, advancements in machine learning and artificial intelligence hold great promise for bioinformatics in genome editing. Deep learning algorithms and neural networks can potentially enhance the accuracy and efficiency of off-target prediction, on-target editing efficiency prediction, and functional annotation of genomic variants. Integrating these advanced computational techniques into existing bioinformatics pipelines will contribute to more precise and reliable genome editing outcomes. Another future direction lies in the development of user-friendly bioinformatics tools and software platforms. Simplifying the accessibility and usability of bioinformatics tools will democratize their use and enable a broader community of researchers to employ these powerful techniques in their genome editing experiments. User-friendly interfaces, intuitive workflows, and comprehensive documentation will enhance the adoption and impact of bioinformatics in the field. Bioinformatics plays a pivotal role in genome editing, but challenges persist. Addressing these challenges and embracing future directions will propel the field forward. By improving off-target prediction, on-target editing efficiency estimation, multi-omics data analysis, and functional interpretation, bioinformatics will continue to revolutionize genome editing technologies, opening new avenues for precision medicine, agriculture, and fundamental biological research.

8.12 Conclusion

Bioinformatics has emerged as a vital discipline in the field of genome editing, facilitating various aspects of the gene-editing process. Through computational analysis of target sequences, researchers can identify suitable target sites for genome editing and assess their potential impact. Designing and optimizing CRISPR systems using bioinformatics tools enables the development of more efficient and precise gene-editing tools. Prediction and evaluation of off-target effects using computational algorithms aid in minimizing unintended modifications and ensuring the safety of genome editing applications. Functional annotation of genomic variants provides insights into the functional consequences of genetic alterations, guiding researchers in understanding their impact on protein function and gene regulation. Comparative genomics and homology analysis help unravel evolutionary relationships and identify conserved elements across species, contributing to our understanding of gene function and evolution. Integration of multi-omics data allows for a comprehensive view of biological systems, enabling the identification of molecular interactions and the discovery of novel biomarkers and therapeutic targets. Challenges, such as accurate prediction of off-target effects, on-target editing efficiency, data management, and functional interpretation, exist in the field of bioinformatics for genome editing. However, these challenges present opportunities for future advancements, including the integration of machine learning and artificial intelligence, development of user-friendly tools, and enhanced data analysis techniques. By addressing these challenges and embracing future directions, bioinformatics will continue to revolutionize genome editing technologies, empowering researchers with powerful tools for precision medicine, agriculture, and biological research.

References

Akram F, Haq IU, Sahreen S, Nasir N, Naseem W, Imitaz M, Aqeel A (2022) CRISPR/Cas9: a revolutionary genome editing tool for human cancers treatment. Technol Cancer Res Treat 21: 15330338221132078

Cai Z, Poulos RC, Liu J, Zhong Q (2022) Machine learning for multi-omics data integration in cancer. Iscience 25:103798

Cancellieri S, Zeng J, Lin LY, Tognon M, Nguyen MA, Lin J, Pinello L (2023) Human genetic diversity alters off-target outcomes of therapeutic gene editing. Nat Genet 55(1):34–43

Fernandez-Recio J, Totrov M, Skorodumov C, Abagyan R (2005) Optimal docking area: a new method for predicting protein–protein interaction sites. Proteins 58(1):134–143

Han P, Teo WZ, Yew WS (2022) Biologically engineered microbes for bioremediation of electronic waste: Wayposts, challenges and future directions. Eng Biol 6(1):23–34

Hatanaka F, Suzuki K, Shojima K, Yu J, Takahashi Y, Sakamoto A, Belmonte JCI (2023) Therapeutic strategy for spinal muscular atrophy by combining gene supplementation and genome editing. bioRxiv 2023:535786

Joshi A, Song HG, Yang SY, Lee JH (2023) Integrated molecular and bioinformatics approaches for disease-related genes in plants. Plants 12(13):2454

Kaushik V, Jain P, Akhtar N, Joshi A, Gupta LR, Grewal RK, Chawla M (2022) Immunoinformatics-aided design and in vivo validation of a peptide-based multiepitope vaccine targeting canine circovirus. ACS Pharmacol Transl Sci 5(8):679–691

Khalil AM (2020) The genome editing revolution. J Genet Eng Biotechnol 18(1):1–16

Kibar G, Vingron M (2023) Prediction of protein–protein interactions using sequences of intrinsically disordered regions. Proteins 91:980

Naeem M, Alkhnbashi OS (2023) Current bioinformatics tools to optimize CRISPR/Cas9 experiments to reduce off-target effects. Int J Mol Sci 24(7):6261

Nagral A, Mallakmir S, Garg N, Tiwari K, Masih S, Nagral N, Aggarwal R (2023) Genomic variations in ATP7B gene in Indian patients with Wilson disease. Indian J Pediatr 90(3): 240–248

Navaridas R, Vidal-Sabanés M, Ruiz-Mitjana A, Perramon-Güell A, Megino-Luque C, Llobet-Navas D, Dolcet X (2023) Transient and DNA-free in vivo CRISPR/Cas9 genome editing for flexible modeling of endometrial carcinogenesis. Cancer Commun 43(5):620

Ratan ZA, Son YJ, Haidere MF, Uddin BMM, Yusuf MA, Zaman SB, Cho JY (2018) CRISPR-Cas9: a promising genetic engineering approach in cancer research. Ther Adv Med Oncol 10: 1758834018755089

Sharma P, Dahiya S, Kaur P, Kapil A (2023) Computational biology: role and scope in taming antimicrobial resistance. Indian J Med Microbiol 41:33–38

Spade G (2023) The revolutionary genome editor: CRISPR-Cas9 systems

Tao X, Xu T, Lin X, Xu S, Fan Y, Guo B, Yue H (2023) Genomic profiling reveals the variant landscape of sporadic parathyroid adenomas in Chinese population. J Clin Endocrinol Metabol 108(7):1768–1775

Tian M, Zhang R, Li J (2023) Emergence of CRISPR/Cas9-mediated bioimaging: a new dawn of in-situ detection. Biosens Bioelectron 232:115302

Yeo HC, Selvarajoo K (2022) Machine learning alternative to systems biology should not solely depend on data. Brief Bioinform 23(6):bbac436

Zhang L, Zhou Q (2014) CRISPR/Cas technology: a revolutionary approach for genome engineering. Sci China Life Sci 57:639–640

Zhang C, Chen Y, Zeng T, Zhang C, Chen L (2022) Deep latent space fusion for adaptive representation of heterogeneous multi-omics data. Brief Bioinform 23(2):bbab600

Zhou H, Arapoglou T, Li X, Li Z, Zheng X, Moore J, Lin X (2023) FAVOR: functional annotation of variants online resource and annotator for variation across the human genome. Nucleic Acids Res 51(D1):D1300–D1311

Bioinformatics in Pathway Identification, Design, Modelling, and Simulation

9

Juveriya Israr, Sahabjada Siddiqui, Sankalp Misra, Indrajeet Singh, and Ajay Kumar

Abstract

Bioinformatics plays a crucial contribution in the study of complex biological systems, particularly in the areas of pathway identification, design, modelling, and simulation. This chapter aims to provide a high-level survey of the applications of bioinformatics in various fields. The methods and algorithms used in bioinformatics to solve problems with pathway identification, design, modelling, and simulation are highlighted. The chapter also includes numerous software examples to illustrate potential applications of bioinformatics in the context of pathway analysis. The materials and information were collected from various data bases and PubMed, Research Gate, Wikipedia, etc. The current comprehensive chapter would lead to develop a foundation for further work in bioinformatics by providing a comprehensive overview of pathway discovery, design, modelling, and simulation.

J. Israr
Faculty of Biosciences, Institute of Biosciences and Technology, Shri Ramswaroop Memorial University, Lucknow, Uttar Pradesh, India

Department of Biotechnology, Era University, Lucknow, Uttar Pradesh, India

S. Siddiqui (✉)
Department of Biotechnology, Era University, Lucknow, Uttar Pradesh, India
e-mail: sahabjada@erauniversity.in

S. Misra
Faculty of Biosciences, Institute of Biosciences and Technology, Shri Ramswaroop Memorial University, Lucknow, Uttar Pradesh, India

I. Singh · A. Kumar
Department of Biotechnology, Rama University, Kanpur, Uttar Pradesh, India

V. Singh, A. Kumar (eds.), *Advances in Bioinformatics*,
https://doi.org/10.1007/978-981-99-8401-5_9

Keywords

Bioinformatics tools · Pathway identification · Pathway discovery · Modelling · Simulation

9.1 Introduction

Bioinformatics is a fast-growing area of biology that uses computation and statistics to interpret and understand massive life science databases. Pathway discovery, design, modelling, and simulation are just a few examples of how bioinformatics is used to better understand biological processes and their underlying principles. Bioinformatics tools allow scientists to locate, generate, model, and simulate pathways with more accuracy and efficiency than ever before (Mubeen et al. 2019; Kim et al. 2012; Park et al. 2009). The role of bioinformatics is crucial in this respect. Scientists can use it to look for patterns in large databases that may shed light on illness mechanisms, treatment options, or metabolic pathways. Through the use of bioinformatics, researchers are able to better comprehend complex biological systems by integrating information from several databases.

Bioinformatics is essential for understanding how living things work. It lays a firm groundwork for identifying pathways, designing them, modelling, and simulating them by merging computer research with biological data. For better disease knowledge, drug discovery, and synthetic biology, bioinformatics-driven pathway analysis yields crucial insights. Bioinformatics is an interdisciplinary field that uses methods from many other fields to analyse and make sense of biological data. Combining the methods of computer science, mathematics, and statistics with data from the life sciences is essential for comprehending the complex biological processes that make up living things. High-throughput techniques, including because methods such as next-generation sequencing, microarrays, and mass spectrometry generate vast amounts of data that require the expertise of bioinformaticians to manage and assess. By identifying key genes, proteins, and pathways in biological processes and developing models and simulations based on these findings, researchers can get a deeper understanding of these processes. This chapter's goal is to evaluate bioinformatics methods for identifying pathways, creating models of those pathways, and simulating their operation. The most widely used techniques, resources, and applications in these fields have been highlighted. Furthermore, by showcasing the current status of pathway bioinformatics research and debating its developing trends and prospective future directions, we intended to contribute to subject development in bioinformatics (Fig. 9.1).

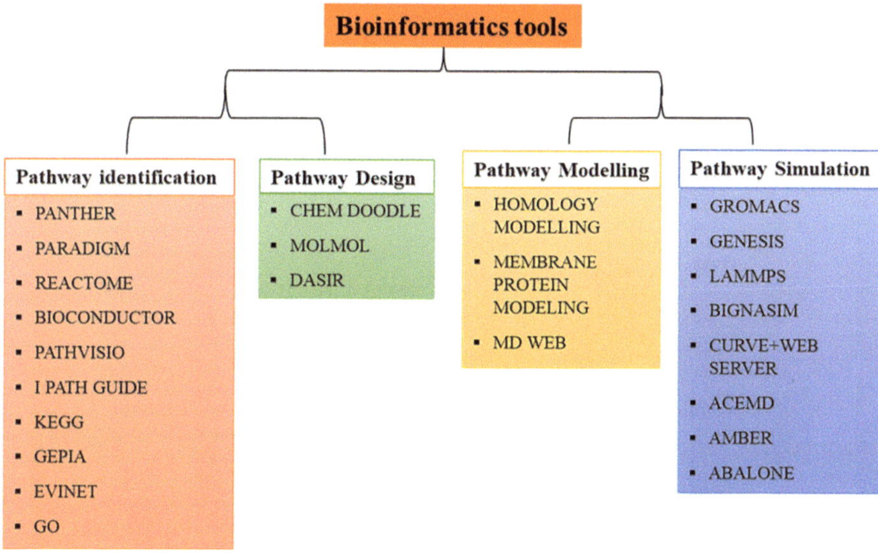

Fig. 9.1 Basic workflow of the tools used in bioinformatics

9.2 Pathway Identification

Discovering the biological pathways or networks responsible for the regulation and maintenance of particular cellular activities is called 'pathway identification'. Bioinformatics methods are used to analyse large data sets in order to identify key genes, proteins, and pathways in these processes (Kanehisa et al. 2016). Bioinformatics has been used, for instance, to chart out the signalling pathways involved in cancer. Gene expression data from tumour samples can shed light on which genes and pathways are aberrantly activated in cancer cells, which would be very useful in drugs design targeting specific pathways (Ashburner et al. 2000). Pathway analysis is an essential part of bioinformatics. Pathway analysis software decodes high-throughput biological data, a common requirement in life science research. Methods of route analysis can be used to construct a novel pathway from genes and proteins previously known to be involved in a given process, or to discover essential genes and proteins within a route in relation to an experiment or pathological condition (García-Campos et al. 2015; Mubeen et al. 2019). The following are some methods for identifying promising avenues (Fig. 9.2a, b):

9.2.1 PANTHER

According to the official website (http://pantherdb.org/) and the research of Mi et al. (2019), to aid with high-throughput analysis, the PANTHER Classification System

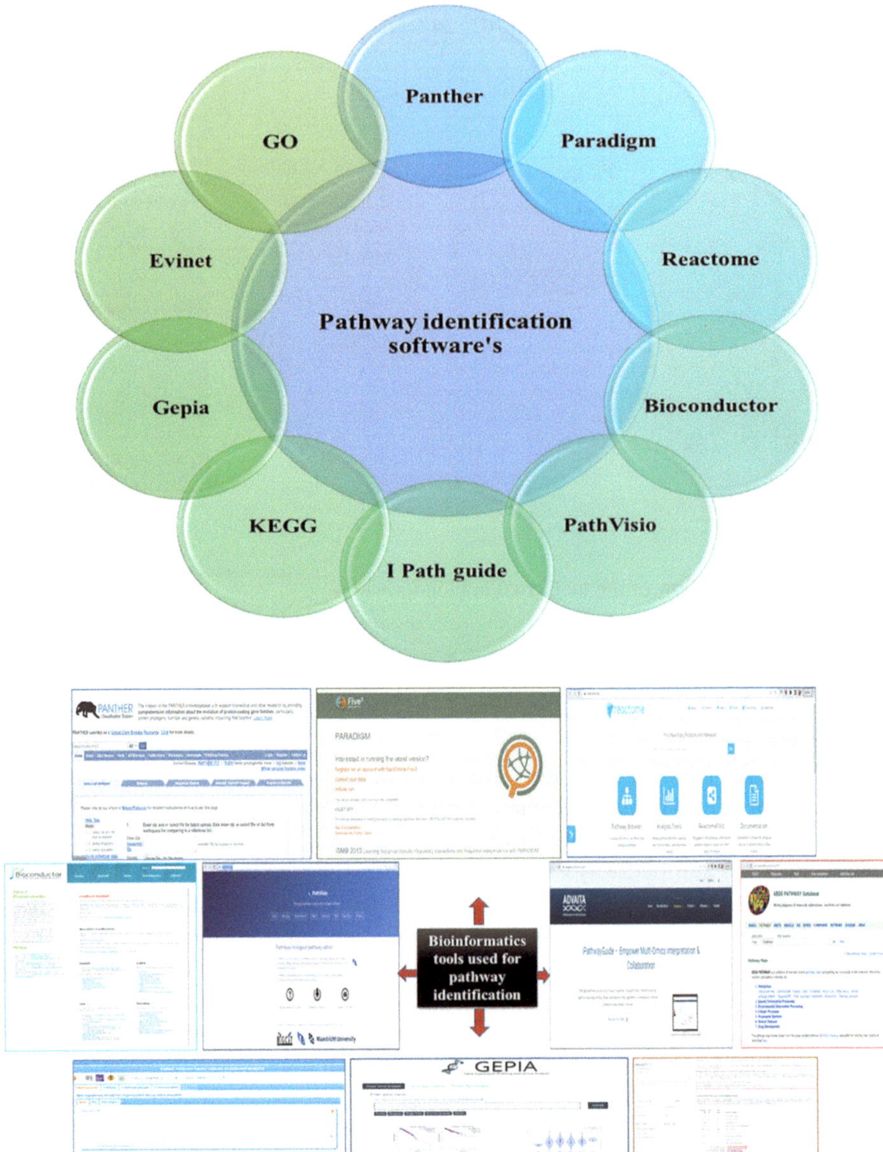

Fig. 9.2 (**a**) List of pathway identification software. (**b**) List of the webpages of different pathway identification software

was developed to categorise proteins and the genes that code for them. The PANTHER program is used to find new pathways by examining the evolutionary connections between proteins. The aforementioned is a well-structured database of information about gene and protein families and the subfamilies that have common functions. This database can be used to efficiently classify gene products and identify their functional characteristics. PANTHER is used in the Gene Ontology Reference Genome Project classifies proteins along with associated genes, enabling more effective high-throughput evaluation. This resource helps researchers classify protein sequences from new sources and analyse gene lists constructed from high-throughput genomic data. The PANTHER software suite allows users to search for routes and their constituents according to a wide variety of parameters, including molecular function, biological process, pathway link, interaction, and PANTHER family. A total of 176 custom pathways built in Cell Designer may be found in the PANTHER database. These routes can be downloaded in either the SBML or SBGN formats, which are utilised in systems biology. Mi et al. (2012, 2016, 2020) have listed the detailed features of PANTHER 16: a revamped family categorisation system, a tree-based classification tool, regions that boost gene expression, and a robust application programming interface.

9.2.2 PARADIGM

Using data from many genetic models, the computer tool PARADIGM may identify genomic pathways. Using functional genomic data and a route diagram, it predicts genetic activity for subsequent investigation. Using multi-dimensional cancer genomes data, the PARADIGM tool is able to infer the functioning of patient-specific pathways. The framework at hand is concerned with factor graphs and their use in direct inference on graphical models for the description and study of routes. In the field of systems biology and genome/pathway informatics, the preferred software platform for constructing a Pathway/Genome Database (PGDB) is the pathway tool. The aforementioned is a production-ready software environment that may be used to build a model-organism database known as a PGDB. Pathway Tools was developed out of a need for all-encompassing biological knowledge resources (http://paradigm.five3genomics.com/) that combine many sources of data (Vaske et al. 2010; Karp et al. 2009, 2021).

9.2.3 Reactome Pathway Database

The Reactome route database is a free, user-friendly, expert-evaluated, curated online tool. User-friendly bioinformatics tools are provided for exploring and making sense of established pathways (https://reactome.org/documentation) (Yu et al. 2023). The aforementioned is a freely available and community-developed database of pathways. The goal of this curated and reviewed set of resources is to make available bioinformatics tools that can be used to better visualise, interpret, and analyse route knowledge. It's meant to aid in areas including education, systems biology, genome analysis, and modelling. Biochemical reactions, regulatory networks, genetic interactions, transport, catalysis, and other related phenomena are all represented in Reactome's all-encompassing model of biology. Experts in the field of biology collaborated with the Reactome editorial staff to write the publication. Many different bioinformatics databases have been used to cross-reference the data. An interactive route browser, an online pathway analysis tool, a gene expression analysis tool, and a cross-species comparison tool are only some of the data analysis instruments available on Reactome. Downloadable versions of the program are available in BioPAX and SBML (Fabregat et al. 2017; Croft et al. 2011; Haw and Stein 2012).

9.2.4 Bioconductor

Pathway and enrichment analysis can be conducted using various tools provided by Bioconductor. SPIA, Path view, Clipper, Gauge, Graphite, and Path net are basics tools available in bioinformatics. Experience with the programming language R is preferred but not required. According to Sepulveda (2020) both the R and Bioconductor software packages can be obtained for free from the official website (https://www.bioconductor.org/). Pathway databases, web-based apps, and software packages are some of the resources available for the purpose of pathway identification. Important genes and proteins in a pre-existing pathway can be identified in relation to a given experimental or pathological setting using the aforementioned methods. Furthermore, they can build a pathway from scratch by assembling known components.

9.2.5 Path Visio

Path Visio is an open-source and free program made specifically for visualising and analysing biological routes. It has some specific features including being able to modify, examine, and create such connections. For further information on this program, check out Fried et al. (2013) or visit the website https://pathvisio.org/.

9.2.6 iPathwayGuide

Advaita's "Impact Analysis" is used by the web-based resource iPathwayGuide to locate useful paths. This method is completely free of the background noise that plagues other approaches. Based on previous research (Ahsan and Drăghici 2017), it takes only a few minutes to get results that are ready to be shared (https://advaitabio.com/ipathwayguide/).

9.2.7 KEGG

Pathway researchers frequently consult this site. This includes routes involved in metabolism and signalling in all major kingdoms of life, from bacteria to humans. Useful search parameters using KEGG's straightforward interface include pathway names, organism names, and gene names. You can make sense of the information in the database with the help of a variety of tools and resources, including KEGG Mapper and KEGG Orthology (https://www.genome.jp/kegg/pathway.html) (Kanehisa 2000).

9.2.8 GEPIA

Expression data from RNA sequencing were analysed using information from the TCGA and GTEx databases, which include 9736 tumours and 8587 normal samples, is made easier with the use of the GEPIA web server. The aforementioned is an online tool for performing interactive gene expression analysis. It is quick and flexible, thanks to the use of TCGA and GTEx data. The GEPIA platform makes it easier for experimental biologists to conduct gene expression analysis without extensive knowledge of computational programming. There are probably around 20,000 coding genes and 25,000 noncoding genes in the genome under consideration. There are also around 14,000 pseudogenes and around 400 T-cell receptor regions. Differential gene analysis, locating the most important survival genes, and other complex analyses based on many genes are simplified with the help of GEPIA. It has been shown that GEPIA (http://gepia.cancer-pku.cn/) may successfully ease the identification of differentially expressed genes across cancer and normal tissues (Tang et al. 2017; Yang et al. 2019), despite not being created as a pathway identification software tool.

9.2.9 EVINET

To do enrichment analysis on a network, you can use the web-based EVINET software. It's a flexible method for identifying groups of genes. The aforementioned is an online resource for doing analyses such as gene set enrichment, exploratory functional analysis, driver vs passenger mutation analysis, and network and pathway enrichment. Pathways and networks can be chosen for analysis using the collection menu. The typical time required to complete the calculation is less than a few minutes. The program will analyse network richness by identifying all available AGS-FGS edges. Both the uploaded data and the results of any analyses are stored in separate locations, password-protected directories for each project. You can visit the EviNet website at https://www.evinet.org/. Pathway Tools is a software platform that makes it simple to create a pathway or genome database (also known as PGDB), and the Pathway/Genome Database (PGDB) is a resource for researchers working in pathway/genome informatics and systems biology. The aforementioned is a production-ready software environment that may be used to build a model-organism database known as a PGDB (Jeggari et al. 2018). The development of Pathway Tools is funded in part the National Institutes of Health (https://www.evinet.org/) funding agency.

9.3 Gene Ontology (GO)

Connections between biological words and specific genes are hand-picked by curators or generated automatically and stored in the GO. The GO framework was created to accurately and consistently portray the well-established relationships between many biological terminologies and the many genes that serve as illustrative examples of these terms. The use of GO has shown to be a very useful tool for organising biological knowledge and analysing genomic data. Some examples of such software are Easy GO (http://bioinformatics.sdstate.edu/go/), go tools (http://bioinformatics.sdstate.edu/go/), and REVIGO (http://bioinformatics.sdstate.edu/go/) (Jeggari et al. 2018; Qi and Chen 2021).

9.3.1 Pathway Design

Creating new biological pathways or modifying existing ones to achieve a certain goal is what's meant by the term "pathway design". The activities of individual components and the entire system can be predicted with the use of bioinformatics software, which is used to design and improve these pathways. The production of biofuels has inspired the use of bioinformatics in the development of metabolic pathways. Researchers can design pathways that maximise the output of target molecules while minimising the production of unwanted by-products by using computer models to predict the actions of different enzymes and metabolic intermediates (Carbonell et al. 2016; Smanski et al. 2016). When it comes to design and modelling, bioinformatics is a powerful tool. There are several examples of the application of bioinformatics to the design and modelling processes. Several programs are used in the bioinformatics community for molecular design (Fig. 9.3a, b).

9.3.1.1 ChemDoodle

ChemDoodle is an online tool for creating chemical diagrams and sketches (https://www.chemdoodle.com/). The chemical sketching program ChemDoodle is an excellent molecular modelling software for chemistry. Everyone from students to chemical engineers can benefit from its straightforward interface. It generates high-resolution 3D pictures of chemical 3D structures and constructs reaction schemes in real time. Software "Widgets" are included in ChemDoodle. These mobile apps are useful for creating molecular diagrams. Tasks like determining molecular masses, analysing elements, researching chemical structures, etc. Additionally, ChemDoodle 3D can make 3D models from 2D chemical structures. The program

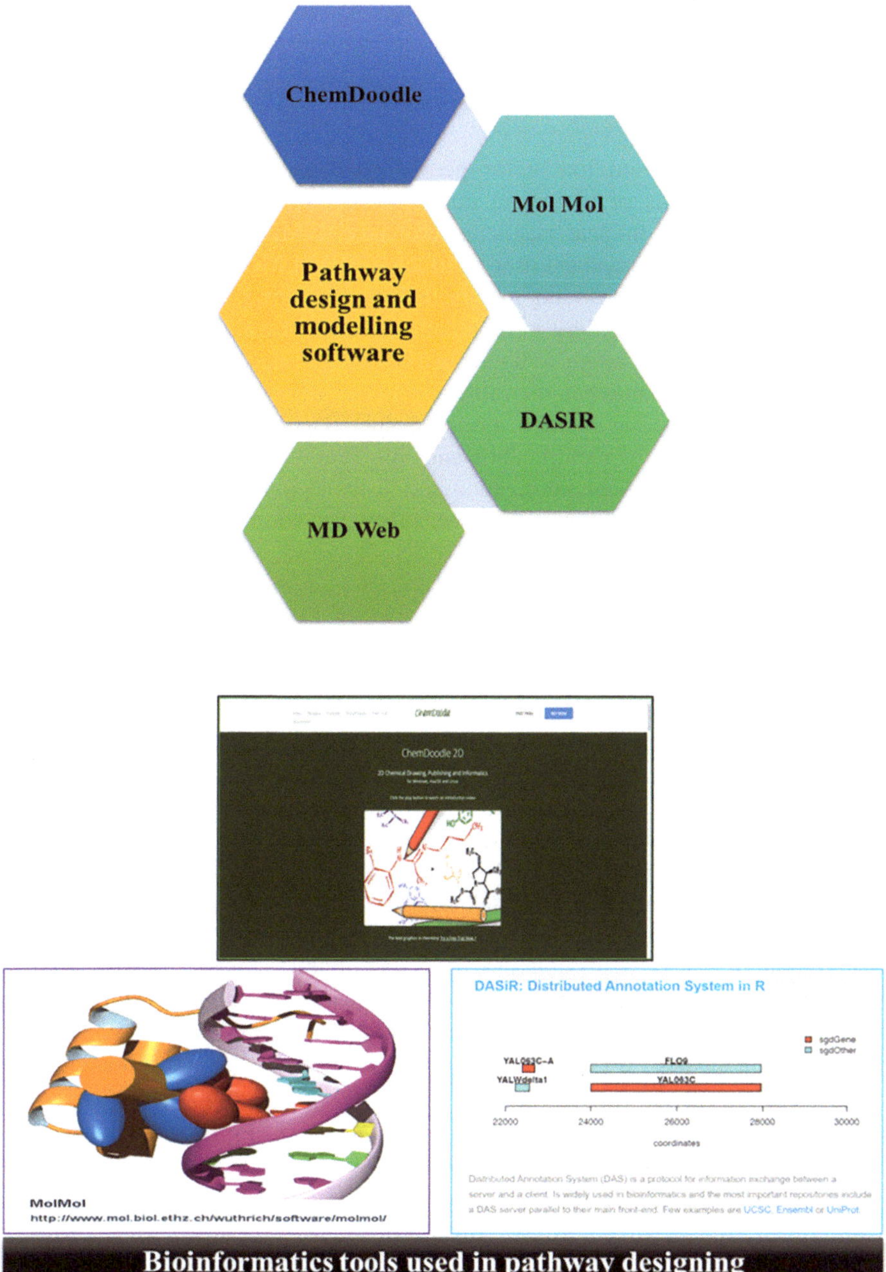

Fig. 9.3 (**a**) List of software used in pathway design and modelling. (**b**) Software webpages used in pathway design and modelling

is downloadable for use on computers running Windows, Mac OS X, and Linux. Smartphones allow for the transfer of 3D models between computers and mobile devices (Burger 2015).

9.3.1.2 MOLMOL

MOLMOL is a program that helps people see, study, and change the three-dimensional structures of biological macromolecules, especially those that have been characterised by nuclear magnetic resonance (NMR) methods, such proteins and nucleic acids. The graphical user interface (GUI) of MOLMOL is comprised of many different sections, including menus, dialogue windows, and documentation. A molecule can be represented by a combination of common and unusual schematics. Changing the number of atoms or bonds in a covalent molecule can cause noticeable structural changes. Changes in the molecule's three-dimensional structure can also result through interaction rotation around bonds. The software has a number of features that allow for the identification and enumeration of short distances between pairs of hydrogen atoms, the verification and visualisation of NMR constraints, the identification of hydrogen bonds, and the superimposition of conformers (Koradi et al. 1996).

9.3.1.3 DASiR

DASiR stands for the Distributed Annotation System in R, Moreover, it is utilised in the process of data analysis for high-throughput sequencing studies. The DAS is a server-client protocol for a distributed annotation system. A DAS server sits alongside the principal front-end at the larger bioinformatics libraries. The University of California, Santa Cruz, Ensembl, and UniProt. DASiR provides an R-DAS interface for programmatically connecting to remote DAS servers across a network. It allows R users an easy interface to a plethora of biological data and supports the DAS 1.6 protocol. DAS uses XML and HTTP protocols, which necessitate less infrastructure and fewer clients than MySQL and BioMart. The web address https://www.dasregistry.org and https://mmb.irbbarcelona.org/www/node/349 provide a directory of more than 1500 accessible DAS servers online. Although the DAS protocol allows queries on a variety of data types, DASiR is optimised for ranges. While you can do queries on genomic sequences and protein structures, you may find more convenient methods of doing so in R (the Biostrings package for genomes or the Bio3dD package for PDB structures, for example) (Dowell et al. 2001).

9.3.2 Pathway Modelling

When we talk about "pathway modelling," we're referring to the process of creating mathematical models that mimic the workings of biological pathways. The aforementioned models can be used to make predictions about how pathways will react to a variety of stimuli or perturbations and can be used to test hypotheses about the underlying mechanisms of biological processes. The circadian clock, a complex biological pathway that regulates the timing of physiological events in living organisms, has been modelled using bioinformatics. Scholars can predict the clock's response to changes in light and other environmental stimuli by developing mathematical models depicting the interplay among the clock's multiple components (Hucka et al. 2003; Klipp et al. 2012). Bioinformatics has several uses in molecular modelling. Here are some specific examples:

9.3.3 Protein Structure Analysis

Biological macromolecules' sequencing, structure, and function can be better understood with the help of bioinformatics tools and resources. According to Schmidt et al. (2014), this method includes investigating the physical and chemical conditions that affect protein structure and function. Briefly said, bioinformatics is used in molecular modelling in a variety of ways, including but not limited to homology modelling, molecular structure simulation, membrane protein modelling, and protein structure analysis. Tools and methods can be used to better predict protein structure and function, create new medications and therapies, and learn how the order, structure, and function of biological macromolecules are interconnected.

9.3.4 Homology Modelling

Homology modelling is often used in bioinformatics to estimate the structure of an uncharacterised protein by comparing it to known structures of homologous proteins (Hoy et al. 2007; Jumper et al. 2021).

9.3.5 Membrane Protein Modelling

Kulp (2010) provide evidence for the use of structural bioinformatics in the modelling of membrane proteins for which structural knowledge is unavailable and in the creation of novel membrane proteins.

9.3.6 MDWeb

Protein structure prediction and analysis are made easier with the use of MDWeb, a web-based platform that incorporates Docking between proteins and ligands, docking between proteins, and the dynamics of proteins. The user-friendly layout makes it simple to submit work and view the outcomes digitally. MDWeb allows users to simulate and model biological networks by providing a pathway modelling method. Although MDWeb has many useful features, including route modelling, it is most commonly used for predicting and analysing protein structures. Pathway

Tools, Visinets, and Paradigm are some examples of software that may be used to model and analyse pathways (Spychala et al. 2015; Karp et al. 2009).

9.3.7 Pathway Simulation

Biological mechanisms are replicated in a number of contexts using computational models. Theoretical frameworks may be tested, new insights can be generated, and prospective interventions can be predicted with the help of simulation models. Bioinformatics is the study of how medicines affect biological processes through the use of computational technologies. Modelling drug–target interactions allows scientists to anticipate how drugs will affect pathways and spot unintended consequences. A computational method that allows for the study of molecular behaviour over time is called molecular dynamics simulation. This contributes to the advancement of research into the structure and function of proteins as well as the development of new therapeutics.

Biomolecular modelling, biochemical networks, data-driven drug discovery, and molecular dynamics simulation are just few of the areas where bioinformatics simulations are being used. It has been suggested that certain technologies can predict protein structure and function, create new pharmaceuticals and therapies, improve research, and speed up the process of drug development (Gillespie 1977; Ciliberto and Novère 2013). Some examples of accessible route simulation software are presented in Fig. 9.4a, b.

9.3.7.1 GROMACS

The GROMACS program is a molecular dynamics suite optimised for modelling macromolecules including proteins, lipids, and nucleic acids. The aforementioned program suite is an open-source, free, and high-performance molecular dynamics and output analysis tool. GROMACS is a popular program because of its lightning-fast processing times. There is support for both CPUs and GPUs. CMake is used exclusively as the build system, and there is plenty of information available on the official website to help with the setup process. Proteins, lipids, and nucleic acids are only some of the biological macromolecules that may be modelled and analysed computationally with the help of GROMACS. Because of its fast computing of non-covalent interactions, it is also used to study them. A pathway modelling mode is included in GROMACS, letting users build and run simulations of biological networks. Although it can be used for pathway modelling (https://www.computabio.com/applications-of-gromacs-software.html), GROMACS was originally developed as a package for molecular dynamics simulations (https://www.gromacs.org/) (Kutzner et al. 2022).

9.3.7.2 GENESIS

The General Neural Simulation System, also known as GENESIS, is a simulation platform that enables users to develop realistic models of neurobiological systems at multiple scales, ranging from individual neurons all the way up to complete brain

Fig. 9.4 (**a**) List of software used in simulation. (**b**) Software webpages used in simulation

circuits. This platform is a flexible simulation tool made to help model brain systems. The GENESIS program serves its purpose by creating virtual worlds that can be used to build models of neurons and other brain systems. This simulation platform has been around for a while, although its original intent was to help with simulating neurological systems. The combination of GENESIS and Yale University's NEURON software facilitates the simulation of neural systems, from subcellular components and biochemical reactions to extensive networks and system-level models. In order to do high-performance molecular dynamics simulations and analyse the resulting data, the GENESIS software package was developed. The main goal of this program is to help with the modelling of biological macromolecules like proteins, lipids, and nucleic acids. While GENESIS can be used for pathway modelling, its primary function is as an environment for simulation that allows for the construction of precise models of neurobiological systems (http://genesis-sim.org/) (Bower et al. 2013).

9.3.7.3 LAMMPS

The Large-scale Atomic/Molecular Massively Parallel Simulator (LAMMPS) was developed at Sandia National Laboratories for use in molecular dynamics simulations. The majority of its applications are in material simulation, where it serves as a conventional molecular dynamics simulator. LAMMPS is capable of simulating a wide variety of systems, from the sub-mesoscopic to the mesoscopic, and from solid-state materials like metals and semiconductors to soft matter like biomolecules and polymers. The relevant software is open-source, which means that its source code can be accessed and modified by anybody. The GNU General Public License (GPL) governs its distribution. To efficiently determine the number of nearby particles, LAMMPS makes use of neighbour lists, specifically Verlet lists. The program can operate in parallel or on separate CPUs thanks to message-passing mechanisms and a geographical segmentation of the simulation region. It was a deliberate design choice to make modifications and additions to the LAMMPS program simple. This tool can be used as a parallel particle simulation technique or to represent atoms at the atomic, meso, and continuum levels. The LAMMPS software package was not designed for use in pathway modelling (https://www.lammps.org/#gsc.tab=0), but rather for molecular dynamics simulations and materials modelling (Chávez Thielemann et al. 2019; Humbert et al. 2019).

9.3.7.4 BIGNASim

Using a NoSQL database, BIGNASim analyses results from simulations of nucleic acids. Next to genomics, molecular dynamics simulation (MD) is the most resource-intensive application running on modern supercomputers. In the course of several months, MD trajectories are calculated, analysed on the spot, and then forgotten. There are efforts to create a database of proteins' stable trajectories, but no such efforts exist for nucleic acids. BIGNASim maintains a one-of-a-kind database for nucleic acid analysis and MD trajectories. In the first batch of data, the new standard for molecular dynamics force fields, parmBSC1, is the clear winner. It takes more than 120 ms to run 156 simulations. New trajectory information is welcome in deposition techniques. Analytical results and simulation data are stored in MongoDB, while trajectories are kept in Cassandra. Mechanical, NMR, helical, and backbone research are also accessible. The portal (https://mmb.irbbarcelona.org/BIGNASim/) provides access to both individual and meta-trajectories.

9.3.7.5 Curves+ Web Server

DNA's three-dimensional structure is analysed through the Curves+ Web Server. Curves+ is a refined approach to analysing the 3D structure of nucleic acids. It is more efficient, provides more recent information, and adheres to standards for analysing nucleic acids. In addition to managing single nucleic acid structures, Curves+ and Canal can analyse molecular dynamics trajectories, create time series, time averaged characteristics, and search for relationships. Canion (https://bisi.ibcp.fr/tools/curves_plus/) enables the investigation of ions or molecules around nucleic acids in helical space (Blanchet et al. 2011).

9.3.7.6 ACEMD

High-performance molecular dynamic simulations can be performed with the help of ACEMD (https://software.acellera.com/acemd/index.html), as stated in Harvey and De Fabritiis (2015).

9.3.7.7 AMBER

AMBER, a molecular dynamics simulation program, has extensive analytic capabilities, as stated by Meyer et al. (2018). Amber, a package of programs for modelling biomolecules. Visit our contributions and history pages for more details. The word "Amber" can be used in two different ways. In the first stage, molecular mechanical force fields that are available to the public are used to simulate biomolecules. The software package for molecular simulation is accompanied by source code and illustrative examples demonstrating its utilisation. The Amber22 and AmberTools23 software packages are currently being released. Unlike Amber22, which is incompatible with AmberTools23, Amber23 can run on its own. Get the Amber toolkit so you can use its code standards. Amber was created through a group effort by Peter Kollman, David Case, Tom Cheatham, Ken Merz, Adrian Roitberg, Carlos Simmerling, Darrin York, Ray Luo, Junmei Wang, Maria Nagan, and others. (https://ambermd.org/) (Case et al. 2005; Salomon-Ferrer et al. 2012; Meyer et al. 2018).

9.3.7.8 Abalone

Abalone is a program designed for biomolecular modelling and simulation. Multiple studies have shown its value when used to simulate protein folding, such as those by Lexei et al. (2014) and Campos and Sanz-Serna (2015). ChemDoodle, MOLMOL, BIGNASim, Curves+ Web Server, DASiR, ACEMD, AMBER, and Abalone are only few of the many software programs used in the field of bioinformatics for molecular modelling. These tools can be used to create replicas of DNA and RNA, simulate biomolecular interactions, predict how proteins will fold and function, and design new drugs and treatments (http://www.biomolecular-modeling.com/Abalone/).

9.4 Conclusion

Bioinformatics techniques are employed to identify potential therapeutic targets within biological pathways. The utilisation of computational methodologies to anticipate the impact of medications or small compounds on pathway activity can facilitate the discovery of novel treatments or enhancements to existing ones. The field of bioinformatics is of utmost importance in the advancement of biological pathway research. It facilitates the integration of various biological data, the formulation of mathematical models, the simulation of pathway dynamics, and the acquisition of knowledge pertaining to intricate biological systems. This tool serves to enhance the progress of scientific comprehension pertaining to biological pathways, their regulation, and their potential applications across diverse fields including medicine, agriculture, and biotechnology.

Acknowledgments Authors acknowledged the Department of Biotechnology, Era University, Lucknow, India, for central computational facilities.

References

Ahsan S, Drăghici S (2017) Identifying significantly impacted pathways and putative mechanisms with iPathwayGuide. Curr Protoc Bioinform 57:7–15. https://doi.org/10.1002/cpbi.24

Ashburner M, Ball CA, Blake JA et al (2000) Gene ontology: tool for the unification of biology. Nat Genet 25:25–29. https://doi.org/10.1038/75556

Blanchet C, Pasi M, Zakrzewska K, Lavery R (2011) CURVES+ web server for analyzing and visualizing the helical, backbone and groove parameters of nucleic acid structures. Nucleic Acids Res 39:W68–W73. https://doi.org/10.1093/nar/gkr316

Bower JM, Cornelis H, Beeman D (2013) Genesis, the general neural simulation system. In: Encyclopedia of computational neuroscience. Springer, New York, pp 1–8

Burger MC (2015) ChemDoodle Web Components: HTML5 toolkit for chemical graphics, interfaces, and informatics. J Chem 7:35. https://doi.org/10.1186/s13321-015-0085-3

Campos CM, Sanz-Serna J (2015) Extra chance generalized hybrid Monte Carlo. J Comput Phys 281:365–374

Carbonell P, Currin A, Jervis AJ et al (2016) Bioinformatics for the synthetic biology of natural products: integrating across the design–build–test cycle. Nat Prod Rep 33:925–932. https://doi.org/10.1039/c6np00018e

Case DA, Cheatham TE, Darden T et al (2005) The Amber biomolecular simulation programs. J Comput Chem 26:1668–1688. https://doi.org/10.1002/jcc.20290

Chávez Thielemann H, Cardellini A, Fasano M et al (2019) From GROMACS to LAMMPS: GRO2LAM. J Mol Model 25:147. https://doi.org/10.1007/s00894-019-4011-x

Ciliberto A, Novère NL (2013) Using COPASI for modeling and simulation of biochemical networks. In: Systems biology. Humana Press, pp 247–276

Croft D, O'Kelly G, Wu G, Haw R, Gillespie M, Matthews L, Caudy M, Garapati P, Gopinath G, Jassal B, Jupe S, Kalatskaya I, Mahajan S, May B, Ndegwa N, Schmidt E, Shamovsky V, Yung C, Birney E, Hermjakob H, D'Eustachio P, Stein L (2011) Reactome: a database of reactions, pathways and biological processes. Nucleic Acids Res 39:691–697

Dowell RD, Jokerst RM, Day A, Eddy SR, Stein L (2001) The distributed annotation system. BMC Bioinform 2:7

Fabregat A, Sidiropoulos K, Viteri G, Forner O, Marin-Garcia P, Arnau V, D'Eustachio P, Stein L, Hermjakob H (2017) Reactome pathway analysis: a high-performance in-memory approach. BMC Bioinform 18(1):142

Fried JY, van Iersel MP, Aladjem MI, Kohn KW, Luna A (2013) PathVisio-Faceted Search: an exploration tool for multi-dimensional navigation of large pathways. Bioinformatics 29(11): 1465–1466

García-Campos MA, Espinal-Enríquez J, Hernández-Lemus E (2015) Pathway analysis: state of the art. Front Physiol 17(6):383

Gillespie DT (1977) Exact stochastic simulation of coupled chemical reactions. J Phys Chem 81(25):2340–2361

Harvey MJ, De Fabritiis G (2015) AceCloud: molecular dynamics simulations in the cloud. J Chem Inf Model 55:909–914. https://doi.org/10.1021/acs.jcim.5b00086

Haw R, Stein L (2012) Using the reactome database. Curr Protoc Bioinform 38(1):8.7.1–8.7.23. https://doi.org/10.1002/0471250953.bi0807s38

Hoy JA, Robinson H, Trent JT, Kakar S, Smagghe BJ, Hargrove MS (2007) Plant hemoglobins: a molecular fossil record for the evolution of oxygen transport. J Mol Biol 371(1):168–179. https://doi.org/10.1016/j.jmb.2007.05.029

Hucka M, Finney A, Sauro HM et al (2003) The systems biology markup language (SBML): a medium for representation and exchange of biochemical network models. Bioinformatics 19(4): 524–531

Humbert MT, Zhang Y, Maginn EJ (2019) PyLAT: python LAMMPS analysis tools. J Chem Inf Model 59(4):1301–1305. https://doi.org/10.1021/acs.jcim.9b00066. Epub 2019 Mar 15

Jeggari A, Alekseenko Z, Petrov I, Dias JM, Ericson J, Alexeyenko A (2018) EviNet: a web platform for network enrichment analysis with flexible definition of gene sets. Nucleic Acids Res 46(W1):W163–W170. https://doi.org/10.1093/nar/gky485

Jumper J, Evans R, Pritzel A, Green T, Figurnov M, Ronneberger O, Tunyasuvunakool K, Bates R, Žídek A, Potapenko A, Bridgland A, Meyer C, Koh SAA, Ballard AJ, Cowie A (2021) Highly accurate protein structure prediction with AlphaFold. Nature 596(7873):583–589. https://doi.org/10.1038/s41586-021-03819-2

Kanehisa M (2000) KEGG: Kyoto encyclopedia of genes and genomes. Nucleic Acids Res 28:27–30. https://doi.org/10.1093/nar/28.1.27

Kanehisa M, Sato Y, Kawashima M, Furumichi M, Tanabe M (2016) KEGG as a reference resource for gene and protein annotation. Nucleic Acids Res 44(D1):D457–D462. https://doi.org/10.1093/nar/gkv1070

Karp PD, Paley SM, Krummenacker M et al (2009) Pathway tools version 13.0: integrated software for pathway/genome informatics and systems biology. Brief Bioinform 11:40–79. https://doi.org/10.1093/bib/bbp043

Karp PD, Midford PE, Billington R, Kothari A, Krummenacker M, Latendresse M, Ong WK, Subhraveti P, Caspi R, Fulcher C, Keseler IM, Paley SM (2021) Pathway tools version 23.0 update: software for pathway/genome informatics and systems biology. Brief Bioinform 22(1): 109–126. https://doi.org/10.1093/bib/bbz104

Kim HU, Sohn SB, Lee SY (2012) Metabolic network modeling and simulation for drug targeting and discovery. Biotechnol J 7(3):330–342. https://doi.org/10.1002/biot.201100159

Klipp E, Liebermeister W, Wierling C, Kowald A (2012) Systems biology: a textbook. Wiley-Blackwell

Koradi R, Billeter M, Wüthrich K (1996) MOLMOL: a program for display and analysis of macromolecular structures. J Mol Graph 14(1):51–55. https://doi.org/10.1016/0263-7855(96)00009-4

Kulp D (2010) Using structural bioinformatics to model and design membrane proteins (Publicly Accessible Penn Dissertations). University of Pennsylvania, p 233. https://repository.upenn.edu/edissertations/233

Kutzner C, Kniep C, Cherian A, Nordstrom L, Grubmüller H, de Groot BL, Gapsys V (2022) GROMACS in the cloud: a global supercomputer to speed up alchemical drug design. J Chem Inf Model 62(7):1691–1711. https://doi.org/10.1021/acs.jcim.2c00044. Epub 2022 Mar 30

Lexei MN, Yury VM, Lyubartsev AP (2014) A new AMBER-compatible force field parameter set for alkanes. J Mol Model 20:2143. https://doi.org/10.1007/s00894-014-2143-6

Meyer F, Hofmann P, Belmann P, Garrido-Oter R, Fritz A, Sczyrba A, McHardy AC (2018) AMBER: assessment of metagenome BinnERs. Gigascience 7(6):giy069. https://doi.org/10.1093/gigascience/giy069

Mi H, Muruganujan A, Thomas PD (2012) Panther in 2013: modeling the evolution of gene function, and other gene attributes, in the context of phylogenetic trees. Nucleic Acids Res 41 (D1):D377–D386. https://doi.org/10.1093/nar/gks1118

Mi H, Huang X, Muruganujan A, Tang H, Mills C, Kang D, Thomas PD (2016) Panther version 11: expanded annotation data from gene ontology and reactome pathways, and data analysis tool enhancements. Nucleic Acids Res 45(D1):D183–D189. https://doi.org/10.1093/nar/gkw1138

Mi H, Muruganujan A, Huang X, Ebert D, Mills C, Guo X, Thomas PD (2019) Protocol update for large-scale genome and gene function analysis with the PANTHER classification system (v.14.0). Nat Protoc 14(3):703–721. https://doi.org/10.1038/s41596-019-0128-8. Epub 2019 Feb 25

Mi H, Ebert D, Muruganujan A, Mills C, Albou L, Mushayamaha T, Thomas PD (2020) Panther version 16: a revised family classification, tree-based classification tool, enhancer regions and extensive API. Nucleic Acids Res 49(D1):D394–D403. https://doi.org/10.1093/nar/gkaa1106

Mubeen S, Hoyt CT, Gemünd A, Hofmann-Apitius M, Fröhlich H, Domingo-Fernández D (2019) The impact of pathway database choice on statistical enrichment analysis and predictive modeling. Front Genet 10:1203. https://doi.org/10.3389/fgene.2019.01203

Park JM, Kim TY, Lee SY (2009) Constraints-based genome-scale metabolic simulation for systems metabolic engineering. Biotechnol Adv 27:979–988

Qi D, Chen K (2021) Bioinformatics analysis of potential biomarkers and pathway identification for major depressive disorder. Comput Math Methods Med 2021:3036741. https://doi.org/10.1155/2021/3036741

Salomon-Ferrer R, Case DA, Walker RC (2012) An overview of the Amber biomolecular simulation package. Wiley Interdiscip Rev Comput Mol Sci 3(2):198–210. https://doi.org/10.1002/wcms.1121

Schmidt T, Bergner A, Schwede T (2014) Modelling three-dimensional protein structures for applications in drug design. Drug Discov Today 19(7):890–897. https://doi.org/10.1016/j.drudis.2013.10.027

Sepulveda JL (2020) Using R and bioconductor in clinical genomics and transcriptomics. J Mol Diagn 22(1):3–20. https://doi.org/10.1016/j.jmoldx.2019.08.006

Smanski MJ, Bhatia S, Zhao D et al (2016) Functional optimization of gene clusters by combinatorial design and assembly. Nat Biotechnol 34(6):638–646

Spychala J, Spychala P, Gomez S, Weinreb GE (2015) Visinets: a web-based pathway modeling and dynamic visualization tool. PLoS One 10(5):e0123773. https://doi.org/10.1371/journal.pone.0123773

Tang Z, Li C, Kang B, Gao G, Li C, Zhang Z (2017) GEPIA: a web server for cancer and normal gene expression profiling and interactive analyses. Nucleic Acids Res 45(W1):W98–W102. https://doi.org/10.1093/nar/gkx247

Vaske CJ, Benz SC, Sanborn JZ, Earl D, Szeto C, Zhu J, Haussler D, Stuart JM (2010) Inference of patient-specific pathway activities from multi-dimensional cancer genomics data using paradigm. Bioinformatics 26(12):i237–i245. https://doi.org/10.1093/bioinformatics/btq182

Yang Y, Sui Y, Xie B, Qu H, Fang X (2019) Glioma DB: a web server for integrating glioma omics data and interactive analysis. Genom Proteom Bioinform 17(4):465–471. https://doi.org/10.1016/j.gpb.2018.03.008

Yu N, Hwang M, Lee Y, Song BR, Kang EH, Sim H, Ahn BC, Hwang KH, Kim J, Hong S, Kim S, Park C, Han JY (2023) Patient-derived cell-based pharmacogenomic assessment to unveil underlying resistance mechanisms and novel therapeutics for advanced lung cancer. J Exp Clin Cancer Res 42(1):37. https://doi.org/10.1186/s13046-023-02606-3

Integration of Metabolomics and Flux Balance Analysis: Applications and Challenges

10

Gholamreza Abdi, Nil Patil, Mukul Jain, and Mukul Barwant

Abstract

This book chapter presents an in-depth analysis of the integration of metabolomics and flux balance analysis (FBA) as powerful tools for understanding metabolic processes and their applications in various scientific disciplines. The potential applications of metabolomics in these fields were discussed, highlighting the valuable insights it offers into metabolic pathways and networks. The subsequent sections delve into the different techniques employed in metabolomics research, including targeted and untargeted approaches using "LC–MS, GC–MS, and NMR". The chapter also explores important tools utilized in flux balance analysis, such as OptKnock, OptGene, OptStrain, COBRA Tools, MetaboAnalyst 4.0, OptFlux, CellNetAnalyzer, SBRT, and Escher-FBA. Furthermore, the chapter discusses metabolomics integration using FBA and highlights the methodologies for identifying and annotating metabolites, including the use of metabolite databases and spectral libraries. The integration of metabolomics data with genome-scale metabolic models was explored, along

G. Abdi (✉)
Department of Biotechnology, Persian Gulf Research Institute, Persian Gulf University, Bushehr, Iran
e-mail: abdi@pgu.ac.ir

N. Patil · M. Jain
Cell and Developmental Biology Lab, Centre of Research for Development, Parul University, Vadodara, Gujarat, India

Department of Life Sciences, Parul Institute of Applied Sciences, Parul University, Vadodara, Gujarat, India

M. Barwant
Department of Botany, Sanjivani Arts, Commerce and Science College, Kopargoan, Ahmednagar, Maharashtra, India

V. Singh, A. Kumar (eds.), *Advances in Bioinformatics*,
https://doi.org/10.1007/978-981-99-8401-5_10

with the estimation of metabolic fluxes from metabolomics data using the "Constraint-Based Reconstruction and Analysis (COBRA) Toolbox". The chapter presents case studies and applications that demonstrate the utility of metabolomics and FBA in various contexts, including therapeutic and diagnostic applications. It explores the application of metabolomics in blood, urine, and saliva, highlighting their potential as non-invasive diagnostic tools. Moreover, the chapter addresses the challenges and limitations associated with integrating metabolomics and FBA, providing insights into future perspectives and directions for further research.

Keywords

Diagnostic · Flux balance analysis · LC–MS · GC–MS · Metabolomics · Metabolic process · NMR · OptKnock · OptGene · OptStrain · COBRA tools strain

10.1 Introduction

Many disorders include hypoxia, which can be extremely harmful to cells. Designing medicines to improve cellular defenses against hypoxia stress is a key objective of medical research, as a prerequisite, additional basic research into the defense systems and processes of hypoxia cell death is needed. The low oxygen level causes a sharp decline in mitochondrial respiratory activity due to the consequence of metabolic (Feala et al. 2009). Due to its-known genome, easily accessible methods for genetic alteration, fecundity, short lifespan, and inherent tolerance to considerable oxygen level changes, *Drosophila melanogaster* has been a popular model organism for systems biology techniques as well as hypoxia investigations. Because of widely recognized genetic makeup, availability of genetic manipulation tools, high reproductive rate, short lifespan, and natural ability to withstand significant variations in oxygen levels, *Drosophila melanogaster* has emerged as a favored model organism for studying systems biology and investigating hypoxia (Krishnan et al. 1997). It is possible for organisms to adapt to shifting environmental conditions while still carrying out vital survival processes from an economic standpoint, it is critical to investigate the effects of these metabolic adaptations, but they can also have major effects on health and disease (Kitano 2004). The modification of fluxes in metabolic networks is one example of molecular adaptation in action. Flux alterations in affected metabolic pathways occur when there are changes in substrate availability. According to Stelling (2004), these flux variations can offer crucial insights into cellular physiology, output, and how organisms react to disruptions. Since intracellular fluxes cannot be observed directly, concentration measurements must be used to quantify them. To achieve this, various experimental and computational methods, including kinetic models, are employed to approximate dynamic fluxes within metabolic networks (Teusink et al. 2000). Dynamic flux balance analysis (DFBA) and ^{13}C-metabolic flux analysis (MFA) have been developed.

Novel methodologies such as ^{13}C-metabolic flux analysis (MFA) and dynamic flux balance analysis (DFBA) have been devise to advance the field (Van Winden et al. 2005). A constraint-based modeling technique called flux balance analysis (FBA) is used to find fluxes in a steady state (Willemsen et al. 2015). The networks of metabolic consist of a larger number of reactions compared to metabolites, resulting in a situation where the stoichiometry of the network imposes mass balance constraints. As a result, an under-determined system of linear equations is created (O'Grady et al. 2012; Wiechert 2001). Capacity constraints, which specify the top and lower bounds of the fluxes, are additionally imposed to condense the solution space. The range of feasible flux values is constrained by these restrictions. The ideal flow distribution within the specified restrictions is then sought after by solving the under-determined linear system as an optimization problem (Förster et al. 2003). The phenotype is described by this objective function as a biological aim like biomass production (e.g., maximal growth yield or energetic efficiency). Each reaction's relative contribution to the phenotype is quantified by the objective function. FBA might be used to calculate fluxes for various steady state conditions in a perturbed system, however this method would not account for transient behavior following a disturbance. By maximizing the objective function over the desired time period, DFBA calculates the transient behavior of the fluxes following a disturbance (Mahadevan et al. 2002). Both capacity constraints and dynamic mass balance constraints were apply to the objective function in flux balance analysis (FBA). The alterations in metabolite concentrations dependent on fluxes, biomass, and other kinetic parameters are described by the dynamic mass balance constraints as differential equations. The maximum rate of change for the fluxes over specific periods of time can be specified using additional restrictions, provided they are available (Varma and Palsson 1994). Dynamic flux balance analysis (DFBA), in contrast to flux balance analysis (FBA), entails evaluating the derivatives (changes) of metabolite concentrations over time. Compared to FBA, where the change in metabolite concentrations is assumed to be zero and only the fluxes are in doubt, this introduces additional unknown parameters. Consequently, the system in DFBA becomes more complex with a higher number of unidentified parameters (Willemsen et al. 2015). To address the dynamic optimization challenge, orthogonal collocation on finite elements is utilized to parameterize the dynamic equations. This approach enables the estimation of dynamic flux profiles. However, it is important to note that this method may be less suitable for modeling extracellular dynamics over extended timeframes or larger metabolic networks. In situations where the driving objective of an organism under varying conditions is known, such as during a diauxic shift, classical dynamic flux balance analysis (DFBA) is employed to estimate the dynamic flux profiles (Mahadevan et al. 2002).

When there are several parameters to estimate, dynamic flux balance analysis (DFBA) is frequently used. Its primary application is for bigger structures to generate dynamic flux profiles, which involve both internal and external fluxes. DFBA is particularly useful when studying the response of an organism to specific perturbations, where external flows through time following the perturbation are of interest. Additionally, DFBA is employ to assess objective functions that influence transient cellular behavior after disturbances. Experimental techniques like mass spectrometry and liquid or gas chromatography (LC/GC–MS) enable direct measurements of metabolite concentration profiles, allowing for the determination of how concentrations change over time. In DFBA, these measured concentration profiles can be used to determine time derivatives and metabolite concentrations, rather than relying solely on estimation. However, measured concentration profiles cannot be directly incorporate into DFBA. A new technique called MetDFBA is created for overcoming this restriction. MetDFBA constructs a system of linear equations via derivatives that directly calculated from observed concentration profiles and then replacing them into the mass balance equation (Willemsen et al. 2015). This method significantly lessens the computational difficulty of DFBA (Schuetz et al. 2007). This paradigm is especially appropriate for larger systems because of the lower complexity and fewer unknowns. We used time-resolved metabolomics data from a feast-famine experiment employing *Penicillium chrysogenum* to produce these estimates. By leveraging this data, we were able to assess and validate the flow changes predicted by MetDFBA against experimental measurements. This comparison serves to demonstrate the accuracy and reliability of our method in capturing the dynamics of metabolic fluxes in response to perturbations (Canelas et al. 2008; Willemsen et al. 2015).

10.1.1 Metabolomics Overview

A scientific activity called "metabolome analysis" focuses on finding and quantifying every metabolite that exists in a biological system. The metabolome encompasses a broad range of metabolites, varying in terms of concentration and physio-chemical characteristics. Due to the extensive nature of the metabolome, it is not feasible for a single technology to enable the simultaneous assessment of all metabolites at once. Researchers use a number of platforms and analytical methods, each with strengths and weaknesses, to address this problem. These methods include gas chromatography (GC), liquid chromatography (LC), nuclear magnetic resonance (NMR) spectroscopy, and others. The metabolome can be covered more thoroughly by the use of several complimentary approaches, which enables researchers to discover and quantify a wider variety of metabolites inside the biological system under study (2015) Töpfer et al. As a result, the word "metabolomics" refers to a group of technologies that study various metabolome components (Redestig et al. 2011). Metabolomics research produces multivariate data, which can make statistical analysis difficult, especially when there are more variables than experimental samples. In such cases, the high dimensionality of the data can make it difficult to extract meaningful insights. To address this issue, principal component analysis (PCA) is commonly employed. The dimensionality of multivariate data can be decreased by using the vector transformation technique known as PCA. By relocating the data "cloud" onto fresh axes in the multivariate space, it does this. These new axes, called principle components, are an orthogonal set of basis vectors and are weighted combinations of the original variables (in this case, metabolites).

The original multivariate data can be represented in a lower-dimensional space using PCA, preserving as much information as feasible. This transformation simplifies the data visualization and analysis, enabling the identification of patterns, clusters, and relationships among the metabolites and samples. PCA serves as a valuable exploratory tool in metabolomics research, aiding in the interpretation and understanding of complex datasets (Coquin et al. 2008). Metabolomics is a supplementary approach to genomics and proteomics for investigating the responses of complex biological systems to environmental, physical, and genetic influences (Griffin and Bollard 2004). Metabolomics studies frequently provide relative quantifications of metabolites, which are evaluated based on the fold-change in peak size between two samples. However, absolute metabolite quantifications must be obtained in order to compare metabolite concentrations precisely. Measurements expressed in moles per unit weight of tissue, such as mol per gram (g) of fresh weight (FW), are the result of calibration curves utilizing standards for each metabolite. There is an increasing emphasis on determining the absolute amounts of metabolites, even if relative changes in metabolite levels are frequently adequate for many uses (Dettmer et al. 2007). In-depth conversations and consultations have been held to provide a thorough understanding of the different metabolic techniques. The complex nature of metabolites, which are integral to a network structure, allows the metabolome to be regarded as a distinct cellular level. Furthermore, the metabolome serves as a crucial bridge connecting genotype and phenotype (Töpfer et al. 2015). The metabolome (Nielsen and Jewett 2007) refers to the full set of metabolites, non-genetically encoded substrates, intermediaries, and products of metabolic pathways that are coupled with a cell as a result of advances in complex network research (Kueger et al. 2012).

Significant advancements have been achieved recently in the creation and application of metabolomics technologies, which make it easier to identify and measure metabolites in their whole. Due to these developments, it is now possible to examine metabolites on a massive scale, which has allowed researchers to better understand metabolic processes and their effects (Töpfer et al. 2015). These tools add to the tried-and-true methodology used in studies on e-nomics, transcriptome, and proteomics, which stand out for a careful analysis of the pertinent cellular components (Romero et al. 2008; Töpfer et al. 2015). Metabolite fluxes were measured during hypoxia and recovery phases in order to look into the processes underlying the age-related drop in hypoxia tolerance. After incorporating these flux estimates into a model, network simulations were run to look at changes in important fluxes including ATP, H, and glucose. This method was used to create theories to explain the observed decline in hypoxia tolerance with aging. The consistency and concordance of these assumptions with the experimental findings was further confirmed by comparing them to transcription patterns seen in young and old flies (Griffin and Bollard 2004). The systematic examination of every metabolite present in a biological sample is the focus of the biological area of metabolomics. It primarily emphasizes the characterization and description of metabolites that are soluble in water. By examining the water-soluble metabolites, metabolomics aims to provide insights into the metabolic processes and pathways occurring within the biological system under investigation. This branch of study plays a crucial role in understanding the biochemical changes and metabolic profiles associated with various biological phenomena, including disease states, drug responses, and environmental interactions (Xia et al. 2013). Metabolomics is widely regarded as a crucial tool in the study of systems biology because of its relationship to genomes and proteomics. The manipulation of gene and protein expression levels controls biological processes, and metabolomics, along with proteomics and transcriptomics, provides a thorough understanding of a system's behavior. Biopsies are taken from two or more experimental groups, and the samples' metabolites are then isolated and evaluated in metabolomics investigations. This makes it possible for researchers to examine and contrast the metabolic profiles of various groups, making it easier to identify important metabolites linked to particular circumstances or experimental variables (Narad et al. 2022). Numerous experimental methods, including NMR (nuclear magnetic resonance), MS (mass spectrometry), and LCMS (liquid chromatography-mass spectrometry), are used to evaluate metabolites. By applying these methods, metabolites are located, and the acquired information is used to build metabolic pathways. In terms of precision, sensitivity, quantification, and dependability, tailored metabolomics is superior to untargeted metabolomics and has a reduced percentage of false positives. Enzymes, which play a crucial role in metabolomics, can be affected by factors like chemical stability and temperature, leading to fluctuations in metabolomics samples. In order to assure reliable results, the sample preparation procedure needs to be adjusted. Metabolomics is a relatively new field that seeks to recognize and measure low-molecular-weight exogenous and endogenous compounds in biological systems. Its close relationship with physiology and genotype enables the exploration of how genotype and environment interact. The field of metabolomics focuses on understanding an organism's full metabolome, which is the collection of tiny compounds that interact inside biological systems and have an impact on diet, genetics, and the environment. It has significant applications in areas such as molecular and personalized medicine, toxicology, and other related disciplines (Narad et al. 2022). The metabolome, which includes variations in gene and protein expression, serves as the organism's final downstream result. It serves as the molecular phenotype reflecting both health and disease states. The "Human Metabolome Database (HMDB)" is a valuable resource containing extensive data on various substances, includes lipid- and water-soluble metabolites, organic acids, nucleotides, lipids, steroids, carbohydrates, and amino acids. Metabolomics can be divided generally into two categories: "targeted" and "untargeted." Targeted metabolomics involves the systematic quantification and identification of specific metabolites based on a predetermined hypothesis or set of target compounds. On the other hand, untargeted metabolomics takes a hypothesis-driven approach, aiming to comprehensively identify and analyze metabolites without predefining specific targets. In metabolomics, consideration is given to both the exo-metabolome (metabolites released outside the cell) and the endo-metabolome (metabolites within the cell). This holistic approach allows for a comprehensive understanding of the metabolic processes and interactions within biological systems. Overall, the field of metabolomics can be divided into targeted and untargeted approaches, with the aim of quantifying and identifying metabolites, including considerations of both exo-metabolome and endo-metabolome, to unravel the molecular complexities of biological systems (Narad et al. 2022). The systematic quantification also includes the exo- and endo-metabolome and metabolite identification (Putri et al. 2013), Untargeted metabolomics are discovered through a hypothesis-driven approach that permits complete metabolome scanning, also known as metabolic fingerprinting, and pattern recognition. According to Toya and Shimizu (2013) and Narad et al. (2022), the main goal of targeted metabolomics, which is based on hypothesis testing, is to confirm the results of the untargeted study.

10.1.2 Applications of Metabolomics

Metabolomics has demonstrated its widespread applicability in the fields of health, synthetic biology, and food sciences, demonstrating its significance in a variety of fields. The following is a discussion of the numerous metabolomics applications: (Putri et al. 2013; Narad et al. 2022).

10.1.2.1 Microbial Science

Microbes are a prominent resource in metabolomics due to their amenability to experimental modifications. However, high-resolution analysis, regulated ambient conditions for metabolite identification, and improved sample preparation methods are required for microbial metabolomics. When preparing samples for microbial metabolomics, extraction and quenching are the two crucial procedures. In order to stop biological reactions in cells, a procedure known as quenching must first be used to collect metabolites from the cells. To ensure accurate and reproducible results, sample quenching is performed at a specific time point during the metabolomics workflow. This quenching step allows for the determination of the actual quantity of metabolites present at that particular moment, enhancing the reliability of the results obtained. Two key aspects—short-term biological reaction halting and minimal metabolite leakage—validate quenching. The microbial cells are subjected to the extraction process depending on the chemical characteristics of the target analyte, the reactivity of the enzymes, the cell characteristics, and the durability of the cell membrane. Depending on how effectively microbial cells can tolerate the demanding environment, high temperature, methanol, chloroform, or free thawing are utilized. Stable isotopes are typically used in microbial metabolomics. For instance, ^{14}C glucose was utilized to investigate the connection between overall control and cellular metabolism in *Saccharomyces cerevisiae*. MFA has also been used to analyze the central metabolism using ^{13}C-labeled intermediates. The application of microbial metabolomics holds significant potential for advancing the study of higher organisms. The knowledge gained from microbial metabolomics can be extrapolated and applied to better understand and analyze the metabolomes of higher organisms. By leveraging the insights and techniques derived from microbial metabolomics, researchers can enhance their understanding of complex metabolic processes in higher organisms, contributing to advancements in fields such as medicine, ecology, and agriculture (Putri et al. 2013; Narad et al. 2022).

10.1.2.2 Plant Science

In order to understand the complicated biological processes and decipher the roles of numerous important genes, metabolomics is crucial for plant science (Toya and Shimizu 2013). Plant metabolic status is significantly influenced by transcriptional control under a variety of developmental and environmental circumstances. It was reported that the, the mechanism behind the regulatory roles in metabolic phenotype and gene expression remains enigmatic. Metabolomics enables botanists to conduct detailed investigations into the dynamic behavior of plant metabolic systems. By analyzing plant metabolomics, researchers can assess both metabolites and gene expression, providing valuable insights into plant physiology. The AtMetExpress dataset has been utilized to study the metabolic profiles and gene expression patterns of the *Arabidopsis thaliana* plant, contributing to a deeper understanding of its metabolic pathways and regulatory mechanisms (Putri et al. 2013). Based on a report, the genome of *Arabidopsis thaliana* contains a substantial number of vital metabolic genes involved in the production of commercially significant plant compounds. The data revealed the presence of 1589 metabolic signals related to various phytochemicals and a total of 167 distinct metabolites within the plant. This information highlights the rich metabolic potential of *Arabidopsis thaliana* and its significance in the production of essential compounds (Putri et al. 2013), the diversity of plants' secondary metabolism and the source of dynamics are determined by the transcription of metabolites and the regulation of those transcripts. It is also admirable how metabolomics is being used in breeding and agriculture sciences. It is important to identify the genetic components that play a crucial role in controlling metabolic levels in order to increase the nutritional value of the crops. Metabolomics is frequently used to explore these linkages since the relationship between biomass/yield and metabolite composition controls plant metabolism (Putri et al. 2013). Metabolomics has emerged as a valuable tool in breeding and crop sciences, particularly for enhancing the nutritional value of crops. By identifying genetic factors that influence metabolic levels, researchers can make targeted improvements. Understanding the intricate relationship between biomass, yield, and metabolite composition is crucial for manipulating plant metabolism. Metabolomics provides a comprehensive approach to studying these relationships, enabling researchers to explore and optimize crop traits related to metabolite composition and overall crop quality (Narad et al. 2022). Metabolomics is important for both defining the necessary level of risk management and for the effective production of genetically modified crops. Researchers can now produce enormous amounts of phytochemicals thanks to improvements in metabolomics technology, which opens up a wide range of potential uses. These developments contribute to the exploration and utilization of plant metabolites for agricultural purposes, including the development of genetically modified crops and effective risk assessment strategies (Narad et al. 2022). For plant metabolomics, three main approaches are required: a method to calculate false discovery rates, a vast mass spectral library of phytochemicals, and MS spectrum data for elucidating metabolite structure (Toya and Shimizu 2013; Narad et al. 2022).

10.1.2.3 Animal Science

With the use of metabolomics technology, it is now possible to investigate the biological processes of important model organisms like fruit flies and zebrafish. These species' metabolism can be thoroughly study to learn a great deal about their pathogenic, physiological, and developmental processes.

A popular model organism for investigating different biological, behavioral, and biomedical processes connected to organogenesis and embryogenesis in vertebrate development is the zebrafish (*Danio rerio*). Its genetic resemblance to humans, quick development, clear embryos, and simplicity of genetic modification all contribute to its popularity. Researchers extensively study zebrafish to gain insights into fundamental developmental processes, investigate disease mechanisms, and screen potential therapeutic compounds. It is a useful tool for enhancing our understanding of vertebrate development and human health due to its usefulness as a model organism. Due to its ease of breeding in high numbers and relatively low maintenance cost compared to other model organisms, it has been extensively utilized for research into the development of drugs and diseases (Riekeberg and Powers 2017). The association between embryogenesis and metabolome can be ascertained with the help of the metabolomics approach, which acts as a fingerprint for analyzing the embryonic process and helps with medication therapies. *Caenorhabditis elegans*, a different model organism, is frequently used to research aging, genetic disorders, physiology, lifespan, and medication toxicity screening (Narad et al. 2022). Due to their short lifespan, similarity to human aging, and ease of breeding, fruit flies, also known as *Drosophila melanogaster*, are extensively employed to study the physiology and genetics of aging.

A useful model organism for metabolomics research on topics like embryonic biology, the effects of phenobarbital, pesticide resistance, oxidative stress, and more is the zebrafish (*Danio rerio*). Its resistance to mutations, hypoxia, and cold shock further enhances its suitability for these research areas (Putri et al. 2013; Narad et al. 2022).

10.1.2.4 Medical Science

Numerous medical fields have utilized metabolomics. It frequently used to examine the biomarkers found in physiological fluids and to ascertain how drugs work. The use of metabolomics, which tracks metabolite changes in biofluids, is common in pharmacological therapy and medical therapy (Toya and Shimizu 2013). Metabolites serve as biomarkers for illnesses, therefore a change in their content in bodily fluids denotes the presence of a disease. As a result, a variety of metabolites provide data on treatment response with a high degree of selectivity and sensitivity (Narad et al. 2022). Additionally, a lot of people utilize metabolomics to forecast how a drug will react to a certain condition and to gauge how the disease may develop in the future. Furthermore, the zebrafish is utilized in predicting personalized treatment options for patients. Precision medicine, single cell metabolic phenotyping, personalized medicine, metabolome-wide association studies (MWAS), and epidemiological population research are a few areas where metabolomics has applications. These diverse applications highlight the value of metabolomics in monitoring and understanding health and disease. The identification and analysis of metabolites, the characterization of tiny molecules, and the high-dimensional profiling of individual cells all depend on metabolomics. It is extensively utilized for the discovery of clinical biomarkers through various approaches such as metabolomics fingerprinting, profiling, foot printing, and metabolome-wide association studies (MWAS). These methodologies enable the identification and utilization of metabolomics signatures for diagnostic, prognostic, and therapeutic purposes in various clinical applications (Putri et al. 2013). It is also used to research a variety of metabolic syndromes, including serious conditions brought on by sugar and lipid metabolism, like cancer, heart disease, and cerebrovascular disease. Through the pathophysiological study of metabolites and biomarkers, it assists in the early detection of fatal diseases (Narad et al. 2022). The pathophysiological investigation of metabolites and biomarkers using metabolomics aims to detect life-threatening disorders early. Using LCMS-based metabolomics techniques, biomarkers such as trimethylamine oxide have been identified as indicators of cardiovascular diseases. This highlights the potential of metabolomics in uncovering valuable biomarkers for timely diagnosis and intervention in critical medical conditions (Toya and Shimizu 2013). It could function as a biomarker for illnesses like myocardial infarction, coronary artery disease, and peripheral artery disease. It is frequently employed in cancer research, early cancer diagnosis, and accurate prognosis. In order to measure metabolic flux in lung cancer cells, the metabolomics method based on ^{13}C stable isotopes has been used. It revealed an excess of alanine, lactate, and glutamine, three substances crucial to the growth of cancer (Putri et al. 2013). In order to comprehend the biochemical alterations in cancer cells, isotopomer-based metabolomics is employed. It is a promising, non-invasive, and extremely sensitive cancer diagnostic technique. Additionally, it is used to understand neurological disorders and psychological issues (Putri et al. 2013). As a potential biomarker for brain metabolomics, cerebral spinal fluid has been studied using 1H NMR-based metabolomics. This method is used to study neurodegenerative conditions including Alzheimer's and Parkinson's. Furthermore, neural metabolomics offers insights into psychiatric conditions like depression and schizophrenia, where alterations in neurotransmitter systems and phospholipids in neuronal membranes are implicated in the pathogenesis of schizophrenia (Narad et al. 2022). Since alterations in lipid metabolism are a contributing factor in schizophrenia, lipidomic analysis is regularly carried out to identify the likely biomarkers underlying pathophysiology. Drug toxicity testing, early diagnosis, therapy, and research of biochemical alterations in mood disorders all make use of metabolomics. Our comprehension of the relationship between pathological diseases and molecular abnormalities in the body is improved by combining metabolomics with other omics approaches. This comprehensive approach offers valuable insights into disease mechanisms and potential therapeutic approaches (Putri et al. 2013; Narad et al. 2022).

10.1.2.5 Food and Herbal Medicines

A promising method to evaluate the safety and quality of food and herbal treatments is metabolomics. Food product quality can be influenced by milling, atmospheric storage, and genetic modifications. The quality control of finished food products and the safety assurance of herbal remedies can be successfully implemented by sensory evaluation and metabolomics (Putri et al. 2013). The five senses—touch, hearing, sight, taste, and smell—are employed in sensory evaluation, a scientific method, to examine, evoke, interpret, and quantify product quality. The preservation of quality standards and cost management are crucial in the food sector. To evaluate the quality of different food products such fruits, cereals, crops, and drinks, MS-based metabolomics approaches are used. The field of food metabolomics encompasses the organization of flavor-active chemicals and the simulation of human senses, contributing to advancements in food quality assessment (Narad et al. 2022). Sensomics, a subfield of food metabolomics, mimics human hearing, taste, sight, smell, and touch to assess the quality of food. In food metabolomics, methods like NMR and GCMS are used. These are also utilized for industrial, pharmaceutical, and research related to herbal medicines. They are employed in the analysis of pharmacological and toxic effects. Consequently, metabolomics is developing into a strong, trustworthy, useful, and promising instrument for quality assurance and sensory chemistry. It provides valuable insights into the chemical composition and sensory characteristics of various products, enabling effective quality assessment and control measures. Metabolomics contributes to enhancing the overall understanding and evaluation of product quality, reinforcing its importance in the field of sensory chemistry (Putri et al. 2013; Narad et al. 2022).

10.2 Flux Balance Analysis

Metabolic networks are modelled using a variety of mathematical techniques that are based on the determination of a single solution specifying all the fluxes via a metabolic network. FBA, or constraint-based analysis, provides a fundamental understanding of how a metabolic network is made up and functions. Mathematical formulas are used to reflect these constraints (Kauffman et al. 2003). The annotated genomic sequences provide information on the enzymes involved in these processes. In order to find the enzymes involved in metabolism, Because of the advancement of homology searches, it is now possible to compare known genes to those that are unknown. Table 10.1 lists a few useful databases for genetic and metabolic information (Narad et al. 2022).

Table 10.1 Database and tools used in metabolic network analysis (Narad et al. 2022)

Sr.	Section	Databases/tools	Description	References
1	Tools	MetaFluxNet	Metabolic flux analysis	Lee et al. (2003)
2		Yana	Network reconstruction, analysis, and visualisation	Schwarz et al. (2005)
3		System Biology Research Tool	Analysis of stoichiometric networks through multiple methods	Wright and Wagner (2008)
4		Constraint Based Reconstruction and Analysis Toolbox	Works with MATLABs for metabolic network analysis, gene deletions, etc., using FBA	Becker et al. (2007)
5		PathwayAnalyzer	Uses MoMA, FBA for gene deletion studies and metabolic networks	Raman and Chandra (2009)
6		BML Software Guide	Model databases	Hucka et al. (2015)
7		CellNetAnalyzer	Databases of functional and structural analysis	Klamt et al. (2007)
8		SNA—Stoichiometric Network Analysis	Mathematic toolbox for metabolic networks	Urbanczik (2006)
1	Database	BRENDA	Information of molecular and biochemical pathways on enzymes	Schomburg et al. (2002)
2		BioCyc	Databases of pathways for several organisms	Karp et al. (2019)
3		Reactome	Curated databases of biological processes in humans	Fabregat et al. (2017)
4		PEDANT	Genome annotations	Riley et al. (2007)
5		Biomodels	Databases of kinetic models of pathways	Li et al. (2010)

10.3 Metabolomics Techniques

In this book chapter we give you, an overview of metabolomics, which is mainly segregated into two main, approaches targeted and untargeted. Below we discuss some frequently used techniques such as, LCMS, GCMS and NMR and their consequent data analysis procedure.

10.3.1 Targeted and Untargeted Metabolomics Techniques

The Targeted approach is mainly focused on identification and quantification of specific class of metabolites or metabolites. These might be substances from a certain class, direct products of a protein, enzyme substrates, or participants in a certain pathway. Normally, the targeted approach is hypothesis-driven, aiming to test specific hypotheses. Another metabolome technique, the untargeted analysis involves measuring metabolites within biological system.

10.3.1.1 LC–MS

LC–MS, a "Liquid Chromatography–Mass Spectrometry" combined the principles of both LC and MS. LC is mainly used to separate the available molecule in liquid mobile phase by using solid stationary phase (Pitt 2009). The high resolving power of LC based analysis is used to determine the structure and quantified degradation of compounds and impurities in different materials. By combining, LC–MS makes it possible to identify and characterize individual components within a complex mixture based on their mass-to-charge ratios (Karpievitch et al. 2010). This yields important information on the make-up and standard of the analyzed materials. An ion source, a mass analyzer, and a detector are the three crucial parts of a mass spectrometer. Sample molecules are turned into ions by the ion generator, and then utilizing an electromagnetic field, the mass analyzer separates the ions according to their mass-to-charge ratios. The detector then captures and measures the separated ions, providing valuable information about their abundance and mass properties (Fuerstenau and Benner 1995). Once these ions have been separated by the mass analyzer, the detector finally measures them. Electrospray ionization (ESI), atmospheric pressure chemical ionization (APCI), atmospheric pressure photoionization (APPI), and fast atom bombardment (FAB) are just a few of the flexible ion sources that can be used with mass spectrometry. These different ion sources provide diverse ionization mechanisms and are selected based on the specific requirements of the analysis or sample type in mass spectrometry (Agarwal and Goyal 2017). ESI stands out among these other ion sources because of its gentle ionization capabilities, which makes it easier to produce plenty of ions by charge exchange in solution. The first identification of analytes is aided by this property. For the measurement of polar and semi-polar metabolites, LC–MS is a flexible analytical method that is often used in metabolomics investigations (Xiao et al. 2012). It works well for profiling small molecules, including amino acids, organic acids, nucleotides, carbohydrates, lipids, and other water-soluble compounds (Sato et al. 2004). LC–MS enables the comprehensive characterization and quantification of these compounds, providing valuable insights into the metabolic profile and composition of biological samples (Fig. 10.1). Metabolites are essential components involved in crucial cellular processes, signaling pathways, energy metabolism, and various disease-related pathways. LC-ESI-MS has emerged as the preferred technique for analyzing and profiling metabolites in complex biological samples. By incorporating chromatographic separation, the complexity of the sample can be reduced, and any potential matrix effects during ionization can be minimized. This approach allows for enhanced sensitivity, specificity, and accuracy in metabolite analysis, making LC-ESI-MS a powerful tool in metabolomics research (Böttcher et al. 2007). Reverse phase liquid chromatography (RPLC) is often employed, with the use of C18 columns, to effectively separate semi-polar compounds such as phenolic acids, flavonoids, glycosylated steroids, alkaloids, and other glycosylated species in LC-ESI-MS (Lu et al. 2008). RPLC is a popular technique for effective semi-polar chemical separation in LC-ESI-MS,

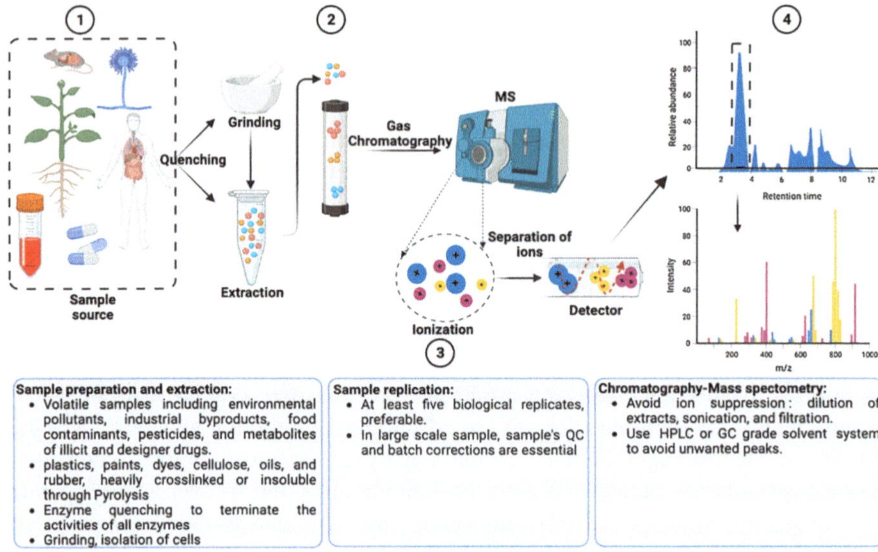

Fig. 10.1 General flowchart of LC–MS for metabolomics

frequently utilizing C18 columns. Different substances, such as phenolic acids, flavonoids, glycosylated steroids, and other glycosylated species, can be separated using this method. These semi-polar metabolites can be thoroughly analyzed and identified in complex biological samples using RPLC and LC-ESI-MS (Zhou et al. 2012). In LC–MS data, variations can be observed not only in the spectra obtained from different instruments but also in the MS/MS spectra generated under different experimental conditions. These discrepancies arise from the use of diverse combinations of ionization sources, collision energies, mass analyzers, and detectors. These variations in instrument types and experimental settings can impact the characteristics and interpretation of the MS/MS spectra, highlighting the need for careful consideration and standardization in data analysis and comparison. As a result, several MS/MS spectra can be seen for the same metabolite, highlighting how experimental variables can affect the final spectrum features.

The raw LC–MS data must undergo a number of preprocessing steps in order to create a peak list that facilitates analysis and comparison of different runs. Outlier identification, peak matching, baseline correction, filtering, outlier screening, and retention time alignment, ion annotation, normalization, transformation, and use of software tools are some of these stages that are listed below (Castillo et al. 2011). Every one of these preprocessing procedures is essential for cleaning up the data and getting it ready for insightful analysis and future comparisons. In order to account for matrix effects and time-dependent changes brought on by instrument sensitivity variations, detected peaks are normalized using an internal reference peak. Gradient elution in LC–MS allows high-resolution spectrum data of metabolites, permitting specialized metabolite study (Griffiths and Wang 2009). Various equipment configurations cause variations in LC–MS and MS/MS spectra, resulting in various spectra for the same metabolite. Currently, manual verification is utilized after mass-based search to identify metabolites in untargeted metabolic research. Because there exist chemicals with extremely similar molecular weights, it has been shown that even with an accuracy of less than 1 ppm, which is substantially more precise than most analytical systems can attain, it is still insufficient for unambiguous metabolite identification (Calderón-Santiago et al. 2017). Secondly, isomers with the same elemental content but distinct structures cannot be distinguished by mass-based metabolite identification. Third, there is a dearth of information in all metabolite databases (Sleno 2012). A significant fraction of the detected ions in a typical LC–MS-based metabolomics experiment is still unidentified or has numerous plausible identifications. Through mass-based searches, less than 30% of these ions may be accurately identified. But the use of QqQ-based LC–MS, LC-SRM-MS, and LC-HRMS full scan analysis has shown how crucial metabolite quantification is for comprehending the response to illnesses, treatments, and environmental factors. These techniques enable accurate and precise measurement of metabolite levels, providing valuable insights into metabolic changes and their implications (Dowling 2017). Additionally, some analytes, including those that show the neutral loss of H_2O or CO_2, may exhibit non-specific transitions that are frequently seen in matrix interferences. This lack of specificity can impact the accuracy of quantification in the selected reaction monitoring (SRM) method, leading to incorrect results. It is important to consider and address these challenges in order to ensure reliable and precise quantification of analytes in metabolomics studies (Pozo et al. 2006). A global MS detection employing HRMS, such as FTICR, Orbitrap, TOF, or QTOF, can get beyond these limitations in SRM analysis (Amer et al. 2023). In full-scan mode, high-resolution mass spectrometry (HRMS) enables the identification of almost all compounds present in a sample. With the advancements in HRMS technology, such as fast scan rates, it is possible to capture an ample number of data points across chromatographic peaks. By producing extracted ion chromatograms (EICs) within a small mass window (e.g., 5–10 mmu) centered on the theoretical *m*/*z* value of each analyte, this permits accurate quantification. This approach enhances the accuracy and sensitivity of quantification in metabolomics studies (Alygizakis et al. 2023). In summary, LC–MS has revolutionized the field of metabolomics, enabling researchers to comprehensively study the dynamic and intricate world of small molecules. With its ability to provide detailed insights into metabolic pathways, biomarker discovery, and understanding of disease mechanisms, LC–MS continues to drive groundbreaking discoveries and advancements in various scientific disciplines. It's potential to transform healthcare, agriculture, environmental studies, and personalized medicine is immense, making it an indispensable tool in the quest to unravel the complexities of the metabolome.

10.3.1.2 GC–MS

By separating molecules based on their volatility, gas chromatography. Its initial use was explained in 1952 (Bartle and Myers 2002). The analytes are initially adsorbed to a GC column's surface at a slightly raised temperature in order to accomplish separation. The GC column can be quickly heated and cooled since it is housed inside an oven. The temperature is increased once the analytes are bonded, which causes them to leave the column surface in decreasing order of volatility (McNair et al. 2019). Once the analytes have been thermally desorbed, they are transported down the column surface toward the detector using a carrier gas (mobile phase, often helium). With its unmatched capabilities for the investigation of tiny molecules, GC–MS has completely transformed the area of metabolomics (Cui et al. 2018). A crucial technique for comprehending the metabolome is gas chromatography with mass spectrometry (GC–MS), which combines the separation power of gas chromatography with the sensitive and focused detection of mass spectrometry (Smart et al. 2010). Chemical derivatization is required to improve the volatility of metabolites containing polar functional groups, such as carboxylic and amino groups, in contrast to LC–MS and NMR-based metabolomics. This modification is essential to enhance their vaporization properties for improved detection and analysis in techniques such as GC–MS (Zeki et al. 2020). Pre- or post-derivatization GC–MS analysis and data collection are performed on the volatile compounds. To understand the complex mass signals, a data processing technique should be used to recognize the real signals, classify the signals into different compounds, and align these compounds from different samples (Ràfols et al. 2018). The metabolic route linked to particular physiological or pathological abnormalities can be discovered after peak annotation. As a result, the complete procedure entails the following steps: collecting the samples, extracting the metabolites, derivation the compounds, analyzing the instruments, analyzing the data, and annotating the metabolites and the pathways. The following graphic illustrates the basic steps of GC–MS-based metabolomics, which can be used to analyze the metabolites in biofluids, tissues, or cell samples (Fig. 10.2).

One of the main factors contributing to the adoption of GC–MS in metabolomics investigations is accurate and repeatable chemical identification. In GC-based metabolomics applications, MS detection with electron ionization (EI) is commonly employed in combination with GC. The use of EI mode allows for non-discriminatory identification of all compounds suitable for GC analysis, as the scan response is generally proportional to the injected compound quantity. Various GC columns are utilized to separate fatty acids, amino acids, sugars, and monosaccharides, with the 5% phenyl, 95% methyl siloxane column being frequently used due to its broad selectivity for untargeted metabolomics applications (Zaikin and Halket 2009). Peaks and retention time were given to the same variable in each sample after being separated from the raw data. Three approaches in particular—target analysis, peak selecting, and deconvolution—have proven to be beneficial for this purpose (Koek et al. 2011). In gas chromatography-mass spectrometry (GC–MS), many detector types can be employed to examine and locate chemicals that have been separated by the gas chromatograph. Four typical GC–MS detectors are listed below:

1. Flame Ionization Detection (FID): In GC–MS, the FID detector is a popular and very sensitive one. It works by creating ions from the chemicals that elute from

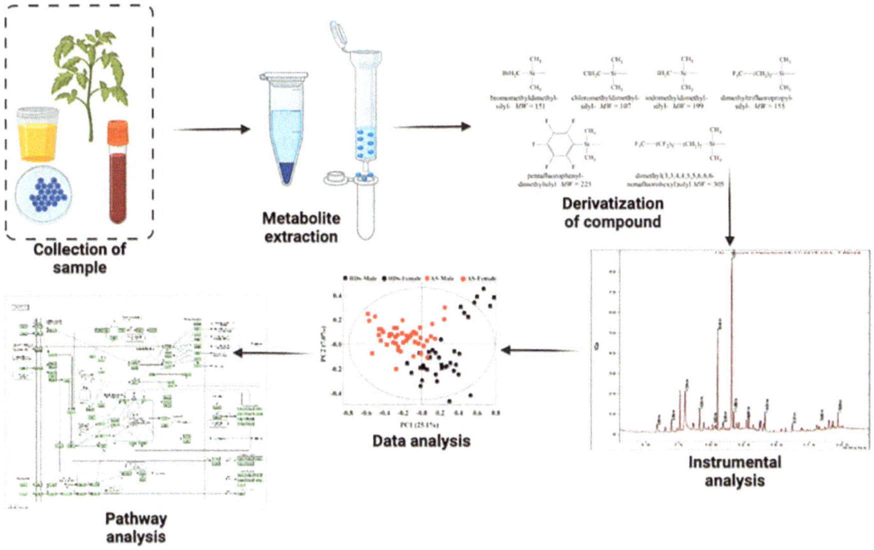

Fig. 10.2 General working flow of GC–MS for metabolomics

the gas chromatograph by burning them in a flame of hydrogen and air (Poole 2015). After being collected and tested, the produced ions reveal how many of the chemicals are present in the sample. The investigation of a variety of chemicals, including hydrocarbons, is made possible by the FID's outstanding sensitivity, large linear dynamic range, and resilience (Jalili et al. 2020).

2. "Single Quadrupole Mass Spectrometry (SQ-MS)": SQ-MS mass spectrometers are frequently utilized in GC–MS. It is made up of a single quadrupole mass filtering ions with a preference for those with a higher mass-to-charge ratio (*m*/*z*) (Morain 2013). SQ-MS can analyze various *m*/*z* values to identify and quantify specific chemicals in a sample (Modisha et al. 2018). It helps with structural elucidation and compound identification and quantification by giving details about the mass fragments created during ionization.
3. "Time-of-Flight Mass Spectrometry (TOF-MS)": An additional mass spectrometer type employed in GC–MS is TOF-MS. It determines the ion's flight duration from the ion source to the detector based on their *m*/*z* values (Guilhaus 1995). The mass resolution, sensitivity, and accuracy of TOF-MS are all quite excellent. It is frequently employed for untargeted analysis and is capable of providing thorough details on the full mass range contained in a sample, enabling the identification of unidentified chemicals and the discovery of trace-level analytes (Hird et al. 2014).
4. "Fourier Transform Ion Cyclotron Resonance Mass Spectrometry (FT-ICR-MS)": It traps ions with a magnetic field and detects the oscillation frequencies, which enables extremely precise mass determinations (Heck et al. 2011). For complicated mixture analysis and isobaric chemical identification, FT-ICR-MS is a great choice due to its high mass resolution, mass accuracy, and

sensitivity. But FT-ICR-MS equipment is expensive and complicated, and it's frequently employed in high-end research settings (Seger et al. 2013).

A target analysis list is created, detailing each metabolite expected to be found in the data file's *m/z* value and defined retention time window. The instrument vendor's software uses the target list to calculate each metabolite's peak area, enabling quantification (Fiehn 2016). In addition, the purity and quality of isolated peaks should be controlled by utilizing internal standards that have been isotopically labeled for extraction and derivatization. In conclusion, GC–MS has become a potent and crucial technique in metabolomics, providing accurate and sensitive investigation of polar and semi-polar metabolites. With its broad selection of detectors, including FID, SQ-MS, TOF-MS, and FT-ICR-MS. GC–MS also facilitates thorough metabolite profiling, identification, and quantification, with greater insights into biological processes, disease causes, and the identification of new biomarkers. An important technique advancing metabolomics research and its applications in areas including personalized medicine, environmental studies, and agriculture, GC–MS is able to handle a variety of sample types and provide high-resolution data.

10.3.1.3 NMR

NMR spectroscopy, a non-destructive analytical method, is a cornerstone of metabolomics because it offers priceless insights into the structure, dynamics, and interactions of metabolites (Cheng et al. 2013). The field of metabolomics can benefit from NMR's major properties, which are listed below (Trimigno et al. 2015).

1. NMR has excellent quantitative and reproducibility.
2. The target list is used by the instrument vendor's software to determine each metabolite's peak area, allowing for quantification.
3. As analytical technological advances have resulted in the detection of an increasing number of signals in complex biological mixtures, many of which remain unidentified, NMR allows for the identification of unknown metabolites, which is crucial.
4. There is no need for sample preparation or separation because to NMR's ability to examine entire biofluids and tissue, which is critical because these operations greatly increase analytical variability.
5. NMR preserves the sample integrity after analysis, allowing for potential reanalysis using NMR or other techniques such as MS in the future.
6. NMR allows for the tracing of metabolic pathways and the measuring of metabolic fluxes by using precursors that have been stable isotope-labeled.
7. Using one or more atomic nuclei, such as ^{1}H, ^{13}C, ^{31}P, or ^{15}N, NMR can detect metabolites.
8. NMR analysis is advantageous for sensitive metabolites like glutamine and coenzymes, as it does not require harsh sampling or ionization voltage treatment.
9. The NMR workflow for metabolomics involves key steps such as signal detection, metabolite identification using 1D and 2D NMR methods, database

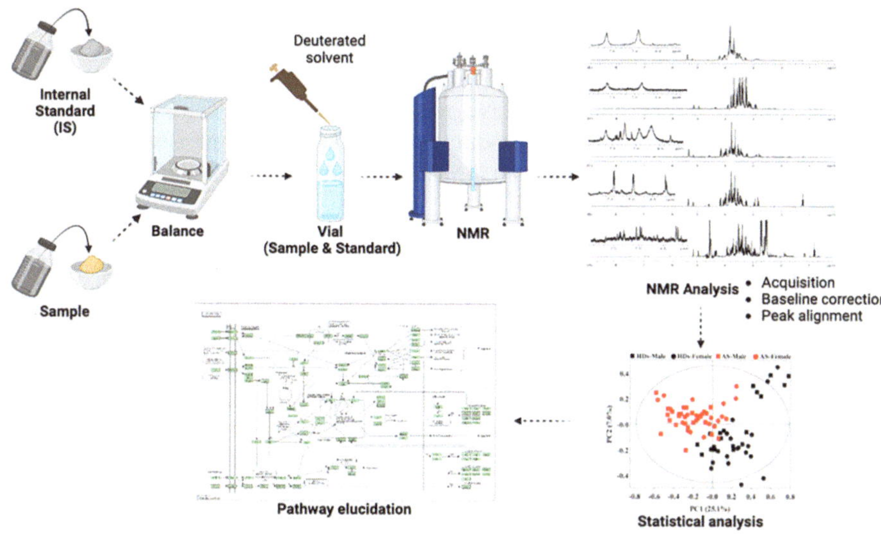

Fig. 10.3 General workflow of NMR for metabolomics

searching and verification, and quantification of identified metabolites. These procedures are essential for the NMR-based metabolomics examination of biological material (Fig. 10.3).

NMR offers some advantages over metabolomics that are unmatched. NMR offers a window into seeing and precisely measuring all of the more prominent compounds found in biological fluids, cell extracts, and tissues without the requirement for time-consuming sample preparation or separation. For compounds that are challenging to ionize or require derivatization for MS analysis, NMR has advantages. It enables the identification of substances, even those with different isotopomer distributions, that have identical masses, providing valuable insights in metabolomics research (Markley et al. 2017). NMR serves as the primary method for elucidating the structures of unknown substances. It enables the investigation of metabolic pathway compartmentalization and provides insights into the kinetics and mechanisms of metabolite conversions through the utilization of stable isotope labels (Fan and Lane 2016). A variety of cutting-edge techniques are being used with NMR-based metabolomics to produce fresh and in-depth data. The next section goes through a few of these techniques.

10.3.1.3.1 Isotope Enhanced NMR to Track Metabolism

This approach utilizes NMR spectroscopy, which possesses the unique ability to identify atom-specific positional isotopomer distributions that arise from the utilization of stable isotope-enriched precursors (Lane et al. 2008). Numerous stable isotope-rich materials, such ^{13}C, ^{15}N, and ^{2}H, have been thoroughly studied (Yu et al. 2023). In order to quantitatively assess the downstream metabolic products of several pathways, including as glycolysis, the tricarboxylic acid (TCA) cycle, and the pentose phosphate pathway (PPP), ^{13}C-labeled glucose is frequently used to follow metabolism (Antoniewicz 2018). This tracking doesn't provide an active metabolic pathway but they also give rates of metabolism step, where labeled substrate gets consumed and their product formed. Therefore, the isotope-enhanced NMR becomes useful in cancer metabolism or cellular metabolism investigation (Tavares et al. 2015).

10.3.1.3.2 Micro-Coil NMR

In this NMR approach, liquid chromatography is used to separate metabolites from complex mixtures, and then direct online detection is used, frequently after pre-concentration online. Micro-coil NMR is a common technique utilized for this purpose (Nagana Gowda and Raftery 2019). This approach is not characterized by high throughput but is better suited for analyzing metabolites in samples with limited mass. Due to the increasing interest and demand for NMR miniaturization, major instrument suppliers now offer micro-coil probes with multinuclear capabilities (Badilita et al. 2012).

When utilizing micro coil NMR, caution must be exercised when concentrating samples, as metabolites with low solubility and sample matrices containing high levels of salt or proteins can negatively impact the relative and absolute quantities of metabolites. It is important to consider these factors to ensure accurate and reliable results during micro coil NMR analysis of samples (Anderson et al. 2012). A recent assessment of the impact of sample concentration on commonly used serum and urine samples revealed that the sensitivity improvement achieved varied for different metabolites and sample matrices. The observed sensitivity enhancement did not follow a linear trend as expected, highlighting the complex relationship between concentration and sensitivity in metabolite analysis.

10.3.1.3.3 Fast NMR Method

High-throughput analysis is yet another crucial requirement in metabolomics, the fastest in terms of data gathering. Numerous developments in 2D NMR have made it possible to obtain data quickly (Croasmun and Carlson 1996). HMQC and HSQC are acronyms for heteronuclear single and multiple quantum coherence and are widely used in 2D NMR experiments in metabolomics. These strategies can greatly speed up data collection by using forward maximum entropy reconstruction and non-uniform sampling, cutting down on the time needed for thorough metabolite profiling (Rouger et al. 2017).

In order to increase the steady-state magnetization, a shorter T1 relaxation time and a better flip angle are combined in the SOFAST NMR (selective optimal flip angle short transient) method. This makes it possible to acquire data quickly, especially for the SOFAST-HMQC 2D experiment, which may be finished in about 10–15 s. Real-time monitoring of metabolism in living cells is made possible by such quick data collecting, giving important insights into cellular processes (Sibille et al. 2012). Single-scan acquisition methods have shown great potential in metabolomics, particularly in 2D experiments. These methods allow for the acquisition of data from a single scan, offering advantages such as reduced acquisition time and improved sensitivity. With their ability to provide valuable information in a shorter time frame, single-scan 2D experiments have proven to be valuable tools in metabolomics research.

10.3.1.3.4 Hyperpolarization Method

There is a significant interest in utilizing hyperpolarization of nuclear spins to enhance sensitivity for real-time in vivo metabolism research. Hyperpolarization techniques enable the generation of highly polarized nuclear spins, resulting in enhanced signal intensity in NMR experiments. This enhanced sensitivity allows for the detection of metabolites in real time, opening up new possibilities for studying dynamic metabolic processes in living organisms (Wang et al. 2019). By introducing hyperpolarized substrates into biological systems, downstream metabolites can be identified with high sensitivity while preserving the polarized nuclear spin state. Unlike the PHIP method, the DNP approach allows for the hyperpolarization of various metabolites without encountering major challenges (Gowda and Raftery 2015). Real-time metabolic investigations benefit most from the dissolution DNP method, which involves quickly melting and injecting a hyperpolarized solid containing the target substrate, a glassing agent, and a polarizing agent into cells, tissue, or organs (Hurd et al. 2012). Overall, although these technologies are still in their infancy, they hold great promise for the sector. In conclusion, NMR spectroscopy is essential to metabolomics research because it provides important information on the intricate metabolic patterns of biological materials (Zhang et al. 2013). It is an essential tool for researching metabolic changes in health and illness since it may offer metabolite identification, quantitative analysis, and information on metabolic pathways. NMR data may be used with other omics methods to help researchers fully comprehend metabolic networks and find new biomarkers for diagnostic and therapeutic uses.

10.3.2 Important Tools of Flux Balance Analysis

10.3.2.1 OptKnock

Genes make up a sizable portion of a standard metabolic model (B1000). Therefore, as the set size increases, it takes more processing resources to do an exhaustive search of knockout sets. OptKnock is based on the duality theory, which claims that there is a single dual LP problem for each primal LP problem that equals the primal's objective function. By setting the objective functions equal to one another, the dual problem is used to increase constraint while maximizing biomass output (Burgard et al. 2003). A single mixed integer linear programming (MILP) problem is used to combine the growth-maximizing problem and the maximum product yield problem. The Optx and OptKnock algorithms can be used independently as a programmed to predict knockouts (Narad et al. 2022).

10.3.2.2 OptGene

OptKnock's disadvantage is that nonlinear objective functions are not optimized. With the MILP problem, it becomes a computationally demanding process when there are many knockouts (Burgard et al. 2003). OptGene gets around these limitations. With the following genetic algorithm:

1. It generates a collection of arbitrary optimal conditions.
2. The metabolic model for each set is solved using MOMA and FBA.
3. Each member of the set is given a score based in part on their metabolic status.
4. The condition with the highest score is chosen as the ideal one.
5. Up until the best score is obtained, step 2 is repeated (Burgard et al. 2003; Narad et al. 2022).

10.3.2.3 OptStrain

Although the OptGene and OptKnock algorithms are thought to be quite effective at predicting knockouts, their range of use is restricted to changes in metabolic processes. By creating a library of biotransformation to enhance the prediction of heterologous routes, OptStrain solves this issue (Burgard et al. 2003). The strategy for using OptStrain is as follows:

- To determine the highest level of product production, LP is used. It serves as the yield's starting point.
- MILP determines the minimum number of heterologous genes required to match the baseline production. It is essential to make the premise that product yield, not growth, should be maximized.
- The stoichiometric model includes the genes that were found in step 2's identification process. OptKnock is a method for optimization (Narad et al. 2022).

10.3.2.4 COBRA Tools

MATLAB's COBRA Toolbox is a collection of software (Becker et al. 2007). It is commonly employed in the MOMA Analysis and growth optimization processes. The benefit of COBRA is that it uses a number of scope functions to optimize the model, including objective functions and solution methods. It is also very flexible and simple to use (Schellenberger et al. 2011; Narad et al. 2022).

10.3.2.5 MetaboAnalyst 4.0

It is a tool created for the analysis, functional interpretation, and visualization of metabolic data. Using the R package, it generates clear and reproducible analyses. It has been observed functional enrichment analysis is utilized to control metabolic pathways. The mummichog algorithm determines untargeted metabolomics data. It facilitates the integration of multiomics data and meta-analysis of biomarkers. It consists of 12 unique modules that are divided into four categories according to their functions. These categories are: (1) data fusion and systems biology; (2) exploratory statistical analysis; (3) data processing and utility functions; and (4) functional analysis. The exploratory statistical analysis accepts data from both targeted and untargeted metabolites. The functional analysis category of MS data includes pathway activity prediction data (Chong et al. 2018; Narad et al. 2022).

10.3.2.6 Opt Flux

The COBRA Toolbox functions similarly to Opt Flux. It does both MOMA optimization and growth maximization. It runs on the JavaScript platform rather than MATLAB and offers an easy-to-use user interface. The OptKnock method and Boolean logic are both used in Opt Flux. However, it cannot be modified for simple one-time alterations (Aurich et al. 2016). It offers consumers access to open-source software applications. It is an open-source platform that gives users freedom of movement while making the representation of background informatics straightforward. It is modular and supports SBML as well as other file formats. It works with ROOM, MOMA, and FBA. The import, export, and visualization of stoichiometric metabolic models, including equations, metabolic processes, and links between gene reactions, are only a few of the services available. It can be used with databases like the BiGG database and BioModels, as well as tools like CellDesigner. Incorporating exogenous metabolites and identifying biomass production reactions require an explicit definition. Opt Flux conducts simulations using three different techniques, including ROOM, FBA, and MOMA. The fluxes of wild-type or mutant strains are calculated using the LP formulation by the FBA approach. ROOM employs MILP and LP while MOMA uses quadratic programming. Opt Knock and the metaheuristic algorithms EA and SA are used for optimization (Aurich et al. 2016; Narad et al. 2022).

10.3.2.7 OpenFlux

It is a simple spreadsheet-based user interface made to operate models based on isotopomers and metabolites. It is used in sensitivity analysis, flux estimation, FBA, and the creation of extensive metabolic models and networks. An isotopomer balance model is produced by OpenFlux using the elementary metabolite units (EMU) decomposition technique. It is more effective computationally. As a productive and versatile instrument for ^{13}C MFA, it is validated against the results. Compared to ^{13}C Flux, it is easier to understand and faster. Statistical analysis makes it simple to identify unknown free fluxes in large-scale metabolic models (Quek et al. 2009; Narad et al. 2022).

10.3.2.8 CellNetAnalyzer

It makes use of COBRA Toolbox for MATLAB. Unlike the MATLAB command window, it uses a straightforward graphical user interface to operate (Cheng 2012). It makes it simple to use numerous interactive and visualization tools by heavily relying on Boolean logic. Other than maximizing growth, it does not employ MOMA or any other sophisticated problem-solving strategies (Cheng 2012; Narad et al. 2022).

10.3.2.9 SBRT

The Systems Biology Research Tool (SBRT), a JavaScript-written piece of software, is used in FBA. It is a plug-in-capable software package that is open source (Wright and Wagner 2008).

10.3.2.10 Escher-FBA

Escher developed Escher-FBA, a flexible visualization tool with an easy-to-use interface, to investigate metabolic pathways. It is a quick and easy way for mapping GEM-containing reactions in the model and displaying both metabolites and reactions. Escher FBA offers users a lot of flexibility in terms of changing parameters and seeing results, including reaction knockouts, flux boundaries, and objective functions. Mobile devices are among the platforms it functions on. This makes it a popular tool in academic labs for visualizing, investigating, and learning FBA models. Users can load, edit, and save their maps that are saved as JSON files using this feature. It offers interactive tooltips for changing the FBA simulation's parameters. It uses the GNU Linear Programming Kit (Rocha et al. 2010; Narad et al. 2022).

10.4 Integration of Metabolomics and FBA

Flux balance analysis (FBA) and metabolomics integration has become a potent strategy for improving our comprehension of cellular metabolism. Metabolomics provides comprehensive information on the metabolite concentrations within a biological system, while FBA is a computational method that predicts metabolic flux distributions based on stoichiometric models. By integrating metabolomics data with FBA, researchers can gain insights into the dynamic behavior and regulation of metabolic pathways. Several studies have demonstrated the successful integration of metabolomics and FBA. Development an integrated approach called FBAwMC (FBA with Metabolomics Constraints) to improve flux predictions in *Saccharomyces cerevisiae*. They combined metabolomics data with FBA by constraining the model with measured metabolite concentrations. This integration improved the accuracy of flux predictions and provided a more realistic representation of the metabolic state of the yeast cells (Lewis et al. 2012). MFA (Metabolomics-assisted Flux Analysis) is another method that integrates metabolomics data into the FBA framework. They applied MFA to investigate the metabolism of *Escherichia coli* and *Saccharomyces cerevisiae* under different conditions. By incorporating metabolomics data as constraints in FBA, they obtained more accurate flux predictions and gained insights into the metabolic response to environmental changes (Volkova et al. 2020). In another study, researchers developed a method called "GIM3E (Gene Inactivation Moderated by Metabolism, Metabolomics, and Expression)" to integrate metabolomics data, gene expression data, and FBA. They applied GIM3E to analyze the metabolism of *Escherichia coli* under various genetic and environmental perturbations (Schmidt et al. 2013). The integration of metabolomics and FBA allowed them to identify metabolic bottlenecks and potential regulatory mechanisms in the system. Moreover, MFAwFBA (Metabolomics-assisted Flux Balance Analysis with confidence intervals) is another integration method that integrates metabolomics data with FBA to estimate fluxes and their uncertainties. They applied MFAwFBA to investigate the metabolism of *Corynebacterium glutamicum* and validated the predicted fluxes using experimental data (Zhang et al. 2017). This integrated strategy made it easier to identify important metabolic pathways and gave more accurate flux calculations. In conclusion, a strong framework for studying and analyzing cellular metabolism is provided by the combination of metabolomics and Flux Balance Analysis (FBA). By incorporating metabolomics data into FBA models, researchers can improve the accuracy of flux predictions and gain insights into the regulation and dynamics of metabolic pathways. The successful integration of metabolomics and FBA has been demonstrated in various studies, highlighting the potential of this approach for advancing our understanding of complex metabolic networks.

10.4.1 Identification and Annotation of Metabolite

An essential challenge in metabolomics research is the identification and annotation of metabolites, as it provides insights into the chemical composition and functional roles of small molecules within biological systems. Due to the complexity of metabolite combinations and the incomplete coverage of current reference databases, this approach might be difficult. For the purpose of improving metabolite identification and annotation, numerous techniques and technologies have been created. One common strategy for metabolite identification is the use of "high-resolution mass spectrometry (HRMS)" coupled with chromatographic techniques such as "liquid chromatography (LC)" or "gas chromatography (GC)". The exact mass measurements and fragmentation patterns offered by HRMS can help identify metabolites. Additionally, "tandem mass spectrometry (MS/MS)" techniques can be employed to obtain fragmentation spectra, which can be matched against spectral libraries or used for de novo identification. Reference databases play a crucial role in metabolite identification, allowing researchers to compare acquired mass spectra and retention times with existing data. In metabolomics research, a number of databases, such as the "Human Metabolome Database (HMDB)" (Wishart et al. 2007), the "Kyoto Encyclopedia of Genes and Genomes (KEGG)" (Kanehisa and Goto 2000), and the Metlin database (Guijas et al. 2018). These databases contain extensive collections of metabolite information, including mass spectra, chemical structures, and associated biological pathways, facilitating the identification and annotation of metabolites. However, it is important to note that the coverage and accuracy of these databases are not exhaustive, and there are limitations in metabolite annotations. The incompleteness of reference databases often leads to unannotated or mis-annotated metabolites. Efforts are being made to address this issue by continuously updating and expanding these databases, incorporating new metabolites and improving annotation accuracy. In addition to spectral matching against databases, complementary approaches have been developed to enhance metabolite identification. These include the use of fragmentation prediction algorithms, such as MetFrag (Ruttkies et al. 2019) and "CFM-ID" (Allen et al. 2014), which generate in silico fragmentation spectra based on metabolite structures and facilitate the annotation process. Additionally, network-based approaches, such as network annotation propagation (NAP) (da Silva et al. 2018), utilize metabolic networks and pathway information to improve metabolite annotation by leveraging the known properties of related compounds. To foster community-driven efforts in metabolite identification and annotation, collaborative platforms have been established. For example, the "Global Natural Products Social Molecular Networking (GNPS)" platform enables the sharing and comparison of MS/MS data and facilitates crowd-sourced annotations. Such platforms encourage data sharing and collaboration among researchers, contributing to the collective knowledge and accuracy of metabolite identification. The identification and annotation of metabolites are fundamental steps in metabolomics research. The integration of HRMS, spectral libraries, and reference metabolic databases enables researchers to identify and annotate metabolites based on mass spectra and retention times.

10.4.1.1 Metabolite Databases

Metabolic databases play a crucial role in organizing and disseminating information related to metabolites, enzymatic reactions, and metabolic pathways. These databases serve as valuable resources for researchers, allowing them to access comprehensive and curated information on metabolite structures, properties, and functions. One widely used metabolic database is the "Kyoto Encyclopedia of Genes and Genomes (KEGG)," which provides a comprehensive collection of metabolic pathways and associated genes for various organisms (Kanehisa et al. 2022). Another notable database is the "Human Metabolome Database (HMDB)," which focuses on human metabolites and contains extensive information on metabolite structures, properties, biofluid concentrations, and associated pathways (Wishart et al. 2022). Furthermore, the MetaboLights database serves as a repository for metabolomics data, enabling researchers to share and access metabolomics datasets along with their associated metadata and analysis results (Haug et al. 2017). These databases support pathway analysis, network modeling, the identification of possible biomarkers and therapeutic targets, in addition to facilitating metabolite identification and annotation. These databases considerably advance our understanding of metabolism and its consequences in numerous disciplines of research by offering a consolidated and curated source of metabolic data.

10.4.1.2 Spectral Libraries

Due to their extensive collection of reference spectra for numerous chemicals, spectral libraries are essential in the discipline of spectroscopy. These libraries serve as valuable resources for researchers, allowing them to compare and identify unknown spectra obtained from experimental analyses. One widely used spectral library is the "National Institute of Standards and Technology (NIST)" Mass Spectral Library, which contains a vast collection of mass spectra for organic compounds (Stein 2012). This library has been widely utilized in fields such as forensic analysis, environmental monitoring, and metabolomics research. Another notable spectral library is the "Human Metabolite Database (HMDB)", which includes reference spectra for a variety of metabolites (Wishart et al. 2022). This resource has proven invaluable in metabolomics studies, enabling researchers to identify and annotate metabolites based on their spectral characteristics. Additionally, the "Protein Data Bank (PDB)" provides a spectral library for proteins, containing information about their structure, function, and associated spectra (Berman et al. 2002). Spectral libraries not only aid in compound identification but also support the development and validation of spectroscopic techniques and analysis methods. By providing a standardized reference for comparison, spectral libraries contribute significantly to advancing research in various scientific disciplines.

10.4.2 Metabolomics Data Integration with Genome-Scale Metabolic Models

The combination of metabolomics data with genome-scale metabolic models is a potent method that enables a thorough examination of cellular metabolism. Metabolomics provides information about the 'Biological systems' small-molecule metabolites are represented by genome-scale metabolic models (GEMs), which depict the intricate web of biochemical processes taking place inside a cell. By integrating metabolomics data with GEMs, researchers can gain insights into the metabolic state of an organism and predict its behavior under different conditions. Several studies have demonstrated the benefits of integrating metabolomics data with GEMs. For example, in a study, researchers used this approach to look into how *Escherichia coli* reacts metabolically to genetic and environmental changes. They were able to pinpoint important metabolic pathways that were impacted by the perturbations by integrating metabolomics data with a genome-scale model of *E. coli* metabolism, (Wang et al. 2021). Similarly, in another study implementation of metabolomics data with a GEM of *Saccharomyces cerevisiae* to study the metabolic changes associated with different growth conditions (Oftadeh et al. 2021). Specific metabolites and metabolic pathways that responded to environmental changes were identified through their investigation. The integration of metabolomics data with GEMs also enables the identification of metabolic biomarkers and the discovery of novel metabolic pathways. Integration of metabolomics data analysis with a GEM of *Arabidopsis thaliana* to identify metabolic biomarkers associated with salt stress, they identified specific metabolites that were significantly altered under salt stress conditions (Awlia et al. 2021). These metabolites served as potential biomarkers for salt stress in plants. In a study of human metabolism to discover a novel pathway for the metabolism of the amino acid methionine. Analysis revealed a previously unknown enzyme-catalyzed reaction that played a role in methionine metabolism (Parkhitko et al. 2019). Furthermore, the integration of metabolomics data with GEMs can be used to improve the accuracy of metabolic flux predictions. Metabolic flux analysis is a technique that quantifies the flow of metabolites through metabolic pathways. By incorporating metabolomics data into GEM-based flux analysis, researchers can refine the predictions of metabolic fluxes and gain a more detailed understanding of cellular metabolism. The integration of metabolomics data with genome-scale metabolic models provides a powerful tool for understanding and predicting cellular metabolism. It enables the discovery of fresh metabolic pathways, the identification of metabolic biomarkers, and the improvement of metabolic flux forecasts. Numerous creatures including bacteria, yeast, plants, and people, have successfully used this integrated method, and it has the potential to contribute to advancements in fields such as biotechnology, medicine, and bioengineering.

10.4.3 Flux Estimation from Metabolomics Data

Flux estimation from metabolomics data is a valuable approach that allows for the quantification of metabolic fluxes within a biological system. Metabolomics provides information about the abundance of metabolites in a cellular environment, while determine the rates at which metabolites are created or consumed in metabolic reactions using flux estimates. By integrating metabolomics data with mathematical models, researchers can infer metabolic fluxes and gain insights into the dynamic behavior of cellular metabolism. Several studies have demonstrated the application of flux estimation from metabolomics data. For example, a method called ^{13}C-assisted metabolite analysis (CAMA) to estimate fluxes in central carbon metabolism. They combined metabolomics data with ^{13}C-labeling experiments and developed a mathematical model to estimate fluxes in *Saccharomyces cerevisiae* (Van Winden et al. 2005). The integration of metabolomics data allowed them to improve the accuracy of gaining knowledge of carbon flux dispersion in yeast metabolism and flux estimation. In another study, a method is being called ^{13}C metabolic flux analysis with multiple labeling experiments (^{13}C-MFA-MLE) to estimate fluxes in microbial systems. They integrated metabolomics data from ^{13}C-labeling experiments with a mathematical model and used maximum likelihood estimation to infer metabolic fluxes. Their method gave information on the control of central carbon metabolism and allowed precise flow estimation in *Escherichia coli* (Yao et al. 2019). Furthermore, there is another method called metabolic flux ratio analysis (Metabolic Flux Ratio Analysis—MeFRA) to estimate relative fluxes from metabolomics data. MeFRA allows for the determination of flux ratios between different metabolic reactions without the need for absolute flux quantification. By integrating metabolomics data with a stoichiometric model, they demonstrated the application of MeFRA in estimating flux ratios in both microbial and mammalian cell cultures (Sauer et al. 1999). Analysis of metabolic flux based on the idea of fundamental flux modes (^{13}C-MFA-EFM) to estimate fluxes in large-scale metabolic networks is another method, in which they integrated metabolomics data from ^{13}C-labeling experiments with a genome-scale metabolic model and used elementary flux modes to calculate flux distributions. This method made it possible to estimate fluxes in intricate metabolic networks and provided insights into pathway usage and regulation (Gerstl et al. 2015). Using flux estimation from metabolomics data is an effective way to comprehend and measure cellular metabolism. By integrating metabolomics data with mathematical models, researchers can infer metabolic fluxes and gain insights into the dynamic behavior of metabolic pathways. Various methods and approaches have been developed to estimate fluxes from metabolomics data, enabling accurate quantification of flux distributions in different biological systems.

10.4.4 Constraint-Based Reconstruction and Analysis (COBRA) Toolbox

The "Constraint-Based Reconstruction and Analysis (COBRA)" Toolbox is a widely used computational tool for modeling and analyzing metabolic networks. COBRA Toolbox employs a constraint-based approach, which leverages the stoichiometry of metabolic reactions, along with physiological and environmental constraints, to predict and analyze metabolic fluxes (Ng et al. 2022). It provides a comprehensive suite of algorithms and functions for tasks such as "flux balance analysis (FBA)", "flux variability analysis (FVA)", and metabolic pathway analysis. The COBRA Toolbox has been applied in numerous studies across different organisms and has contributed to advancements in various fields. For instance, in the field of bioengineering, The COBRA Toolbox is being used to create microbial strains that will produce certain chemicals. They used FBA to find genetic alterations that could increase *Escherichia coli*'s production of desired metabolites. By integrating the COBRA Toolbox with experimental data, they successfully engineered strains with improved production capabilities (O'Brien et al. 2015). The application of COBRA Toolbox to research systems biology the metabolic adaptations of *Mycobacterium tuberculosis* during infection. They reconstructed a genome-scale metabolic model of *M. tuberculosis* and used FBA to predict the metabolic fluxes under different conditions. The flow distributions between the in vivo and in vitro environments are contrasted, they identified metabolic pathways that were upregulated or downregulated during infection, providing insights into the metabolic strategies of the pathogen (Colijn et al. 2009). Furthermore, the study of human metabolism has used COBRA Toolbox. To analyze the metabolic rewiring in cancer cells. They reconstructed a genome-scale metabolic model of human metabolism and integrated it with gene expression data from cancer cells. By applying FBA, they identified metabolic alterations that were specific to cancer cells and could potentially be targeted for therapeutic interventions (Jerby et al. 2010). The COBRA Toolbox continues to evolve, with new features and functionalities being added over time, COBRApy package, which is a Python implementation of the COBRA Toolbox. Constraint-based modeling and analysis are performed using COBRApy, which also offers extra features for visualization and integration with other Python packages (Ebrahim et al. 2013). "The Constraint-Based Reconstruction and Analysis (COBRA)" Toolbox is a powerful computational tool for the analysis and modeling of metabolic networks. It has been extensively used in various fields, including bioengineering, systems biology, and human metabolism. Using the

COBRA Toolbox, metabolic engineering strategies may be designed and optimized while also being able to anticipate and analyze metabolic fluxes, providing insights into cellular metabolism.

10.5 Case Studies and Applications

In order to gather knowledge about the metabolic pathways and activities that occur in cells, tissues, and organisms, metabolomics involves the identification and measurement of these metabolites. Metabolomics has a significant impact on therapeutics and diagnostic in a number of ways:

10.5.1 Therapeutic

Early disease detection plays a crucial role in effective patient care, and the focus on biomarker discovery has intensified with advancements in technology. Changes in metabolites within biofluids serve as indicators of physiological or pathological variations. The quantitative and qualitative examination of metabolites in biological systems is the focus of the rapidly developing discipline of metabolomics (Zhang et al. 2015). Biomarker discovery, reliability relies on quantitative detection, high sensitivity, and specificity in reflecting biological states utilizing analytical technology, metabolomics enables the characterization of metabolites in clinical samples (blood, urine, feces, and tumor tissue) such as NMR, GC/MS, and LC/MS. By applying multivariate statistical methods, significant metabolite markers can be identified to differentiate between different groups (Zeki et al. 2020). As the Warburg effect suggests, many metabolites would be found in glycolysis pathway is associated with cell proliferation, and metastasis (Johar et al. 2021). AML, breast cancer, renal cancer, intrahepatic cholangiocarcinoma, and papillary thyroid carcinoma are just a few of the cancers that have been shown in numerous studies to include 2-hydroxyglutarate (2-HG), a byproduct of IDH1/IDH2 mutations (Wang et al. 2016). Breast cancer is linked to Omega 3-fatty acids, eicosapentaenoic acid (EPA), and docosahexaenoic acid (DHA), according to fatty acid metabolomics therefore, these metabolite can be act as biomarker for breast cancer (Fabian et al. 2015). It is anticipated that more and more metabolomics discoveries will become clinical cancer biomarkers as profiling technologies continue to advance and standardize. Studies conducted in vitro have demonstrated that while reduced glycolysis slows the growth of AML cells and increases the cytotoxicity of Ara-C, increased glycolysis confers less susceptibility to the “anti-leukemic drug” arabinofuranosyl cytidine (Ara-C) (Liu et al. 2019).

10.5.2 Diagnostic

In diagnostic medicine, metabolomics has become a potent tool that opens up new possibilities for illness diagnosis, monitoring, and stratification.

10.5.2.1 Metabolomics in Blood

Due to the varied physicochemical properties of metabolites, a multiplatform metabolomics method is strongly advised for untargeted metabolic fingerprinting in order to thoroughly investigate metabolites and capture the complexity of biological systems. Blood is one of the biological samples most frequently used in metabolomics research. As the main metabolite carrier in the body, a specific biological system's pathological and physiological state can be inferred from blood serum and plasma at any given time (Zhang et al. 2012). A new study used metabolomics to analyze blood samples from patients with advanced metastatic breast cancer (MBC) and localized early breast cancer (EBC). To create and test a model, patients with EBC and MBC were used as an outside test group for accurately distinguishing between the two conditions, achieving a sensitivity of 89.8% and specificity of 79.3% (Bujak et al. 2015). Histidine, acetoacetate, glycerol, and glutamate were among the statistically significant metabolites that suggested the possibility of an breast cancer patients' diagnosis, prognosis, and therapy using an NMR-based metabolomics strategy (Jobard et al. 2014). In recent study of Kobayashi et al. (2013), of analysis of serum based metabolomics through GC–MS for pancreatic cancer suggests that, xylitol, 1,5-anhydro-D-glucitol, histidine and inositol are having high specificity (88.1%) and sensitivity (86%) (Kobayashi et al. 2013). Through the use of DIMS and RP-UHPLC, Alzheimer's disease patients' sera's phospholipid profile revealed elevated levels of sphingo-phospholipid in cognitively normal condition, MCI, and finally AD stage patients (González-Domínguez et al. 2017). Therefore, the elevated level of sphingo-phospholipids metabolite can act as an early Alzheimer's disease indicator.

10.5.2.2 Metabolomics in Urine

Urine, similar to saliva, is considered an ideal biological sample for biomarker analysis in urogenital cancer due to its non-invasive collection process and easy storage. A targeted approach utilizing LC-QqQ/MS with phenylboronic acid gel as a selective medium for *cis*-diol compounds has been used to determine urinary nucleosides as potential biomarkers for urogenital cancer (Struck-Lewicka et al. 2014).

Out of the 12 nucleosides that were quantitatively measured, five of them (inosine, 3-methyluridine, *N*2-methylguanosine, 6-methyladenosine, and *N*,*N*-dimethyl guanosine) exhibited significant differences between cancer patients ($n = 61$) and healthy controls ($n = 68$) with statistical significance ($p < 0.05$). These results imply that these nucleosides might function as potential biomarkers for detecting or monitoring cancer (Struck et al., 2013). PLS-DA and the k-NN approach were used in multivariate statistical analyses for the statistically relevant metabolites, yielding sensitivity levels of 62–89% and specificities of 28–50%

(Sugimoto et al. 2012). Sarcosine, alanine, leucine, and proline were quantitatively analyzed in a distinct research by Schamsipur et al. from several cancer patients. They suggested DDLLME (Dispersive Derivatization Liquid-Liquid Micro-extraction) by LC, GC, and paired with "GC–MS" and "LC–MS" as a novel method for sample pretreatment (Shamsipur et al. 2013). The measurement of these four metabolites is low in a wider population of prostate cancer patients. All together, these metabolites have potential as prostate cancer biomarkers. On the other hand, "GC-IT/MS" was also used to determine some of the volatile urine metabolites (Monteiro et al. 2014). The study by Stephens et al. used NMR's OPLS-DA approach to analyze metabolites. As a consequence, TCA-related metabolites such as succinate, *trans*-aconitate, and citrate, amino acids (1-methylhistidine, lysine, and asparagine), and it was discovered that other metabolites, including taurine and creatine, may be linked to inflammatory bowel disease (Stephens et al. 2013).

10.5.2.3 Metabolomics in Saliva

Salivary metabolites play an important part in shedding light on the molecular mechanisms behind a variety of diseases, making it perfectly suited for the early identification of a number of diseases, including periodontal and oral cancer (Freire et al. 2021). Many diagnostic kits, such as the Oral Fluid Nano Sensor Test (OFNASET) for oral cancer, my PerioPath (OralDNA Labs) for periodontal disease, and HPV test for detecting the severity of the human papillomavirus in oral cancer, have recently undertaken novel detection of biomarker from saliva (Cova et al. 2015). The experimental results indicate that tumor exosomes or tumor-specific proteins, miRNA, or mRNA may be detected in plasma and saliva. As metabolomics' wide-ranging potential for early detection and specialized diagnostic responses revolutionizes the area of diagnostics and gives medical sciences a new focus for the early identification of numerous disorders.

10.6 Challenges, Future Perspectives and Conclusion

Our understanding of cellular metabolism has greatly benefited through metabolomics and flux-based analyses and they have opened up new avenues for studying complex biological systems. However, these approaches also come with challenges and present exciting opportunities for future advancements. The detection and annotation of metabolites is one of the main difficulties in metabolomics. Metabolomics experiments generate vast amounts of data, and accurately identifying and quantifying metabolites from complex mixtures is still a significant hurdle. Standardized databases and improved analytical techniques are needed to enhance metabolite identification and annotation, enabling more robust and reproducible analyses.

References

Agarwal P, Goyal A (2017) Ionization sources used in mass spectroscopy: a review

Allen F, Pon A, Wilson M, Greiner R, Wishart D (2014) CFM-ID: a web server for annotation, spectrum prediction and metabolite identification from tandem mass spectra. Nucleic acids Res 42(W1):W94–W99

Alygizakis N, Lestremau F, Gago-Ferrero P, Gil-Solsona R, Arturi K, Hollender J, Thomaidis NS (2023) Towards a harmonized identification scoring system in LC-HRMS/MS based non-target screening (NTS) of emerging contaminants. TrAC Trends Anal Chem 159:116944

Amer B, Deshpande RR, Bird SS (2023) Simultaneous quantitation and discovery (SQUAD) analysis: combining the best of targeted and untargeted mass spectrometry-based metabolomics. Meta 13(5):648

Anderson R, Groundwater PW, Todd A, Worsley A (2012) Antibacterial agents: chemistry, mode of action, mechanisms of resistance and clinical applications. Wiley, New York

Antoniewicz MR (2018) A guide to ^{13}C metabolic flux analysis for the cancer biologist. Exp Mol Med 50(4):1–13

Aurich MK, Fleming RM, Thiele I (2016) MetaboTools: a comprehensive toolbox for analysis of genome-scale metabolic models. Front Physiol 7:327

Awlia M, Alshareef N, Saber N, Korte A, Oakey H, Panzarová K, Julkowska MM (2021) Genetic mapping of the early responses to salt stress in *Arabidopsis thaliana*. Plant J 107(2):544–563

Badilita V, Meier RC, Spengler N, Wallrabe U, Utz M, Korvink JG (2012) Microscale nuclear magnetic resonance: a tool for soft matter research. Soft Matter 8(41):10583–10597

Bartle KD, Myers P (2002) History of gas chromatography. TrAC Trends Anal Chem 21(9–10): 547–557

Becker SA, Feist AM, Mo ML, Hannum G, Palsson BØ, Herrgard MJ (2007) Quantitative prediction of cellular metabolism with constraint-based models: the COBRA toolbox. Nat Protoc 2(3):727–738

Berman HM, Battistuz T, Bhat TN, Bluhm WF, Bourne PE, Burkhardt K, Zardecki C (2002) The protein data bank. Acta Crystallogr D Biol Crystallogr 58(6):899–907

Böttcher C, Roepenack-Lahaye EV, Willscher E, Scheel D, Clemens S (2007) Evaluation of matrix effects in metabolite profiling based on capillary liquid chromatography electrospray ionization quadrupole time-of-flight mass spectrometry. Anal Chem 79(4):1507–1513

Bujak R, Struck-Lewicka W, Markuszewski MJ, Kaliszan R (2015) Metabolomics for laboratory diagnostics. J Pharm Biomed Anal 113:108–120

Burgard AP, Pharkya P, Maranas CD (2003) Optknock: a bilevel programming framework for identifying gene knockout strategies for microbial strain optimization. Biotechnol Bioeng 84(6): 647–657

Calderón-Santiago M, López-Bascón MA, Peralbo-Molina A, Priego-Capote F (2017) MetaboQC: a tool for correcting untargeted metabolomics data with mass spectrometry detection using quality controls. Talanta 174:29–37

Canelas AB, van Gulik WM, Heijnen JJ (2008) Determination of the cytosolic free NAD/NADH ratio in *Saccharomyces cerevisiae* under steady-state and highly dynamic conditions. Biotechnol Bioeng 100(4):734–743

Castillo S, Gopalacharyulu P, Yetukuri L, Orešič M (2011) Algorithms and tools for the preprocessing of LC–MS metabolomics data. Chemom Intell Lab Syst 108(1):23–32

Cheng Q (ed) (2012) Microbial metabolic engineering: methods and protocols, vol 834. Humana Press

Cheng JH, Dai Q, Sun DW, Zeng XA, Liu D, Pu HB (2013) Applications of non-destructive spectroscopic techniques for fish quality and safety evaluation and inspection. Trends Food Sci Technol 34(1):18–31

Chong J, Soufan O, Li C, Caraus I, Li S, Bourque G, Xia J (2018) MetaboAnalyst 4.0: towards more transparent and integrative metabolomics analysis. Nucleic Acids Res 46(W1):W486–W494

Colijn C, Brandes A, Zucker J, Lun DS, Weiner B (2009) Interpreting expression data with metabolic flux models: predicting *Mycobacterium tuberculosis* mycolic acid production. PLoS Comput Biol 5:e1000489

Coquin L, Feala JD, McCulloch AD, Paternostro G (2008) Metabolomic and flux-balance analysis of age-related decline of hypoxia tolerance in Drosophila muscle tissue. Mol Syst Biol 4(1):233

Cova MAMN, Castagnola M, Messana I, Cabras T, Ferreira RMP, Amado FML, Vitorino RMP (2015) Salivary omics. In: Advances in salivary diagnostics. Springer, pp 63–82

Croasmun WR, Carlson RM (eds) (1996) Two-dimensional NMR spectroscopy: applications for chemists and biochemists, vol 15. Wiley, New York

Cui L, Lu H, Lee YH (2018) Challenges and emergent solutions for LC–MS/MS based untargeted metabolomics in diseases. Mass Spectrom Rev 37(6):772–792

da Silva RR, Wang M, Nothias LF, van der Hooft JJ, Caraballo-Rodríguez AM, Fox E, Dorrestein PC (2018) Propagating annotations of molecular networks using in silico fragmentation. PLoS Comput Biol 14(4):e1006089

Dettmer K, Aronov PA, Hammock BD (2007) Mass spectrometry-based metabolomics. Mass Spectrom Rev 26(1):51–78

Dowling G (2017) Analysis of bitterness compounds by mass spectrometry. In: Bitterness: perception, chemistry and food processing. Wiley, New York, pp 161–194

Ebrahim A, Lerman JA, Palsson BO, Hyduke DR (2013) COBRApy: constraints-based reconstruction and analysis for python. BMC Syst Biol 7:1–6

Fabian CJ, Kimler BF, Hursting SD (2015) Omega-3 fatty acids for breast cancer prevention and survivorship. Breast Cancer Res 17(1):1–11

Fabregat A, Sidiropoulos K, Viteri G, Forner O, Marin-Garcia P, Arnau V, Hermjakob H (2017) Reactome pathway analysis: a high-performance in-memory approach. BMC Bioinform 18(1): 1–9

Fan TWM, Lane AN (2016) Applications of NMR spectroscopy to systems biochemistry. Prog Nucl Magn Reson Spectrosc 92:18–53

Feala JD, Coquin L, Zhou D, Haddad GG, Paternostro G, McCulloch AD (2009) Metabolism as means for hypoxia adaptation: metabolic profiling and flux balance analysis. BMC Syst Biol 3(1):1–15

Fiehn O (2016) Metabolomics by gas chromatography–mass spectrometry: combined targeted and untargeted profiling. Curr Protoc Mol Biol 114(1):30–34

Förster J, Famili I, Fu P, Palsson BØ, Nielsen J (2003) Genome-scale reconstruction of the *Saccharomyces cerevisiae* metabolic network. Genome Res 13(2):244–253

Freire M, Nelson KE, Edlund A (2021) The oral host–microbial interactome: an ecological chronometer of health? Trends Microbiol 29(6):551–561

Fuerstenau SD, Benner WH (1995) Molecular weight determination of megadalton DNA electrospray ions using charge detection time-of-flight mass spectrometry. Rapid Commun Mass Spectrom 9(15):1528–1538

Gerstl MP, Ruckerbauer DE, Mattanovich D, Jungreuthmayer C, Zanghellini J (2015) Metabolomics integrated elementary flux mode analysis in large metabolic networks. Sci Rep 5(1):1–8

González-Domínguez R, Sayago A, Fernández-Recamales Á (2017) Metabolomics in Alzheimer's disease: the need of complementary analytical platforms for the identification of biomarkers to unravel the underlying pathology. J Chromatogr B 1071:75–92

Gowda GN, Raftery D (2015) Can NMR solve some significant challenges in metabolomics? J Magn Reson 260:144–160

Griffin JL, Bollard ME (2004) Metabonomics: its potential as a tool in toxicology for safety assessment and data integration. Curr Drug Metab 5(5):389–398

Griffiths WJ, Wang Y (2009) Mass spectrometry: from proteomics to metabolomics and lipidomics. Chem Soc Rev 38(7):1882–1896

Guijas C, Montenegro-Burke JR, Domingo-Almenara X, Palermo A, Warth B, Hermann G, Siuzdak G (2018) METLIN: a technology platform for identifying knowns and unknowns. Anal Chem 90(5):3156–3164

Guilhaus M (1995) Special feature: tutorial. Principles and instrumentation in time-of-flight mass spectrometry. Physical and instrumental concepts. J Mass Spectrom 30(11):1519–1532

Haug K, Salek RM, Steinbeck C (2017) Global open data management in metabolomics. Curr Opin Chem Biol 36:58–63

Heck M, Blaum K, Cakirli RB, Rodríguez D, Schweikhard L, Stahl S, Ubieto-Díaz M (2011) Dipolar and quadrupolar detection using an FT-ICR MS setup at the MPIK Heidelberg. Hyperfine Interact 199:347–355

Hird SJ, Lau BPY, Schuhmacher R, Krska R (2014) Liquid chromatography-mass spectrometry for the determination of chemical contaminants in food. TrAC Trends Anal Chem 59:59–72

Hucka M, Bergmann FT, Dräger A, Hoops S, Keating SM, Le Novere N, Wilkinson DJ (2015) Systems biology markup language (SBML) level 2 version 5: structures and facilities for model definitions. J Integr Bioinform 12(2):731–901

Hurd RE, Yen YF, Chen A, Ardenkjaer-Larsen JH (2012) Hyperpolarized ^{13}C metabolic imaging using dissolution dynamic nuclear polarization. J Magn Reson Imaging 36(6):1314–1328

Jalili V, Barkhordari A, Ghiasvand A (2020) Solid-phase microextraction technique for sampling and preconcentration of polycyclic aromatic hydrocarbons: a review. Microchem J 157:104967

Jerby L, Shlomi T, Ruppin E (2010) Computational reconstruction of tissue-specific metabolic models: application to human liver metabolism. Mol Syst Biol 6(1):401

Jobard E, Pontoizeau C, Blaise BJ, Bachelot T, Elena-Herrmann B, Trédan O (2014) A serum nuclear magnetic resonance-based metabolomic signature of advanced metastatic human breast cancer. Cancer Lett 343(1):33–41

Johar D, Elmehrath AO, Khalil RM, Elberry MH, Zaky S, Shalabi SA, Bernstein LH (2021) Protein networks linking Warburg and reverse Warburg effects to cancer cell metabolism. Biofactors 47(5):713–728

Kanehisa M, Goto S (2000) KEGG: Kyoto encyclopedia of genes and genomes. Nucleic Acids Res 28(1):27–30

Kanehisa M, Sato Y, Kawashima M (2022) KEGG mapping tools for uncovering hidden features in biological data. Protein Sci 31(1):47–53

Karp PD, Billington R, Caspi R, Fulcher CA, Latendresse M, Kothari A, Subhraveti P (2019) The BioCyc collection of microbial genomes and metabolic pathways. Brief Bioinform 20(4): 1085–1093

Karpievitch YV, Polpitiya AD, Anderson GA, Smith RD, Dabney AR (2010) Liquid chromatography mass spectrometry-based proteomics: biological and technological aspects. Ann Appl Stat 4(4):1797–1823

Kauffman KJ, Prakash P, Edwards JS (2003) Advances in flux balance analysis. Curr Opin Biotechnol 14(5):491–496

Kitano H (2004) Biological robustness. Nat Rev Genet 5(11):826–837

Klamt S, Saez-Rodriguez J, Gilles ED (2007) Structural and functional analysis of cellular networks with CellNetAnalyzer. BMC Syst Biol 1(1):1–13

Kobayashi T, Nishiumi S, Ikeda A, Yoshie T, Sakai A, Matsubara A, Yoshida M (2013) A novel serum metabolomics-based diagnostic approach to pancreatic CancerSerum Metabolomic analysis of pancreatic cancer. Cancer Epidemiol Biomark Prev 22(4):571–579

Koek MM, Jellema RH, van der Greef J, Tas AC, Hankemeier T (2011) Quantitative metabolomics based on gas chromatography mass spectrometry: status and perspectives. Metabolomics 7: 307–328

Krishnan SN, Sun YA, Mohsenin A, Wyman RJ, Haddad GG (1997) Behavioral and electrophysiologic responses of *Drosophila melanogaster* to prolonged periods of anoxia. J Insect Physiol 43(3):203–210

Kueger S, Steinhauser D, Willmitzer L, Giavalisco P (2012) High-resolution plant metabolomics: from mass spectral features to metabolites and from whole-cell analysis to subcellular metabolite distributions. Plant J 70(1):39–50
Lane AN, Fan TWM, Higashi RM (2008) Isotopomer-based metabolomic analysis by NMR and mass spectrometry. Methods Cell Biol 84:541–588
Lee DY, Yun H, Park S, Lee SY (2003) MetaFluxNet: the management of metabolic reaction information and quantitative metabolic flux analysis. Bioinformatics 19(16):2144–2146
Lewis NE, Nagarajan H, Palsson BO (2012) Constraining the metabolic genotype–phenotype relationship using a phylogeny of in silico methods. Nat Rev Microbiol 10(4):291–305
Li C, Donizelli M, Rodriguez N, Dharuri H, Endler L, Chelliah V, Laibe C (2010) BioModels database: an enhanced, curated and annotated resource for published quantitative kinetic models. BMC Syst Biol 4(1):1–14
Liu T, Peng XC, Li B (2019) The metabolic profiles in hematological malignancies. Indian J Hematol Blood Transfus 35:625–634
Lu X, Zhao X, Bai C, Zhao C, Lu G, Xu G (2008) LC–MS-based metabonomics analysis. J Chromatogr B 866(1–2):64–76
Mahadevan R, Edwards JS, Doyle FJ (2002) Dynamic flux balance analysis of diauxic growth in *Escherichia coli*. Biophys J 83(3):1331–1340
Markley JL, Brüschweiler R, Edison AS, Eghbalnia HR, Powers R, Raftery D, Wishart DS (2017) The future of NMR-based metabolomics. Curr Opin Biotechnol 43:34–40
McNair HM, Miller JM, Snow NH (2019) Basic gas chromatography. Wiley, New York
Modisha PM, Jordaan JH, Bösmann A, Wasserscheid P, Bessarabov D (2018) Analysis of reaction mixtures of perhydro-dibenzyltoluene using two-dimensional gas chromatography and single quadrupole gas chromatography. Int J Hydrog Energy 43(11):5620–5636
Monteiro M, Carvalho M, Henrique R, Jeronimo C, Moreira N, de Lourdes Bastos M, de Pinho PG (2014) Analysis of volatile human urinary metabolome by solid-phase microextraction in combination with gas chromatography–mass spectrometry for biomarker discovery: application in a pilot study to discriminate patients with renal cell carcinoma. Eur J Cancer 50(11): 1993–2002
Morain BÉV (2013) In-situ and operando infrared investigations on supported ionic liquid-and ionic liquid crystal-based catalytic materials. Friedrich-Alexander-Universitaet Erlangen-Nuernberg (Germany)
Nagana Gowda GA, Raftery D (2019) Overview of NMR spectroscopy-based metabolomics: opportunities and challenges. In: NMR-based metabolomics: methods and protocols. Springer, pp 3–14
Narad P, Naresh G, Sengupta A (2022) Metabolomics and flux balance analysis. In: Bioinformatics. Academic Press, pp 337–365
Ng RH, Lee JW, Baloni P, Diener C, Heath JR, Su Y (2022) Constraint-based reconstruction and analyses of metabolic models: open-source python tools and applications to cancer. Front Oncol 12:914594
Nielsen J, Jewett MC (eds) (2007) Metabolomics: a powerful tool in systems biology, vol 18. Springer Science & Business Media, Berlin
O'Brien EJ, Monk JM, Palsson BO (2015) Using genome-scale models to predict biological capabilities. Cell 161(5):971–987
O'Grady J, Schwender J, Shachar-Hill Y, Morgan JA (2012) Metabolic cartography: experimental quantification of metabolic fluxes from isotopic labelling studies. J Exp Bot 63(6):2293–2308
Oftadeh O, Salvy P, Masid M, Curvat M, Miskovic L, Hatzimanikatis V (2021) A genome-scale metabolic model of *Saccharomyces cerevisiae* that integrates expression constraints and reaction thermodynamics. Nat Commun 12(1):4790
Parkhitko AA, Jouandin P, Mohr SE, Perrimon N (2019) Methionine metabolism and methyltransferases in the regulation of aging and lifespan extension across species. Aging Cell 18(6):e13034

Pitt JJ (2009) Principles and applications of liquid chromatography–mass spectrometry in clinical biochemistry. Clin Biochem Rev 30(1):19–34

Poole CF (2015) Ionization-based detectors for gas chromatography. J Chromatogr A 1421:137–153

Pozo ÓJ, Sancho JV, Ibáñez M, Hernández F, Niessen WM (2006) Confirmation of organic micropollutants detected in environmental samples by liquid chromatography tandem mass spectrometry: achievements and pitfalls. TrAC Trends Anal Chem 25(10):1030–1042

Putri SP, Nakayama Y, Matsuda F, Uchikata T, Kobayashi S, Matsubara A, Fukusaki E (2013) Current metabolomics: practical applications. J Biosci Bioeng 115(6):579–589

Quek LE, Wittmann C, Nielsen LK, Krömer JO (2009) OpenFLUX: efficient modelling software for ^{13}C-based metabolic flux analysis. Microb Cell Factories 8:1–15

Ràfols P, Vilalta D, Brezmes J, Cañellas N, Del Castillo E, Yanes O, Correig X (2018) Signal preprocessing, multivariate analysis and software tools for MA (LDI)-TOF mass spectrometry imaging for biological applications. Mass Spectrom Rev 37(3):281–306

Raman K, Chandra N (2009) Flux balance analysis of biological systems: applications and challenges. Brief Bioinform 10(4):435–449

Redestig H, Szymanski J, Hirai MY, Selbig J, Willmitzer L, Nikoloski Z, Saito K (2011) Data integration, metabolic networks and systems biology. Annu Plant Rev Biol Plant Metabol 43: 261–316

Riekeberg E, Powers R (2017) New frontiers in metabolomics: from measurement to insight. F1000Research 6:1148

Riley ML, Schmidt T, Artamonova II, Wagner C, Volz A, Heumann K, Frishman D (2007) PEDANT genome database: 10 years online. Nucleic Acids Res 35(suppl_1):D354–D357

Rocha I, Maia P, Evangelista P, Vilaça P, Soares S, Pinto JP, Rocha M (2010) OptFlux: an open-source software platform for in silico metabolic engineering. BMC Syst Biol 4(1):1–12

Romero R, Espinoza J, Gotsch F, Kusanovic JP, Friel LA, Erex O, Tromp G (2008) The use of high-dimensional biology (genomics, transcriptomics, proteomics, and metabolomics) to understand the preterm parturition syndrome. BJOG 113(Suppl. 3):118–135

Rouger L, Gouilleux B, Nantes FPG (2017) Fast *n*-dimensional data acquisition methods

Ruttkies C, Neumann S, Posch S (2019) Improving MetFrag with statistical learning of fragment annotations. BMC Bioinform 20(1):1–14

Sato S, Soga T, Nishioka T, Tomita M (2004) Simultaneous determination of the main metabolites in rice leaves using capillary electrophoresis mass spectrometry and capillary electrophoresis diode array detection. Plant J 40(1):151–163

Sauer UWE, Lasko DR, Fiaux J, Hochuli M, Glaser R, Szyperski T, Bailey JE (1999) Metabolic flux ratio analysis of genetic and environmental modulations of *Escherichia coli* central carbon metabolism. J Bacteriol 181(21):6679–6688

Schellenberger J, Que R, Fleming RM, Thiele I, Orth JD, Feist AM, Palsson BØ (2011) Quantitative prediction of cellular metabolism with constraint-based models: the COBRA Toolbox v2.0. Nat Protoc 6(9):1290–1307

Schmidt BJ, Ebrahim A, Metz TO, Adkins JN, Palsson BØ, Hyduke DR (2013) GIM3E: condition-specific models of cellular metabolism developed from metabolomics and expression data. Bioinformatics 29(22):2900–2908

Schomburg I, Chang A, Schomburg D (2002) BRENDA, enzyme data and metabolic information. Nucleic Acids Res 30(1):47–49

Schuetz R, Kuepfer L, Sauer U (2007) Systematic evaluation of objective functions for predicting intracellular fluxes in *Escherichia coli*. Mol Syst Biol 3(1):119

Schwarz R, Musch P, von Kamp A, Engels B, Schirmer H, Schuster S, Dandekar T (2005) YANA—a software tool for analyzing flux modes, gene-expression and enzyme activities. BMC Bioinform 6:1–12

Seger C, Sturm S, Stuppner H (2013) Mass spectrometry and NMR spectroscopy: modern high-end detectors for high resolution separation techniques–state of the art in natural product HPLC-MS, HPLC-NMR, and CE-MS hyphenations. Nat Prod Rep 30(7):970–987

Shamsipur M, Naseri MT, Babri M (2013) Quantification of candidate prostate cancer metabolite biomarkers in urine using dispersive derivatization liquid–liquid microextraction followed by gas and liquid chromatography–mass spectrometry. J Pharm Biomed Anal 81:65–75
Sibille N, Bellot G, Wang J, Déméné H (2012) Low concentration of a Gd-chelate increases the signal-to-noise ratio in fast pulsing BEST experiments. J Magn Reson 224:32–37
Sleno L (2012) The use of mass defect in modern mass spectrometry. J Mass Spectrom 47(2): 226–236
Smart KF, Aggio RB, Van Houtte JR, Villas-Bôas SG (2010) Analytical platform for metabolome analysis of microbial cells using methyl chloroformate derivatization followed by gas chromatography–mass spectrometry. Nat Protoc 5(10):1709–1729
Stein S (2012) Mass spectral reference libraries: an ever-expanding resource for chemical identification. Anal Chem 84:7274
Stelling J (2004) Mathematical models in microbial systems biology. Curr Opin Microbiol 7(5): 513–518
Stephens NS, Siffledeen J, Su X, Murdoch TB, Fedorak RN, Slupsky CM (2013) Urinary NMR metabolomic profiles discriminate inflammatory bowel disease from healthy. J Crohns Colitis 7(2):e42–e48
Struck W, Siluk D, Yumba-Mpanga A, Markuszewski M, Kaliszan R, Markuszewski MJ (2013) Liquid chromatography tandem mass spectrometry study of urinary nucleosides as potential cancer markers. J Chromatogr A 1283:122–131
Struck-Lewicka W, Kaliszan R, Markuszewski MJ (2014) Analysis of urinary nucleosides as potential cancer markers determined using LC–MS technique. J Pharm Biomed Anal 101:50–57
Sugimoto M, Kawakami M, Robert M, Soga T, Tomita M (2012) Bioinformatics tools for mass spectroscopy-based metabolomic data processing and analysis. Curr Bioinform 7(1):96–108
Tavares LC, Jarak I, Nogueira FN, Oliveira PJ, Carvalho RA (2015) Metabolic evaluations of cancer metabolism by NMR-based stable isotope tracer methodologies. Eur J Clin Investig 45: 37–43
Teusink B, Passarge J, Reijenga CA, Esgalhado E, Weijden CC, van der Schepper M, Walsh MC, Bakker BM, van Dam K, Westerhoff HV, Snoep JL (2000) Can yeast glycolysis be understood in terms of in vitro kinetics of the constituent enzymes? Testing biochemistry. Eur J Biochem 267(17):5313–5329
Töpfer N, Kleessen S, Nikoloski Z (2015) Integration of metabolomics data into metabolic networks. Front Plant Sci 6:49
Toya Y, Shimizu H (2013) Flux analysis and metabolomics for systematic metabolic engineering of microorganisms. Biotechnol Adv 31(6):818–826
Trimigno A, Marincola FC, Dellarosa N, Picone G, Laghi L (2015) Definition of food quality by NMR-based foodomics. Curr Opin Food Sci 4:99–104
Urbanczik R (2006) SNA—a toolbox for the stoichiometric analysis of metabolic networks. BMC Bioinform 7(1):1–4
Van Winden WA, Van Dam JC, Ras C, Kleijn RJ, Vinke JL, Van Gulik WM, Heijnen JJ (2005) Metabolic-flux analysis of *Saccharomyces cerevisiae* CEN.PK113-7D based on mass isotopomer measurements of ^{13}C-labeled primary metabolites. FEMS Yeast Res 5(6–7): 559–568
Varma A, Palsson BO (1994) Metabolic flux balancing: basic concepts, scientific and practical use. Bio/Technology 12(10):994–998
Volkova S, Matos MR, Mattanovich M, Marín de Mas I (2020) Metabolic modelling as a framework for metabolomics data integration and analysis. Metabolites 10(8):303
Wang X, Chen S, Jia W (2016) Metabolomics in cancer biomarker research. Curr Pharmacol Rep 2: 293–298
Wang ZJ, Ohliger MA, Larson PE, Gordon JW, Bok RA, Slater J, Vigneron DB (2019) Hyperpolarized ^{13}C MRI: state of the art and future directions. Radiology 291(2):273–284

Wang CY, Lempp M, Farke N, Donati S, Glatter T, Link H (2021) Metabolome and proteome analyses reveal transcriptional misregulation in glycolysis of engineered *E. coli*. Nat Commun 12(1):4929

Wiechert W (2001) ^{13}C metabolic flux analysis. Metab Eng 3(3):195–206

Willemsen AM, Hendrickx DM, Hoefsloot HC, Hendriks MM, Wahl SA, Teusink B, van Kampen AH (2015) MetDFBA: incorporating time-resolved metabolomics measurements into dynamic flux balance analysis. Mol Biosyst 11(1):137–145

Wishart DS, Tzur D, Knox C, Eisner R, Guo AC, Young N, Querengesser L (2007) HMDB: the human metabolome database. Nucleic Acids Res 35(suppl_1):D521–D526

Wishart DS, Guo A, Oler E, Wang F, Anjum A, Peters H, Gautam V (2022) HMDB 5.0: the human metabolome database for 2022. Nucleic Acids Res 50(D1):D622–D631

Wright J, Wagner A (2008) The systems biology research tool: evolvable open-source software. BMC Syst Biol 2:1–6

Xia J, Broadhurst DI, Wilson M, Wishart DS (2013) Translational biomarker discovery in clinical metabolomics: an introductory tutorial. Metabolomics 9:280–299

Xiao JF, Zhou B, Ressom HW (2012) Metabolite identification and quantitation in LC–MS/MS-based metabolomics. TrAC Trends Anal Chem 32:1–14

Yao R, Li J, Feng L, Zhang X, Hu H (2019) ^{13}C metabolic flux analysis-guided metabolic engineering of *Escherichia coli* for improved acetol production from glycerol. Biotechnol Biofuels 12(1):1–13

Yu D, Zhou L, Liu X, Xu G (2023) Stable isotope-resolved metabolomics based on mass spectrometry: methods and their applications. TrAC Trends Anal Chem 116985:116985

Zaikin V, Halket JM (2009) A handbook of derivatives for mass spectrometry. IM Publications

Zeki ÖC, Eylem CC, Reçber T, Kır S, Nemutlu E (2020) Integration of GC–MS and LC–MS for untargeted metabolomics profiling. J Pharm Biomed Anal 190:113509

Zhang A, Sun H, Wang X (2012) Serum metabolomics as a novel diagnostic approach for disease: a systematic review. Anal Bioanal Chem 404:1239–1245

Zhang AH, Sun H, Qiu S, Wang XJ (2013) NMR-based metabolomics coupled with pattern recognition methods in biomarker discovery and disease diagnosis. Magn Reson Chem 51(9):549–556

Zhang A, Sun H, Yan G, Wang P, Wang X (2015) Metabolomics for biomarker discovery: moving to the clinic. Biomed Res Int 2015:354671

Zhang Y, Cai J, Shang X, Wang B, Liu S, Chai X, Wen T (2017) A new genome-scale metabolic model of *Corynebacterium glutamicum* and its application. Biotechnol Biofuels 10:1–16

Zhou B, Xiao JF, Tuli L, Ressom HW (2012) LC–MS-based metabolomics. Mol BioSyst 8(2):470–481

Bioinformatics in Drug Discovery

11

Ngo Anh Dao, Thuy-Duong Vu, and Dinh-Toi Chu

Abstract

Drug discovery requires high cost and is a time-consuming process, and the facilitation of computer-based drug design methods is one of the most potential approaches to change this challenging situation. In fact, along with the current advancement of science and technology, especially in the field of bioinformatics, the stages of drug discovery can be significantly shortened while the cost is reduced and the efficacy of treatment increases. Bioinformatics tools and platforms can not only advance drug target identification and screening, but also support drug candidate selection and evaluate effectiveness of drug candidates. In recent years, bioinformatics tools have often been used to screen the sequences of gene fragments, uncovering potential binding sites for therapeutic drugs or also known as drug targets. Besides, the high-throughput screen method is a popular method for drug candidate identification for detecting potential small molecules among a large amount of information in available data libraries. Since the early years of the twenty-first century, research has applied bioinformatics to screen targeted molecules using the high-throughput screening model. Bioinformatics also has a huge contribution in virtual screening through the early elimination of substances with undesirable properties through computers and in silico screening, thereby finding the closest compounds to the

N. A. Dao · T.-D. Vu
Center for Biomedicine and Community Health, International School, Vietnam National University, Hanoi, Vietnam

D.-T. Chu (✉)
Center for Biomedicine and Community Health, International School, Vietnam National University, Hanoi, Vietnam

Faculty of Applied Sciences, International School, Vietnam National University, Hanoi, Vietnam
e-mail: toicd@vnu.edu.vn

V. Singh, A. Kumar (eds.), *Advances in Bioinformatics*,
https://doi.org/10.1007/978-981-99-8401-5_11

desired drug. Based on these tools and techniques, the efficacy of drug candidates can be easily and quickly determined, especially in individuals, which revolutionarily benefits drug validation and personalized pharmacological therapies.

Keywords

Bioinformatics · Drug development · Drug screening · Drug validation · Pharmacology

11.1 Introduction

Drug discovery normally starts with the discovery or diagnosis of a novel disease or pathogen that impair the quality of life. Consequently, researchers look for a desirable chemical (which could be a simple molecule or a complex protein) with a therapeutic effect that can benefit patients' health and develop a new drug based on that valuable substance. A potential drug to develop on an industrial scale also requires limited severe and long-term side effects (Xia 2017), low possibility of drug resistance, affordability for patients and profitability for pharmaceutical companies (David et al. 2009; Drews and Ryser 1997) along with minor damage to the environment (Boxall et al. 2012).

The basic process of drug discovery pipeline includes target identification and study, hit discovery, hit to lead generation, lead optimization, candidate identification as well as preclinical and clinical trials (Zhong et al. 2018). Estimates recently suggest that in order to bring a novel prescription drug to the market, the mean expenditure before tax is approximately 3 billion USD (DiMasi et al. 2016) and it takes roughly 13 years (Paul et al. 2010). Nevertheless, only 13% of potential medicinal chemicals are estimated to be successfully approved after clinical phases, which shows a significantly high risk of failure (Zhong et al. 2018). Possible reasons for this low approval success rate includes unexpected toxicity, the inability to successfully compete in the market and most importantly, lack of clinical efficacy (Kola and Landis 2004). The facilitation of computer-based drug design methods is one of the most potential approaches to tackle this challenging situation (Baig et al. 2016).

Bioinformatics is an interdisciplinary science that includes proteomics, genomics, transcriptomics and molecular phylogenetics (Xia 2017). Bioinformatics facilitates drug discovery by using high-throughput molecular data to examine the difference between symptom-carriers such as between cell lines, animal models or patients and the controlled group (Xia 2017). Such comparison is aimed to (1) find the association between diseases and genetic and epigenetic factors as well as other environmental factors affecting gene expression, (2) screen drug targets related to cellular malfunction elimination or function improvement, (3) predict or modulate drug candidates in order to get the desirable outcome and minimize toxicities, (4) measure the influence on the environment and the possibility of drug resistance (Xia 2017).

Symptom-based bioinformatics in drug development depends on the disease types among infectious, genetic diseases and cancer (van Driel and Brunner 2006). Bioinformatics support drug discovery in genetic disorders mainly through identifying noninvasive tools for genetic diagnosis and prognosis (Wooller et al. 2017). For infectious diseases, this science examines the impact of bacterial or viral presence on gene expression and compares it with those of other pathogens or drug-induced results to explore new potentials of existing drugs (Wooller et al. 2017). Bioinformaticians can also identify the main genetic causes of cancer in individual patients and hence, personalize cancer treatment and facilitate the discovery of a novel drug or repurpose the existing drugs (Wooller et al. 2017; Zhang et al. 2009). Regarding drug screening, bioinformatics is able to benefit such processes by using high-throughput screening for library screening related to the drug target and for other secondary assays (Fox et al. 2006; Nemmani 2021). It also contributes to the early elimination of potential candidates with undesirable properties (Smith 2002a). In the next step of the drug discovery pipeline, bioinformatics software and platforms are applied in the process of drug validation. Pharmaceutical companies have gained a better understanding of how human genomes impact the efficacy of drug candidates thanks to bioinformatics tools and techniques (Chang 2005).

One of the earliest and most well-known contributions of bioinformatics to the pharmaceutical industry is the discovery of sequence homology between a platelet-derived growth factor (PDGF) and an oncogene named v-sis from sarcoma virus using simple string matching (Doolittle et al. 1983; Waterfield et al. 1983). This important finding has opened two novel lines of thinking in cancer biology. First, growth factors could be targets for anti-cancer drugs; for example, PDGF (Pietras et al. 2003). Second, cancer can be a final result of any regulatory factors of gene expression. We can say that this bioinformatics-induced finding led to a whole new conceptual framework that enhances the development of anti-cancer drug development (Moffat et al. 2014). That is just one outstanding example of how bioinformatics can facilitate the development of a novel drug. Therefore, this book chapter will focus on how bioinformatics supports the discovery of a potential drug, including the role of this novel area on the process of drug development, drug screening, drug validation and some notable achievements.

11.2 Bioinformatics in Drug Development

Drug development is the work of researching and finding suitable new drug molecules from the early stages to phase III clinical practice and the process of bringing drugs to market as well as testing afterward (Chen et al. 2021). The drug discovery process takes a long time of research and costs a lot of money to find a suitable one (Preziosi 2007); today, along with the development of science and technology, especially in the field of bioinformatics, the stages of drug development can be significantly shortened while the cost is reduced and the effectiveness of treatment increases (Chen et al. 2021; Moore and Allen 2019). Currently, the high-throughput screen method is a popular method for detecting potential small

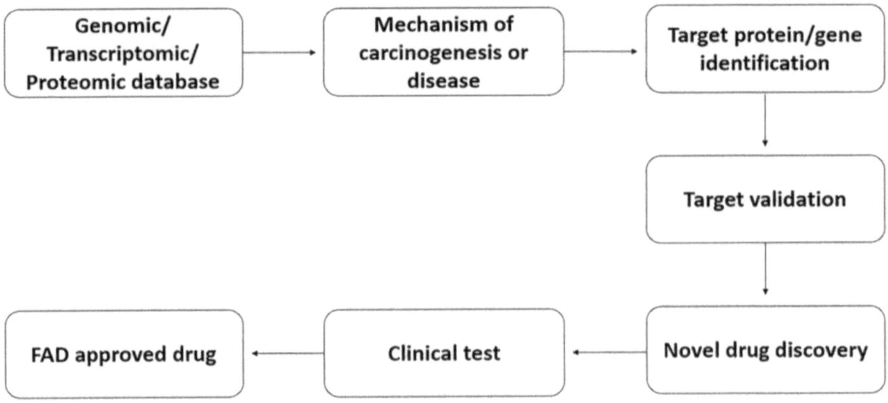

Fig. 11.1 Bioinformatics in drug development

molecules among a large amount of information in available data libraries (McLean 2015). The molecules are then tested for their ability to bind to their target or work in vivo, and if appropriate, can be used as a starting point for drug testing in animals. In addition, bioinformatics also helps scientists study disease symptoms with genetic mutations, identify drugs capable of restoring or eliminating damaged cells, predict the effectiveness and side effects of drugs, as well as assessing drug resistance (Xia 2017).

With the vast amount of data available from gene libraries, reports on mutations or epigenetics, proteomic or biological processes, bioinformatics has greatly contributed to the discovery of potential drugs (Fig. 11.1). Through genome analysis, biologists and pharmacologists can find drugs capable of treating genetic diseases or pathogens. Bioinformatics tools are often used to screen the sequences of gene fragments, thereby uncovering potential binding sites for therapeutic drugs (Xia 2012). For example, bioinformatics research shows that potential LXR response elements (LXREs) regulate the human ADFP gene and they have great implications for the treatment of fatty liver (Kotokorpi et al. 2010). The study of the genomes of pathogens, such as bacteria, has revealed specific genetic sites of disease-causing species, and this is a huge target for the treatment of infections that limit the ability to drug resistance (Gal-Mor and Finlay 2006). Metabolic pathways in pathogenic microorganisms are also explored by bioinformatics, and drugs that target metabolic pathways in pathogens may be developed in the future (Bhatia et al. 2014). Bioinformatics can also reduce the cost of drug discovery by repurposing existing drugs to treat new pathogens (Ding et al. 2014). From the understanding of the genome, about the components of the surface structure of many pathogenic microorganisms, such as Galactofuranose—an important component of pathogenic bacteria but not found in humans, many studies provided potential drug development targets targeting such ingredients (Pedersen and Turco 2003). Bioinformatics also provides a great source of data on epigenetic changes, genes, metabolic processes and related substances, thereby helping to develop more effective drugs (Kanehisa 2013). Bioinformatics not only provides genetic data and mutations, but also provides a large amount of information about transcription, helping drug development through phenotypic screening to identify potential drug candidates and drug target determination (Xia 2017). The information on gene expression or metabolism patterns obtained from bioinformatics databases will play an important role in discovering drugs such as anti-cancer drugs or curing metabolic diseases (Wishart 2016; Xia et al. 2009). Based on the database, Li et al. calculated natural compounds with great potential in drug development against COVID-19 (Muhseen et al. 2021). A series of reports on the use of quantitative structure–activity relationship (QSAR), machine learning, and deep learning have yielded surprising results for the potential development of anti-aging drugs or the treatment of infections (Yeh et al. 2021; Araujo et al. 2020).

11.3 Bioinformatics in Drug Screening

Drug screening is the process of identifying and selecting drugs with great potential, safety, and efficacy before they go into clinical trials. This work needs to work with a large amount of information about the library of medicinal herbs and chemicals to be able to create the best medicine, so bioinformatics has a great application in this process (Table 11.1) (Nature 2023). After going through biochemical screening steps, potential compounds ("hit") will continue to undergo tests to check that they have the appropriate physicochemical and pharmacological properties for development into drugs or not, if passed, it will be considered a "lead". Before entering clinical trials, the "lead" will be chemically and biologically screened and eventually has the potential to develop into a drug. Since the early years of the twenty-first century, research has applied bioinformatics to screen targeted molecules, one of

Table 11.1 Screening models (Hughes et al. 2011)

Name	Explanation	Application
High-throughput screen	A large variety of analytically selected substances designed to run in dishes with 384 wells or more	Compound libraries have been discovered, researched, and updated. In addition, the powerful computer support helps to analyze suitable compounds and increase screening efficiency
Virtual screen	Screening of suitable compounds on the library by X-ray to match the selected molecule and is the basis for further research on the structure and binding ability of the drug molecule	It can be possible to provide the initial structure for a focused screen at a more economical cost, or it may be space to detect new molecular structures from known molecules
Physiological screen	A screening method for the effectiveness of drugs on body tissues	The lower throughput; however, more closely simulates the effect of the drug on the tissues. Screening for fewer drug molecules, results in molecules most relevant to the treatment of the disease

which is high-throughput screening. This model involves screening libraries close to the drug target and then secondary assays for the site of action or ability to function in the target protein (Fox et al. 2006; Nemmani 2021). Bioinformatics also has a huge contribution in virtual screening through the early elimination of substances with undesirable properties through computers and silico screening and thereby finding the closest compounds to the desired drug (Smith 2002b). With technological advancements and the ability to share data, virtual screening programs are exhibiting a higher percentage of "hit" screenings than in the past.

Bioinformatics has been applied in the screening and selection of potential drugs to treat diseases of unknown pathogenesis. By bioinformatics analysis, genes involved in Rheumatoid arthritis expression were discovered (Shi et al. 2020). Compounds with therapeutic potential for this disease were screened through disease-specific gene interaction (*LILRB1*), which resulted in the kaempferol 3-*O*-β-D-glucosyl-(1 molecule) molecule. →2)-β-D-glucoside can inhibit the pathological process of Rheumatoid arthritis. A 2019 study has shown the positive signals of bioinformatics application in the screening of potential compounds that help to proliferate cardiac muscle cells while ensuring physiological activity to heal damaged heart muscle tissue (Mills et al. 2019). This study shows that, from about 5000 compounds in the library, through the screening steps, the research has shown that two compounds have high applicability in myocardial proliferation and have the least side effects. The profiling relative inhibition simultaneously in mixtures (PRISM) method has been developed to increase the ability to test drugs, thereby uncovering potential compounds against cancer cell lines (Corsello et al. 2020). The study also showed unexpected results when drugs that do not treat cancer but also have the ability to inhibit cancer cell lines, allowing further research into the molecular characteristics of these cell lines and the direction of treatment. The development in recent years of bioinformatics has greatly contributed to the screening of drug molecules targeting RNA to fight cancer or infection (Manigrasso et al. 2021). The drug molecular structures are not only studied, calculated for pharmacological activity or virtual screening, but also stored in data libraries for in-depth studies and future prediction (Martin et al. 2021).

11.4 Bioinformatics in Drug Validation

According to FDA (U.S. Food and Drug Administration), drug validation can be understood as the process of collecting and evaluating the effectiveness of a drug from the time it is designed through the time it enters experiments and commercial production, thence, establish a system of reliable, scientific evidence for product quality (Center for Devices and Radiological Health and Center for Biologics Evaluation and Research 2002). With the strong development of science and technology today, the application of technological advances to testing the effectiveness of drugs is also of great interest (Hoffmann et al. 1998). Accordingly, bioinformatics software and platforms have been applied to determine the effectiveness of drug targeting genes to optimize the effects of disease therapies.

Table 11.2 Programs are used to determine the potability of a drug (Wooller et al. 2017)

Name	Searching method	Drug analysis
fPocket	Geometric criteria based on distances to predefined points	Based on the chemical and physical properties of the drug molecule such as hydrophobicity and local hydrophobic density
DoGSScorer	Geometric structure based on 3D image increment technique	Based on penetrability, bulk, or association of amino acids
SiteMap	Structure and energy from 3D grids	Based on hydrophilicity, sequestration ability, binding ability

Bioinformatics can be said to have revolutionized the evaluation of drug efficacy through bioinformatics techniques and tools (Table 11.2). Based on these tools and techniques, drug development companies have been able to better understand how the human genome affects the effectiveness of therapeutic drugs (Chang 2005). In addition, also from the knowledge of the patient's genome, personalized pharmacological therapies will be developed, and prescriptions will also be made to suit his or her drug metabolism. The application of bioinformatics to drug development, such as DNA microarray, has been developed to show the correlation between metabolic pathways and drug side effects, and also to evaluate new potential targets for treatment (Meloni et al. 2004). Working on the application of machine learning and synthesizing data on the relationship between genes and drugs, Wang and colleagues identified 96 drugs that target 10 target genes, which are biomarkers for atherosclerosis (Wang et al. 2022). Some of them have been found to be effective for stroke or atherosclerosis. Using the advantages of machine learning and data mining, pharmacologists can evaluate the pharmacological effects of drugs, make adjustments to the 3D structure or develop drug combination treatment strategies to achieve the best treatment effect with the fewest side effects (Agamah et al. 2020). Using genetic data and bioinformatics analysis, scientists have demonstrated that some drugs such as Echinacea, Omeprazole, Ibudilast are effective in treating periodontitis in type 2 diabetic animals, in which Echinacea and Ibudilasts deserve more research because of their amazing medicinal properties (Pan et al. 2022).

11.5 Conclusion

In this chapter, we have presented the applications of bioinformatics for drug development, screening, and validation. Thereby, providing an overview of the achievements that bioinformatics has been used in the field of pharmacology. However, this report still has some limitations. Machine learning, deep learning models in drug response prediction are often assemblies of information that neglects the biological pathways underlying the prediction; therefore, they often have low predictive accuracy and require much fine-tuning by experts (Ching et al. 2018; Murdoch et al. 2019). In addition, bioinformatics-based predictions and analyzes are often still only models, and so they require clinical trials in animals and humans to draw the most accurate conclusions about safety and efficacy in real situations (Shi et al. 2020; Wang et al. 2022; Papillon-Cavanagh et al. 2013).

The development of science and technology is booming, bioinformatics technologies and software are increasingly perfected and have higher accuracy. As a result, the new era of personalized medicine will receive more research attention to personalize methods and prescriptions to treat diseases, in which bioinformatics will play an important role in helping pharmacists and doctors take advantage of the huge resources available (Bayat 2002). Furthermore, developments in bioinformatics have shown the ability to shorten the search time and cost of producing new drugs and utilize natural sources of medicinal herbs (Agamah et al. 2020; Tutone and Almerico 2021). Advances in biotechnology have opened up the understanding of the characteristics of oncogenes and the biomarkers to detect them, thereby developing potential treatment models or drugs-targeted genes (Nguyen and Caldas 2021). Bioinformatics also has enormous application opportunities in the development of software or models in predictive medicine, increasing the success rate of clinical trials (Kuenzi et al. 2020).

The pathogenesis of diseases of great interest such as cancer will be discovered through genomics, proteomics, and transcriptomics libraries. Accordingly, drug companies will identify the target gene or target protein to treat the disease based on the database of gene interactions, gene sequencing, and related articles (Thomford et al. 2018). *After the process of selecting potential drug molecules based on bioinformatics tools, the interaction effect between the drug and the knock-out gene will be studied* in vivo *and* in vitro *to yield novel drug discovery results. This is followed by preclinical trials or drug efficacy models and then clinical trials to determine the actual safety and effectiveness of the drug in real situations.*

References

Agamah FE et al (2020) Computational/in silico methods in drug target and lead prediction. Brief Bioinform 21(5):1663–1675

Araujo PHF et al (2020) Identification of potential COX-2 inhibitors for the treatment of inflammatory diseases using molecular modeling approaches. Molecules 25(18):4183

Baig MH et al (2016) Computer aided drug design: success and limitations. Curr Pharm Des 22(5): 572–581

Bayat A (2002) Science, medicine, and the future: bioinformatics. BMJ 324(7344):1018–1022

Bhatia B et al (2014) Identification of glutamate ABC-transporter component in *Clostridium perfringens* as a putative drug target. Bioinformation 10(7):401–405

Boxall AB et al (2012) Pharmaceuticals and personal care products in the environment: what are the big questions? Environ Health Perspect 120(9):1221–1229

Center for Devices and Radiological Health and Center for Biologics Evaluation and Research (2002) General principles of software validation. FDA-1997-D-0029. https://www.fda.gov/regulatory-information/search-fda-guidance-documents/general-principles-software-validation

Chang PL (2005) Clinical bioinformatics. Chang Gung Med J 28(4):201–211

Chen Z et al (2021) Applications of artificial intelligence in drug development using real-world data. Drug Discov Today 26(5):1256–1264

Ching T et al (2018) Opportunities and obstacles for deep learning in biology and medicine. J R Soc Interface 15(141):20170387
Corsello SM et al (2020) Discovering the anti-cancer potential of non-oncology drugs by systematic viability profiling. Nat Cancer 1(2):235–248
David E, Tramontin T, Zemmel R (2009) Pharmaceutical R&D: the road to positive returns. Nat Rev Drug Discov 8(8):609–610
DiMasi JA, Grabowski HG, Hansen RW (2016) Innovation in the pharmaceutical industry: new estimates of R&D costs. J Health Econ 47:20–33
Ding H et al (2014) Similarity-based machine learning methods for predicting drug–target interactions: a brief review. Brief Bioinform 15(5):734–747
Doolittle RF et al (1983) Simian sarcoma virus onc gene, v-sis, is derived from the gene (or genes) encoding a platelet-derived growth factor. Science 221(4607):275–277
Drews J, Ryser S (1997) The role of innovation in drug development. Nat Biotechnol 15(13): 1318–1319
Fox S et al (2006) High-throughput screening: update on practices and success. J Biomol Screen 11(7):864–869
Gal-Mor O, Finlay BB (2006) Pathogenicity islands: a molecular toolbox for bacterial virulence. Cell Microbiol 8(11):1707–1719
Hoffmann A et al (1998) Computer system validation: an overview of official requirements and standards. Pharm Acta Helv 72(6):317–325
Hughes JP et al (2011) Principles of early drug discovery. Br J Pharmacol 162(6):1239–1249
Kanehisa M (2013) Molecular network analysis of diseases and drugs in KEGG. Methods Mol Biol 939:263–275
Kola I, Landis J (2004) Can the pharmaceutical industry reduce attrition rates? Nat Rev Drug Discov 3(8):711–715
Kotokorpi P et al (2010) The human ADFP gene is a direct liver-X-receptor (LXR) target gene and differentially regulated by synthetic LXR ligands. Mol Pharmacol 77(1):79–86
Kuenzi BM et al (2020) Predicting drug response and synergy using a deep learning model of human cancer cells. Cancer Cell 38(5):672–684 e6
Manigrasso J, Marcia M, De Vivo M (2021) Computer-aided design of RNA-targeted small molecules: a growing need in drug discovery. Chem 7(11):2965–2988
Martin WJ, Grandi P, Marcia M (2021) Screening strategies for identifying RNA- and ribonucleoprotein-targeted compounds. Trends Pharmacol Sci 42(9):758–771
McLean L (2015) 49—Drug development. In: Hochberg MC et al (eds) Rheumatology, 6th edn. Mosby, Philadelphia, pp 395–400
Meloni R, Khalfallah O, Biguet NF (2004) DNA microarrays and pharmacogenomics. Pharmacol Res 49(4):303–308
Mills RJ et al (2019) Drug screening in human PSC-cardiac organoids identifies pro-proliferative compounds acting via the mevalonate pathway. Cell Stem Cell 24(6):895–907.e6
Moffat JG, Rudolph J, Bailey D (2014) Phenotypic screening in cancer drug discovery—past, present and future. Nat Rev Drug Discov 13(8):588–602
Moore H, Allen R (2019) What can mathematics do for drug development? Bull Math Biol 81(9): 3421–3424
Muhseen ZT et al (2021) Computational determination of potential multiprotein targeting natural compounds for rational drug design against SARS-COV-2. Molecules 26(3):674
Murdoch WJ et al (2019) Definitions, methods, and applications in interpretable machine learning. Proc Natl Acad Sci U S A 116(44):22071–22080
Nature (2023) Drug screening articles from across Nature Portfolio. Nature
Nemmani KVS (2021) Pharmacological screening: drug discovery. In: Poduri R (ed) Drug discovery and development: from targets and molecules to medicines. Springer Singapore, Singapore, pp 211–233
Nguyen LV, Caldas C (2021) Functional genomics approaches to improve pre-clinical drug screening and biomarker discovery. EMBO Mol Med 13(9):e13189

Pan S et al (2022) Identification of cross-talk pathways and ferroptosis-related genes in periodontitis and type 2 diabetes mellitus by bioinformatics analysis and experimental validation. Front Immunol 13:1015491

Papillon-Cavanagh S et al (2013) Comparison and validation of genomic predictors for anticancer drug sensitivity. J Am Med Inform Assoc 20(4):597–602

Paul SM et al (2010) How to improve R&D productivity: the pharmaceutical industry's grand challenge. Nat Rev Drug Discov 9(3):203–214

Pedersen LL, Turco SJ (2003) Galactofuranose metabolism: a potential target for antimicrobial chemotherapy. Cell Mol Life Sci 60(2):259–266

Pietras K et al (2003) PDGF receptors as cancer drug targets. Cancer Cell 3(5):439–443

Preziosi P (2007) 2.06—Drug development. In: Taylor JB, Triggle DJ (eds) Comprehensive medicinal chemistry II. Elsevier, Oxford, pp 173–202

Shi YQ, Qi WF, Kong CY (2020) Drug screening and identification of key candidate genes and pathways of rheumatoid arthritis. Mol Med Rep 22(2):986–996

Smith A (2002a) Screening for drug discovery: the leading question. Nature 418(6896):453–459

Smith A (2002b) Screening for drug discovery: the leading question. Nature 418(6896):453–455

Thomford NE et al (2018) Natural products for drug discovery in the 21st century: innovations for novel drug discovery. Int J Mol Sci 19(6):1578

Tutone M, Almerico AM (2021) Computational approaches: drug discovery and design in medicinal chemistry and bioinformatics. Molecules 26(24):7500

van Driel MA, Brunner HG (2006) Bioinformatics methods for identifying candidate disease genes. Hum Genom 2(6):429–432

Wang J et al (2022) Identification of immune cell infiltration and diagnostic biomarkers in unstable atherosclerotic plaques by integrated bioinformatics analysis and machine learning. Front Immunol 13:956078

Waterfield MD et al (1983) Platelet-derived growth factor is structurally related to the putative transforming protein p28sis of simian sarcoma virus. Nature 304(5921):35–39

Wishart DS (2016) Introduction to cheminformatics. Curr Protoc Bioinform 53(1):14.1.1–14.1.21

Wooller SK et al (2017) Bioinformatics in translational drug discovery. Biosci Rep 37(4): BSR20160180

Xia X (2012) Position weight matrix, Gibbs sampler, and the associated significance tests in motif characterization and prediction. Scientifica (Cairo) 2012:917540

Xia X (2017) Bioinformatics and drug discovery. Curr Top Med Chem 17(15):1709–1726

Xia J et al (2009) MetaboAnalyst: a web server for metabolomic data analysis and interpretation. Nucleic Acids Res 37(Web Server issue):W652–W660

Yeh SJ, Lin JF, Chen BS (2021) Multiple-molecule drug design based on systems biology approaches and deep neural network to mitigate human skin aging. Molecules 26(11):3178

Zhang J, Yang PL, Gray NS (2009) Targeting cancer with small molecule kinase inhibitors. Nat Rev Cancer 9(1):28–39

Zhong F et al (2018) Artificial intelligence in drug design. Sci China Life Sci 61(10):1191–1204

Use of Bioinformatics in High-Throughput Drug Screening

12

Tanya Waseem, Mustafeez Mujtaba Babar, Gholamreza Abdi, and Jayakumar Rajadas

Abstract

Bioinformatics has emerged as a vital component of almost all the fields of biological sciences. Its ability to quickly generate, analyze, and interpret large amounts of data has enabled researchers to integrate it into drug discovery and development. The traditional process of drug discovery had its limitations including increased time consumption and cost, low success rates, inaccurate drug target selection, regulatory and ethical concerns, and lack of personalization. To overcome these challenges, bioinformatics has gained much interest in different stages of drug discovery. In this chapter, we summarize the role of bioinformatics in the high-throughput drug screening process involving both ligand-based and structure-based

The original version of this chapter was revised. A correction to this chapter can be found at https://doi.org/10.1007/978-981-99-8401-5_18

T. Waseem
Department of Pharmaceutical Chemistry, Shifa College of Pharmaceutical Sciences, Shifa Tameer-e-Millat University, Islamabad, Pakistan

M. M. Babar (✉)
Department of Basic Medical Sciences, Shifa College of Pharmaceutical Sciences, Shifa Tameer-e-Millat University, Islamabad, Pakistan

Advanced Drug Delivery and Regenerative Biomaterials, Stanford University School of Medicine, Stanford University, Palo Alto, CA, USA
e-mail: mustafeez.babar@fulbrightmail.org

G. Abdi
Department of Biotechnology, Persian Gulf Research Institute, Persian Gulf University, Bushehr, Iran

J. Rajadas
Advanced Drug Delivery and Regenerative Biomaterials Laboratory, Cardiovascular Institute and Pulmonary and Critical Care Medicine, Stanford University School of Medicine, Stanford University, Palo Alto, CA, USA

V. Singh, A. Kumar (eds.), *Advances in Bioinformatics*,
https://doi.org/10.1007/978-981-99-8401-5_12

screening strategies. The most prominent advantage it offers is the ability to handle large amounts of data within seconds to minutes. This chapter also provides a brief account of various bioinformatics tools and databases which have a prominent role in the drug screening process. Although bioinformatics has been proven beneficial, it still has some limitations in terms of the complexity of data that is handled. With the advent of artificial intelligence and machine learning, it is expected that it would definitely strengthen the biomedical field.

Keywords

High-throughput screening · Virtual screening · Bioinformatics · Drug discovery · Drug development

12.1 Introduction

The continuous developments in the field of bioinformatics have paved ways for scientists to discover its applications in disease genetics, exploration of drug targets and drug design and discovery processes (Xia 2017). The complexity and high cost associated with the drug discovery process has led to the use of in-silico approaches to ease the drug development in connection with the experimental techniques. However, the process of drug development starting with the target discovery or identification and then designing and synthesizing drugs to modify the pathological processes is highly expensive due to the associated cost and time constraints of the drug development pipeline (Pereira et al. 2020).

Bioinformatics has been the mainstay of all scientific research for quite some time now. It is a collection of biological data that can be accessed and analyzed using computational tools and algorithms (Jawdat 2006). The major contributions towards the development of bioinformatics are the whole genome sequencing and the progress in the fields of proteomics, genomics and transcriptomics generating high-quality data which is retained in several databases and accessible for further developments (Jin et al. 2021). Deep learning has gained much attention due to its seemingly flawless performance in tasks of the machine learning including structure prediction, biological sequence analysis, protein interactions, biological diagnosis and image processing along with the prediction of biological properties and features (Li et al. 2019a). The data from the field of metabolomics has enabled researchers to study the biological processes to develop and understand the pathways and factors which are essential in physiological responses. Omics fields has influenced not only medical sciences but also other fields such as it provides considerable input in determining the factors necessary for plant growth and its molecular processes. All this is possible merely due to the availability of data which is a fruit of bioinformatics (Ambrosino et al. 2020). Era of omics has expanded the understanding of systems biology by providing valuable insights at all levels necessary for the understanding of biological systems such as proteins, transcripts, metabolites, and genes, requiring complex data processing and computations which can be done by bioinformatic approaches and tools (Waseem et al. 2020). Translational bioinformatic approaches have gained much interest for the advancements in precision medicine and they are focused on the patient specific needs by researching on the pharmacogenomics with the aid of artificial intelligence techniques (Ritchie et al. 2019). The future of bioinformatics in health sciences is quiet promising as it aims to discover modern approaches for direct clinical practices.

12.2 High-Throughput Drug Screening

Drug discovery involves the detection or identification of drug candidates against a biological target. High-throughput screening (HTS) enables us to screen a large library of drug candidates against a selected target in multiple well plates which ultimately leads to the discovery of a novel lead compound. The major advantage of HTS is that the inactive compounds are eliminated at an early stage before pre-clinical or clinical testing of the drugs, thus saving the cost of analysis of inactive compounds. With the advancements in the area of computational chemistry and genetic biology, a number of new druggable targets have emerged and the library of synthetic and semi-synthetic drugs has grown exponentially (Kainkaryam and Woolf 2009; Hsu et al. 2021). The HTS allows us to examine the effect of a library of compounds on different targets using a single compound per well technique. In this way a library of molecules is screened in-vitro against a specific target and the inhibitors/stimulators are identified for further processing. However, the market output of the drug discovery still faces some challenges due to the unexpected and undesirable pharmacological and toxicological profile of the screened compounds in clinical trials (Scannell et al. 2012). The pharmacokinetic profile of the screened compounds also poses a limitation in the successful drug development process which is identified at a later stage and hence results in the loss of considerable time and resources. Cell-based assays although are relatively slow and expensive as compared to biochemical HTS methods but provide data related to the toxicity profile as well as the kinetic properties. These assays are used in the initial stages of drug development HTS of the library of compounds with the desirable characteristics. Cell-based HTS methods are employed while considering the quality control, and the automation techniques are carefully regulated to optimize the outcome and the development of a new chemical entity (NCE) (Schaduangrat et al. 2020).

Traditional HTS methods employ a single drug per well technique in which large number of chemical resources are wasted in exploring the active compounds as the library only have a few hits among the thousands of compounds (Volochnyuk et al. 2019). The inactive compounds are in large number and utilize great deal of time and resources even after automation. Moreover, sometimes the data results in false positive and negatives leading to the miscalculation and hence polluting the overall drug development process. Miniaturization techniques are used to overcome some of the limitations in which small amount of testing reagents and compounds are utilized and are also time efficient (Wilson et al. 2020; Wölcke and Ullmann 2001). Another strategy used for boosting the efficiency of the HTS process is the use of pooling strategy in which compounds are primarily screened as a mixture and then secondary screening is done for the compounds with positive results from primary pooled mixtures (Kainkaryam and Woolf 2009). Hence it optimizes the resources, and reduces the error and cost. Nevertheless, the choice of pooling design and its development and implementation are some of the limitations. Combinatorial pooling strategy finds its applications in disease diagnostics as well such as one recent study identified the use of HTS for SARS-CoV-2 diagnostic testing for asymptomatic patients who are a carrier of the virus and pose a significant threat to the disease spread (Shental et al. 2020).

12.3 Bioinformatics in High-Throughput Drug Screening

With the advent of bioinformatic tools that complement the drug discovery process, we have overcome many of the limitations and problems encountered in late 90 s in the processes of HTS. Bioinformatics has emerged as a multidisciplinary field which is a vital part of drug discovery and development process such as screening of compound libraries, identification of biological targets, proteomics, genomics, biological, chemical and virtual screening of compounds. Cheminformatics along with the bioinformatics has resolved various problems encountered in the drug development and screening process (Parikh et al. 2023). Bioinformatic techniques are employed for the identification of novel drug targets, modelling of target proteins, designing of druggable compounds, determining their interactions with the target and prediction of physicochemical properties and toxicology profiling. Machine learning techniques and algorithms are being developed to aid these processes (Chavda et al. 2021).

Various data mining tools are being used which provide large datasets for the identification of potential targets as well as the compounds that will bind with those targets and produce a response. These tools also enable us to establish the effectiveness of the drug candidate and its binding interactions (Yang et al. 2012; Patel et al. 2020).

Molecular modelling tools are used to generate models of the target proteins and biological systems using different techniques which enable us to virtually analyze the target structure and predict the binding site as well as the binding interactions or the groups needed for potential binding (Haghighatlari and Hachmann 2019).

Virtual high-throughput screening tools are developed to overcome the limitations of traditional HTS systems which results in loss of reagents and resources. When the HTS assays are complex and tedious, virtual drug screening approaches are used to complement the HTS. The drug libraries are screened using in-silico experimentation to determine their binding interactions with biological targets using molecular modelling tools (Mcintosh-Smith et al. 2015). Such techniques enable scientists to screen large libraries in an efficient manner without the expenditure of viable resources. However, these tools require considerable expertise to operate, operate using complex algorithms and are not always error-free. One of the major advantages of using virtual HTS is that it is economical and less time consuming as compared to the traditional experimentation; using large

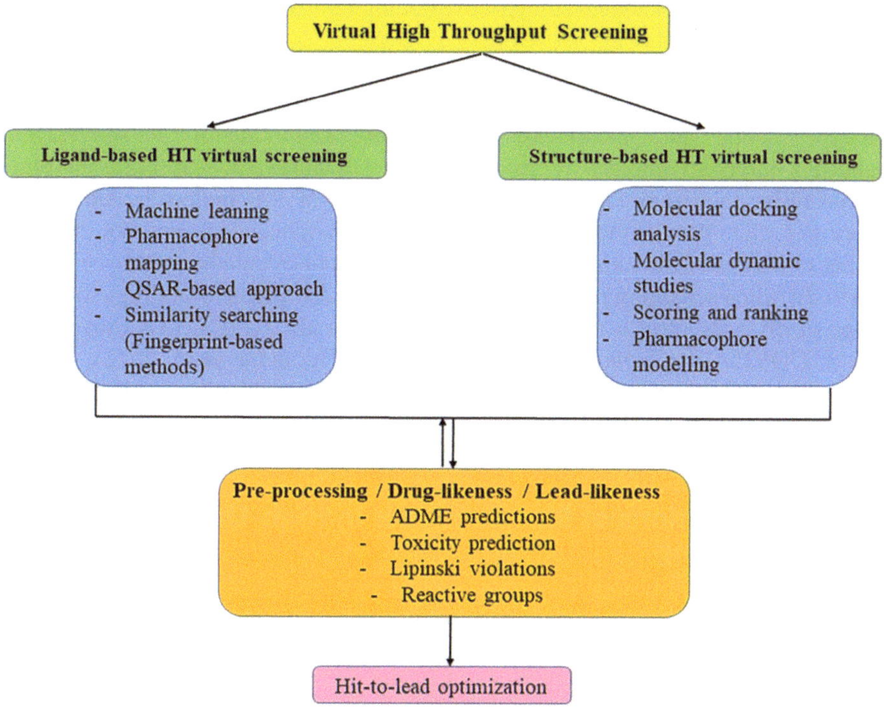

Fig. 12.1 Virtual high-throughput screening strategies for lead identification (Stumpfe and Bajorath 2020; Guterres and Im 2020; Da Silva Rocha et al. 2019)

volume of reagents to screen for a potential active agent from millions of compounds (Mohammad et al. 2021). Structure-based and ligand-based virtual screening are two strategies for hit-to-lead discovery and optimization (Fig. 12.1). These approaches are purely theoretical when compared to the HTS which is purely experimental technique but both aim at the generation of a lead compound for the successful drug discovery process. A combination of both approaches can help in the efficient delivery of successful drug candidates without incurring resource wastage and added costs of analyzing thousands of compounds in HTS experiments (Zhang et al. 2022).

One of the methods for virtual screening is molecular docking which involves the study of interaction between the drug molecule and the biological target followed by the analysis of binding energies and interacting amino acid residues of the binding pocket. The spatial arrangement of drug with its target is based on the induced fit theory and results in the identification of its mechanism of action (Lin et al. 2020). Pharmacophore modelling is another method which is used to design basic structural model of the drug candidate from which lead compound is generated, followed by the screening of databases. The structural features of a pharmacophore are based on its complementary target and by adjusting these features compounds with desirable activity can be designed. Another way is to proceed to the screening of small molecules based on chemical similarity searching on various databases. ZINC is one such database (Seidel et al. 2017; Lin et al. 2020).

Quantitative Structure-activity relationship (QSAR) is a technique in which some quantifiable property of a compound is correlated with its biological activity based on experimental data. Quantitative descriptors are used to identify the active agents against a target of interest by comparing their features such as lipid solubility, permeability, electronic features, size and shape of the molecule and ADME properties. 3D QSAR modelling is still used in pharmaceutical industry due to its ability of accurate structural predictions using minimal calculations (Vucicevic et al. 2019). In-silico screening methods are used for the development of drugs for a wide range of diseases such as tuberculosis (Macalino et al. 2020), CVDs (Savoji et al. 2019), COVID-19 (Gupta et al. 2023), hepatitis (Hdoufane et al. 2022), diabetes (Akhtar et al. 2019), neurodegenerative diseases (Aldewachi et al. 2021) and cancer therapy (Vougas et al. 2019).

12.4 Applications of Bioinformatics in High-Throughput Drug Screening

Omics technology has emerged as a turning point in the health sciences which provides data related to the biological systems and includes proteomics, genomics, metabolomics and transcriptomics. The first step in the HTS is the target identification and various drug targets have been discovered and identified with the help of bioinformatic approaches (Martis et al. 2011). Data mining approaches include high-throughput chemogenomic and proteomics. A wide range of data mining sources are available which have all the necessary information needed for the identification of a biological target such as structural databases (UniPort, PubMed, InterPro), text mining tools (GeneWays, Texrpresso, BioRat), microarray databases (SMD, Oncomine, caArray), clustering database (GenePattern, ArrayMiner, Genecluster), supervised analysis platform (SAM) and interactome and pathway databases (KEGG, PathwayExplorer, Pathguide) (Yang et al. 2012; Agamah et al. 2020). One study reported the use of various bioinformatic tools and databases to develop a human-virus interactome for ZIKA virus using an algorithm OralInt, potentially highlighting various druggable targets against ZIKA virus (Fig. 12.2) (Esteves et al. 2017).

Assay development is a crucial step for the success of screening process. Specificity and sensitivity of assay is the basis of the whole experimentation and bioinformatic techniques have been utilized to develop highly sensitive screening assays. Virtual screening assays are developed as a complementary approach to the HTS and can be regarded as a basic simulation of the HTS assays using the knowledge of biophysics and computer sciences. These simulations are also conducted to optimize the conditions needed to run an assay. In the simulation models, various parameters such as temperature, reagents and time duration can be adjusted leading to the highly sensitive assay. MolMind is one such tool which combines the laboratory based assays and in-silico methods (Szymański et al. 2012). In-silico toxicological analysis

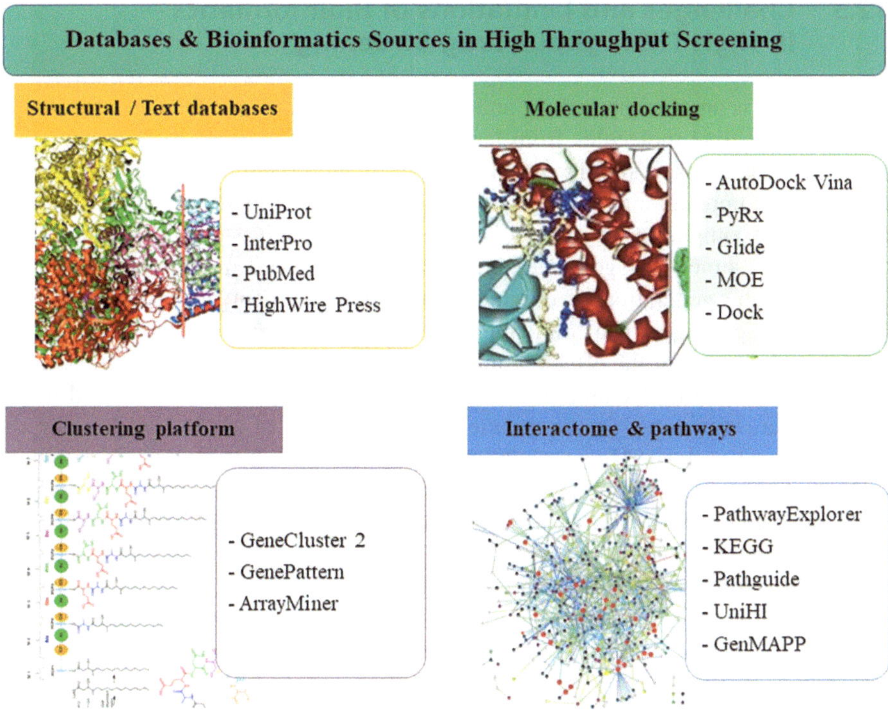

Fig. 12.2 Databases and bioinformatic sources in high-throughput drug screening (Yang et al. 2012)

is a preliminary assay which results in the filtering of potential toxicological compound while virtual screening. ADME-T methods are being used along with computational toxicology methods (i-drug discovery, ToxScope, OncoLogic, MetaDrug, HazardExpert, and e-TOX) for determining the toxicological profile of the drug candidates (Szymański et al. 2012). In one study, imaging techniques and florescent-based methods were combined to create a high-throughput drug screening assay using 3D organoids to assess the organoid growth and the effects of drugs (Li et al. 2022).

Data mining approaches and microarray techniques are utilized for HTS. One study reported the use of microarray analysis for the identification of micro RNAs and genes as biomarkers for the treatment and diagnosis of atrial fibrillation using different databases (Li et al. 2019b). The study of biological pathways utilizing bioinformatic tools has made it possible to identify the disease biomarkers and drug targets. One study reported the involvement of multiple RNAs expression in the regulation and progression of preeclampsia using different bioinformatics tools and databases. It also reported that the activation of JAK-STAT signaling pathway is related to the progression of preeclampsia (Liu et al. 2019).

12.5 Challenges and Limitations of Bioinformatics in High-Throughput Drug Screening

Virtual HTS is an efficient, robust and cost-effective technique for the screening of biologically active molecules from large datasets but it does not replace the traditional HTS methods, it simply complements it by narrowing down the possible hits and leads. Computational analysis, although, enables us to screen for a library of thousands of compounds in a day but it faces some challenges due to the complexity of data and sometimes generates erroneous results. The main focus, however, remains on the generation of efficient leads for subsequent optimization in drug development pipelines. The complexity of data generated by computational analysis is also challenging for the effective interpretation and requires highly skilled analysis. Machine learning techniques such as decision tree models and artificial neural networks are developed to overcome the complexity of data available for the computational analysis (Han et al. 2008; Butkiewicz et al. 2012). The computations and equations of QSAR models are highly complex and requires careful analysis, validations and may sometimes be impractical (Spiegel and Senderowitz 2020). The computational complexity of various multilayered techniques is a hindrance. The method development requires validation of data and sometimes the results are not reproducible raising a question on the validity of the data obtained (Stumpfe and Bajorath 2020). One of the major challenges highlighted and mentioned by multiple researchers in the field is the accuracy of data obtained from virtual screening methods. Sometimes its impractical to translate the outcome in human patients although significant evidence of activity is obtained from computational analysis. In structure-based drug screening, the binding energies of actives and inactive are closely related showing inaccuracy and sometimes putative interactions are generated for inactive leads which like HTS results in the generation of ineffective lead compounds identified at a later stage of testing (Jasial et al. 2016).

Current debate is on the ligand promiscuity of the biological targets which may points towards the inaccuracy of the binding interactions generated through the virtual screening. With the prior knowledge of drug-target binding interactions, virtual screening methods also faces a certain bias in the selection of screening library which leads to the high hit rates confused with the accuracy of the prediction (Stumpfe and Bajorath 2020).

12.6 Future of Bioinformatics in High-Throughput Drug Screening

Bioinformatics has emerged as an indispensable field in the drug discovery and screening processes. The traditional high-throughput screening requires the experimentation of large library of compounds having millions of drugs comprising of large number of inactive candidates. This resulted in wastage of resources, time and money. Bioinformatic tools and techniques enable us to shrink down the chemical library before high-throughput screening assays by ruling out the possible inactive agents in in-silico or virtual screening steps. These virtual screening methods enable us to identify and select only those compounds which show promising results in virtual screening assays (Stumpfe et al. 2012; Stumpfe and Bajorath 2020). Hence these techniques save cost, resources and time by providing highly specific and nearly accurate predictions. Virtual screening era is promising and is predicted to progress further mainly due to its screening efficiency and enormous data handling capacity. Nevertheless, the virtual screening problems need to be encountered in the future to continue an integrative approach towards drug screening. The number one problem which requires attention is the generation of inaccurate binding energies and similarity hits; which require rigorous post-analysis to interpret the accuracy of results. Scientists are working to overcome this problem and have made some progress. In this post-genomic era, the field of molecular and chemical biology remain potential areas of growth that will enhance our understanding as well as the applications of virtual drug screening (Heikamp and Bajorath 2012; Sabe et al. 2021).

The advancements in the field of artificial intelligence are a turning point for the pharmaceutical and health sciences as it is a step forward towards overcoming the limitations encountered in drug discovery (Zhong et al. 2018). Virtual screening is indispensable in drug discovery and development process. Sequential screening which is a widely known concept; computational screening integrated with experimental screening, should be practically incorporated in order to avoid problems at a later stage and overcome the limitations of both techniques (Achary 2020).

12.7 Conclusions and Future Perspectives

The field of bioinformatics has significantly contributed to the drug discovery and development process by providing an avenue though virtual high-throughput screening. However, the vast amount of unverified data available on genetic and protein repositories makes it essential for the bioinformaticians to pre-process it before its integration and interpretation can actually begin. In addition, biological complexity of available data hinders its wider usage. Experimental limitations and lack of availability and accessibility to a variety of user-friendly computer applications also appears to slow-down the HTS process. The advent of publicly available machine learning and artificial intelligence platforms can address some of the identified limitations. Moreover, the inter-disciplinary collaborative research can facilitate the drug development process.

References

Achary PGR (2020) Applications of quantitative structure-activity relationships (QSAR) based virtual screening in drug design: a review. Mini Rev Med Chem 20:1375–1388

Agamah FE, Mazandu GK, Hassan R, Bope CD, Thomford NE, Ghansah A, Chimusa ER (2020) Computational/in silico methods in drug target and lead prediction. Brief Bioinform 21:1663–1675

Akhtar A, Amir A, Hussain W, Ghaffar A, Rasool N (2019) In silico computations of selective phytochemicals as potential inhibitors against major biological targets of diabetes mellitus. Curr Comput Aided Drug Des 15:401–408

Aldewachi H, Al-Zidan RN, Conner MT, Salman MM (2021) High-throughput screening platforms in the discovery of novel drugs for neurodegenerative diseases. Bioengineering 8:30

Ambrosino L, Colantuono C, Diretto G, Fiore A, Chiusano ML (2020) Bioinformatics resources for plant abiotic stress responses: state of the art and opportunities in the fast evolving-omics era. Plan Theory 9:591

Butkiewicz M, Lowe EW, Mueller R, Mendenhall JL, Teixeira PL, Weaver CD, Meiler J (2012) Benchmarking ligand-based virtual high-throughput screening with the PubChem database. Molecules 18:735–756

Chavda V, Sheta S, Changani D, Chavda D (2021) New bioinformatics platform-based approach for drug design. In: Balamurugan S, Krishnan A, Goyal D, Chandrasekaran B, Pandi B (eds) Computation in bioinformatics: multidisciplinary applications. Wiley, pp 101–120

Da Silva Rocha SFL, Olanda CG, Fokoue HH, Sant'anna, C. M. R. (2019) Virtual screening techniques in drug discovery: review and recent applications. Curr Top Med Chem 19:1751–1767

Esteves E, Rosa N, Correia MJ, Arrais JP, Barros M (2017) New targets for *Zika* virus determined by human-viral interactomic: a bioinformatics approach. Biomed Res Int 2017:1734151

Gupta Y, Savytskyi OV, Coban M, Venugopal A, Pleqi V, Weber CA, Chitale R, Durvasula R, Hopkins C, Kempaiah P (2023) Protein structure-based in-silico approaches to drug discovery: guide to COVID-19 therapeutics. Mol Asp Med 91:101151

Guterres H, Im W (2020) Improving protein-ligand docking results with high-throughput molecular dynamics simulations. J Chem Inf Model 60:2189–2198

Haghighatlari M, Hachmann J (2019) Advances of machine learning in molecular modeling and simulation. Curr Opin Chem Eng 23:51–57

Han L, Wang Y, Bryant SH (2008) Developing and validating predictive decision tree models from mining chemical structural fingerprints and high–throughput screening data in PubChem. BMC Bioinform 9:401

Hdoufane I, Bjij I, Oubahmane M, Soliman ME, Villemin D, Cherqaoui D (2022) In silico design and analysis of NS4B inhibitors against hepatitis C virus. J Biomol Struct Dyn 40:1915–1929

Heikamp K, Bajorath J (2012) The future of virtual compound screening. Chem Biol Drug Des 81:33–40

Hsu MN, Tay ZM, Lin WN, Wei S-C (2021) Screening of antigen-specific antibody-secreting cells. Handbook of single-cell technologies. Springer

Jasial S, Hu Y, Vogt M, Bajorath J (2016) Activity-relevant similarity values for fingerprints and implications for similarity searching [version 2; peer review: 3 approved]. F1000Res 5

Jawdat D (2006) The era of bioinformatics. In 2006 2nd international conference on information & communication technologies. IEEE, p 1860–1865

Jin S, Zeng X, Xia F, Huang W, Liu X (2021) Application of deep learning methods in biological networks. Brief Bioinform 22:1902–1917

Kainkaryam RM, Woolf PJ (2009) Pooling in high-throughput drug screening. Curr Opin Drug Discov Devel 12:339

Li Y, Huang C, Ding L, Li Z, Pan Y, Gao X (2019a) Deep learning in bioinformatics: introduction, application, and perspective in the big data era. Methods 166:4–21

Li Y, Tan W, Ye F, Xue F, Gao S, Huang W, Wang Z (2019b) Identification of microRNAs and genes as biomarkers of atrial fibrillation using a bioinformatics approach. J Int Med Res 47:3580–3589

Li X, Fu G, Zhang L, Guan R, Tang P, Zhang J, Rao X, Chen S, Xu X, Zhou Y, Deng Y, Lv T, He X, Mo S, Mu P, Gao J, Hua G (2022) Assay establishment and validation of a high-throughput organoid-based drug screening platform. Stem Cell Res Ther 13:219
Lin X, Li X, Lin X (2020) A review on applications of computational methods in drug screening and design. Molecules 25:1375
Liu S, Xie X, Lei H, Zou B, Xie L (2019) Identification of key circRNAs/lncRNAs/miRNAs/mRNAs and pathways in preeclampsia using bioinformatics analysis. Med Sci Monit 25:1679–1693
Macalino SJY, Billones JB, Organo VG, Carrillo MCO (2020) In silico strategies in tuberculosis drug discovery. Molecules 25:665
Martis E, Radhakrishnan R, Badve R (2011) High-throughput screening: the hits and leads of drug discovery-an overview. J Appl Pharm Sci:02–10
Mcintosh-Smith S, Price J, Sessions RB, Ibarra AA (2015) High performance in silico virtual drug screening on many-core processors. Int J High Perform Comput Appl 29:119–134
Mohammad T, Mathur Y, Hassan MI (2021) InstaDock: a single-click graphical user interface for molecular docking-based virtual high-throughput screening. Brief Bioinform 22:bbaa279
Parikh PK, Savjani JK, Gajjar AK, Chhabria MT (2023) Bioinformatics and cheminformatics tools in early drug discovery. In: Bioinformatics tools for pharmaceutical drug product development, pp 147–181
Patel L, Shukla T, Huang X, Ussery DW, Wang S (2020) Machine learning methods in drug discovery. Molecules 25:5277
Pereira SA, Dyson PJ, Saraiva MLM (2020) Miniaturized technologies for high-throughput drug screening enzymatic assays and diagnostics—a review. TrAC Trends Anal Chem 126:115862
Ritchie MD, Moore JH, Kim JH (2019) Translational bioinformatics: biobanks in the precision medicine era. Pacific symposium on biocomputing 2020. World Scientific, p 743–747
Sabe VT, Ntombela T, Jhamba LA, Maguire GEM, Govender T, Naicker T, Kruger HG (2021) Current trends in computer aided drug design and a highlight of drugs discovered via computational techniques: a review. Eur J Med Chem 224:113705
Savoji H, Mohammadi MH, Rafatian N, Toroghi MK, Wang EY, Zhao Y, Korolj A, Ahadian S, Radisic M (2019) Cardiovascular disease models: a game changing paradigm in drug discovery and screening. Biomaterials 198:3–26
Scannell JW, Blanckley A, Boldon H, Warrington B (2012) Diagnosing the decline in pharmaceutical R&D efficiency. Nat Rev Drug Discov 11:191–200
Schaduangrat N, Lampa S, Simeon S, Gleeson MP, Spjuth O, Nantasenamat C (2020) Towards reproducible computational drug discovery. J Chem 12:1–30
Seidel T, Bryant SD, Ibis G, Poli G, Langer T (2017) 3D pharmacophore modeling techniques in computer-aided molecular design using LigandScout. In: Tutorials in chemoinformatics, pp 279–309
Shental N, Levy S, Wuvshet V, Skorniakov S, Shalem B, Ottolenghi A, Greenshpan Y, Steinberg R, Edri A, Gillis R (2020) Efficient high-throughput SARS-CoV-2 testing to detect asymptomatic carriers. Sci Adv 6:eabc5961
Spiegel J, Senderowitz H (2020) Evaluation of QSAR equations for virtual screening. Int J Mol Sci 21:7828
Stumpfe D, Bajorath J (2020) Current trends, overlooked issues, and unmet challenges in virtual screening. J Chem Inf Model 60:4112–4115
Stumpfe D, Ripphausen P, Bajorath J (2012) Virtual compound screening in drug discovery. Future Med Chem 4:593–602
Szymański P, Markowicz M, Mikiciuk-Olasik E (2012) Adaptation of high-throughput screening in drug discovery—toxicological screening tests. Int J Mol Sci 13:427–452
Volochnyuk DM, Ryabukhin SV, Moroz YS, Savych O, Chuprina A, Horvath D, Zabolotna Y, Varnek A, Judd DB (2019) Evolution of commercially available compounds for HTS. Drug Discov Today 24:390–402

Vougas K, Sakellaropoulos T, Kotsinas A, Foukas G-RP, Ntargaras A, Koinis F, Polyzos A, Myrianthopoulos V, Zhou H, Narang S, Georgoulias V, Alexopoulos L, Aifantis I, Townsend PA, Sfikakis P, Fitzgerald R, Thanos D, Bartek J, Petty R, Tsirigos A, Gorgoulis VG (2019) Machine learning and data mining frameworks for predicting drug response in cancer: an overview and a novel in silico screening process based on association rule mining. Pharmacol Ther 203:107395

Vucicevic J, Nikolic K, Mitchell JB (2019) Rational drug design of antineoplastic agents using 3D-Qsar, cheminformatic, and virtual screening approaches. Curr Med Chem 26:3874–3889

Waseem T, Zargaham MK, Shahid F, Rajput TA, Ibrahim B, Babar MM (2020) New approaches to antimicrobial discovery: current development and future prospects. In: New and future developments in microbial biotechnology and bioengineering. Elsevier, pp 67–77

Wilson BA, Thornburg CC, Henrich CJ, Grkovic T, O'keefe, B. R. (2020) Creating and screening natural product libraries. Nat Prod Rep 37:893–918

Wölcke J, Ullmann D (2001) Miniaturized HTS technologies–uHTS. Drug Discov Today 6:637–646

Xia X (2017) Bioinformatics and drug discovery. Curr Top Med Chem 17:1709–1726

Yang Y, Adelstein SJ, Kassis AI (2012) Target discovery from data mining approaches. Drug Discov Today 17:S16–S23

Zhang Y, Luo M, Wu P, Wu S, Lee T-Y, Bai C (2022) Application of computational biology and artificial intelligence in drug design. Int J Mol Sci 23:13568

Zhong F, Xing J, Li X, Liu X, Fu Z, Xiong Z, Lu D, Wu X, Zhao J, Tan X, Li F, Luo X, Li Z, Chen K, Zheng M, Jiang H (2018) Artificial intelligence in drug design. Sci China Life Sci 61: 1191–1204

Bioinformatics in Precision Medicine and Healthcare

13

Mai-Anh Nguyen, Chia-Ching Wu, and Dinh-Toi Chu

Abstract

In today's healthcare industry, we prefer methods that are highly effective and minimize risks. Precision medicine is a new field that utilizes algorithms in bioinformatics to provide precise treatment for individuals. Bioinformatics tools not only assist doctors find the most appropriate therapeutic solutions for each patient but also aid in uncovering vast amounts of life science and clinical data for healthcare development. Scientists have found that single nucleotide polymorphism is the primary agent of genetic modification and a potential tool for genetic mapping. Bioinformatics tools combine genetic information with phenotypes and drug responses to help doctors choose the most appropriate treatment for a patient. In addition, algorithms in bioinformatics also help doctors reduce their workload, both in the field of diagnosis and the field of treatment. Besides, the development of bioinformatics also helps scientists create comparative models of

M.-A. Nguyen
Center for Biomedicine and Community Health, International School, Vietnam National University, Hanoi, Vietnam

C.-C. Wu
Department of Cell Biology and Anatomy, College of Medicine, National Cheng Kung University, Tainan, Taiwan

International Center for Wound Repair and Regeneration, National Cheng Kung University, Tainan, Taiwan

Department of Biomedical Engineering, National Cheng Kung University, Tainan, Taiwan

D.-T. Chu (✉)
Center for Biomedicine and Community Health, International School, Vietnam National University, Hanoi, Vietnam

Faculty of Applied Sciences, International School, Vietnam National University, Hanoi, Vietnam
e-mail: toicd@vnu.edu.vn

V. Singh, A. Kumar (eds.), *Advances in Bioinformatics*,
https://doi.org/10.1007/978-981-99-8401-5_13

biomolecular sequences or molecular mechanisms and generate medical examination records of patients. By optimizing the treatment method based on the potential of bioinformatics, we can minimize the cost of drugs, medical equipment, and the patient's treatment time.

Keywords

Bioinformatics · Biomedical informatics · Healthcare · Comparative models · Precision medicine

Abbreviations

AI	Artificial intelligence
BMI	Biomedical informatics
BMII	Biomedical imaging informatics
PM	Precision medicine

13.1 Introduction

Bioinformatics is a combination of both biology and information technology covering four areas (Biology, Computer Science, Medicine, Mathematics/Physics) (Fig. 13.1). The main goal of bioinformatics is the analysis of biological data and

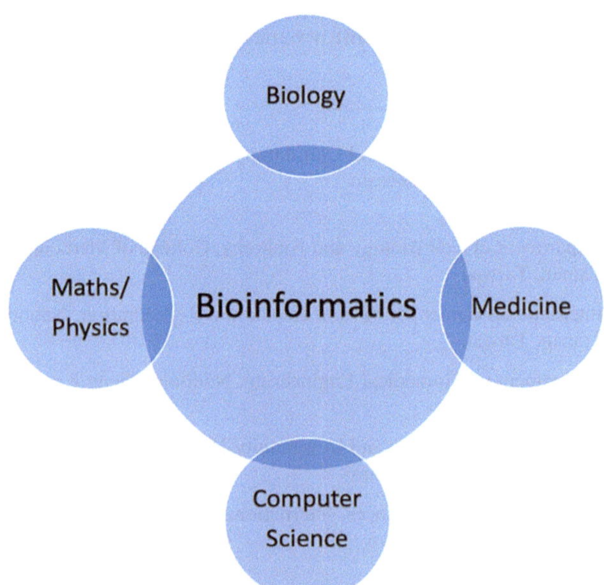

Fig. 13.1 Fields in bioinformatics

Table 13.1 Some free websites for bioinformatics

Names	Applications
Dotplot	Using the dot plot method, compare protein and DNA sequences
BLAST	Based on DNA and protein sequence database, search for similar sequences
DNA sequence analysis	Using biological tools, DNA sequence analysis
Protein sequence analysis	Using biological tools, protein sequence analysis
Modeling	Structure prediction and three-dimensional structure analysis of proteins

the utilization of biological tools to develop software (Bayat 2002). The National Center for Biotechnology Information, a branch of the National Library of Medicine and the National Institutes of Health has defined bioinformatics as the emerging field that deals with the application of computers to the collection, organization, analysis, manipulation, presentation, and sharing of biological data. Through the development of algorithms and software, bioinformatics can extract knowledge from biological data to increase the understanding of biological processes (Pool and Esnayra 2000).

There are many application areas of bioinformatics, such as genomics, image analysis, drug design, and many more. In the field of biometric analytics, bioinformatics is used for identification and access control, thereby facilitating remediation and improvement of crop production and pest control. In addition, bioinformatics plays an important role in the field of precision medicine and preventive medicine, spearheading the development of measures to prevent, control and cure infectious diseases (Bayat 2002). Thanks to funding from the scientific community, there are many freely available bioinformatics tools on the internet. Therefore, anyone can learn about the composition of biomolecules with only basic tools. The three major bioinformatics centers (NCBI, ExPASy and EBI) are the most popular ones, which develop, collect and provide online services on their websites (Luo 2013). Some of the free websites used in bioinformatics are shown in Table 13.1.

In line with the development of society and science, current clinical care needs to adopt therapeutic methods that are optimal in terms of effectiveness and minimal in terms of toxicity (Akhoon 2021). Therefore, in 2015 at the Precision Medicine Initiative launched by Barack Obama, precision medicine (PM) was defined as "providing the right treatment at the right time to the right person and taking into account patients' health history, genes, environments, and lifestyles" (Stone 2016). PM requires a range of tools such as Big Data, artificial intelligence (AI), pharmacodynamics, and omics. In addition, environmental and social factors, as well as the integration of PM with preventive health and population also need to be carefully considered in PM (Naithani et al. 2021a). The purpose of PM is to predict, prevent, diagnose, and treat effectively through the patient's genetic and genomic information. As a result, doctors can choose the most effective treatment methods or prescribe drugs accordingly (Wang et al. 2016). With the aim of optimizing treatment methods and minimizing medical costs, PM is widely applied from the field of diagnosis to the field of treatment (Naithani et al. 2021b).

Bioinformatics plays an important role in the discovery of large amounts of clinical and life science data for medical development. Therefore, combining the steps of bioinformatics research with machine learning breakthroughs is the key to achieving the purpose of PM. Through the use of molecular and digital data, doctors can better understand a patient's current situation, thereby enabling them to make appropriate diagnoses and treatments (Tang et al. 2022). A specific case in the US shows that the "All of Us" research program has been leveraged by subsequent studies to explore the prevalence of eczema in various groups of people in society (Delavar et al. 2022). In addition, this data was also extracted to study glaucoma patients, assessing their knowledge of family health history as well as their ability to purchase medication (Hull and Natarajan 2022). Regarding the healthcare sector, bioinformatics is demonstrating its strong role by providing AI systems. Through these systems, doctors can make reliable diagnoses thanks to algorithms that cannot be grasped by the human mind or eye (Gujar et al. 2020). As a result, technology companies are actively investing to harness the potential of healthcare data, enabling them to manage and utilize digital data more effectively (Arute et al. 2019). Not only that, some technology companies are also aiming to gain a deep understanding of genetics for healthcare, and the combination of AI surgical robots with 5G technology is expected to revolutionize medicine. In the United Kingdom, the government has released the genetic sequences of all newborns with the purpose of laying the foundation for PM and healthcare policies (Gujar et al. 2020). Thus, we see that bioinformatics has great potential in PM as well as healthcare. Therefore, this chapter mainly focuses on discussing the applications of bioinformatics in these two areas.

13.2 Bioinformatics in Precision Medicine

Thanks to significant advances in molecular science, scientists have identified single nucleotide polymorphism as the primary agent of genetic variation and a valuable tool for genetic mapping (Collins et al. 1997). The combination of genetic information with phenotype and drug response will help physicians select the appropriate therapy according to the genotype of each particular patient (Fig. 13.2). The current trend is for doctors to have access to the patient's genome to diagnose diseases and select treatments (Fernald et al. 2011). However, only a small portion of the genome is usable (Collins et al. 1997). Problems arise when scientists conduct association studies, as they often identify variations with small effect sizes and limiting applications in healthcare (Moore et al. 2010). At this time, bioinformatics serves as an effective tool for tailoring PM care, including patient characterization and therapeutic delivery through digital and readily available data (Moore et al. 2010). For example, a scientist might observe an abnormal A1C test result to diagnose a patient with diabetes. They would then notice a prescription for metformin and ultimately observe an improvement in the patient's condition. As a result, large

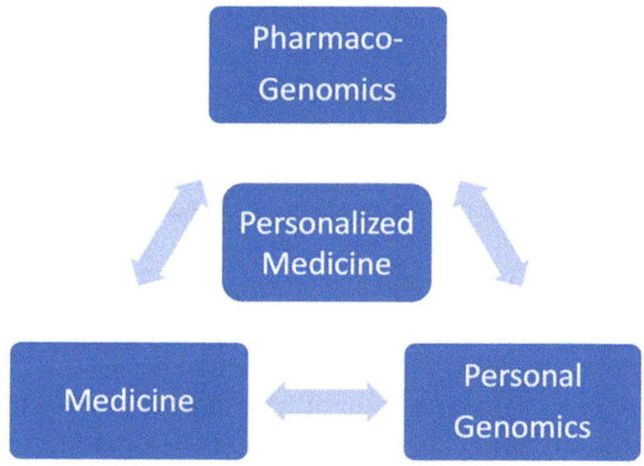

Fig. 13.2 Combination of pharmacogenomics, medicine, and personal genomics in precision medicine

patient groups can be formed from these clinical analyses. In California, scientists identified 97,231 type 2 diabetes patients who were treated with different approaches across five health systems (Moore et al. 2010). In economic and social structural models, another potential application of bioinformatics is the contextualization of research questions (Tang et al. 2022). For example, a study conducted in the UK on children with diabetes revealed that economic status and racial discrimination were associated with patients' treatment regimen (Catherine et al. 2021).

Biomedical informatics (BMI) is a cross-disciplinary field that aims to effectively utilize data and knowledge related to the biomedical field in order to support the improvement of human health. Biomedical imaging informatics (BMII) is a branch of BMI that is emerging due to its significant impact, encompassing diagnostic imaging and imaging information that various disciplines depend on (Hsu et al. 2013). About more than 40 years ago, the need for increased use of radiographic imaging techniques necessitated the creation of new digital methods. These innovative efforts formed the basis of BMII, including image acquisition, image quality control, disease detection, and diagnosis (Sinha et al. 2002; Geis 2007). A case in point is diffuse MRI, which provides structural and functional information that can be used to describe complex diseases or assess the effectiveness of treatments for individual patients (Bui and Taira 2009). The development of BMII will provide accurate pixel data as well as approaches for the most efficient use of image information in the future (Hsu et al. 2013). In addition, BMI is also used and filter data to analyze data, enabling the discovery of new knowledge about neurodegenerative diseases (Miller et al. 2018). The hallmark of neurochemistry is the disruption of complex neural networks (Dennis and Thompson 2014; Collins and Riley 2016), even in the early stages of the disease (Miller and Barr 2017). Diagnosing this disease requires the accurate identification of pathological changes in the patient's brain. This implies the need for reliable biomarkers, as well as imaging data and genetic and phenotypic information (Lista et al. 2015). Therefore, the application of computational network analysis modeling helps integrate different data sources and distributed mappings, creating relationships between them. This enables scientists to map the connection of both brain structure and function (Rubinov and Sporns 2010) and identify common genetic pathways (Talwar et al. 2014). As a result, one can model disease progression over time and predict the subsequent course of the disease, which is one of the important factors of PM (Oxtoby et al. 2017).

13.3 Bioinformatics in Healthcare

With the development of current biomedical applications, comparative modeling of biomolecular sequences or molecular mechanisms has become possible, offering great potential for healthcare (Kuznetsov et al. 2013). However, the creation of medical records is fraught with challenges due to the large storage capacity required and the ability to find the genotype-phenotype association. Other challenges also arise from social and ethical issues related to genetic discrimination (Sethi and Theodos 2009). Besides the challenges, the role of bioinformatics in human healthcare cannot be denied. One of the remarkable achievements was the discovery of the genome sequence of the influenza virus. Every time a new strain of flu appears, there is always the possibility of a large-scale outbreak, so it is essential to understand the characteristics of the new virus. We can remember a case in point in 2009 when the swine flu pandemic was well controlled by scientists, thanks to precise computational methods that contributed to understanding the initial molecular characteristics and process of virus mutations (Garten et al. 2009; Maurer-Stroh et al. 2009; Smith et al. 2009) Modern sequencing technology has made molecular sequence data of the samples readily available. Bioinformatics can rapidly screen for specific mutations from influenza sequences, such as plotting disease patterns over time through the comparison of genomes, structural models and available literature (Kuznetsov et al. 2013). Bioinformatics is also used to discover new influenza mutations, including marker mutations of novel variants (Maurer-Stroh et al. 2010) and novel mutations in neuraminidase that alter drug efficacy (Hurt et al. 2011; van der Vries et al. 2011; Nguyen et al. 2012).

Currently, computing technology has been developed thanks to improvements in image processing and pattern recognition. In particular, imaging is supported by computers, which means that the doctor's workload will be reduced. In addition, the integration of a database of patient medical records has aided physicians in making more accurate diagnoses (Kuznetsov et al. 2013). For example, a disease with a relatively high incidence worldwide is prostate cancer. It is the most common male skin cancer in the US (Jemal et al. 2010) and the third most common in Singapore (Seow et al. 2004). Improving the diagnosis of this disease requires objective computer algorithms to assess the pathology. Many methods have been developed for standard hematoxylin/eosin stain image analysis. The most commonly used techniques include leveling (Naik et al. 2008), machine learning (Teverovskiy et al. 2004; Doyle et al. 2006; Hafiane et al. 2008), and fractal analysis (Naik et al. 2008) with the aim of segmenting routes (Naik et al. 2007) and multiply (Hafiane et al. 2008; Muhammad and Rajpoot 2007) or identify areas of malignancy directly (Doyle et al. 2006).

13.4 Conclusion

In this chapter we discussed the application of bioinformatics in precision medicine and healthcare, thereby providing an overview of the achievements and applications of bioinformatics in the healthcare field of each patient. Bioinformatics holds great potential as it helps scientists explore the molecular structure of genes, their complex interactions, and their role in diseases. Based on that foundation, a new field such as precision medicine is developing more and more and showing its position in healthcare. Bioinformatics is a tool that helps clinical researchers take advantage of the benefits offered by algorithms. However, it still has some limitations. First, protecting the privacy and security of patients' genetic information and advancing research to enhance patient care, are challenging. Failure to pay attention to the protection of patient privacy can lead to the possibility of discrimination. Additionally, advancing the field of translational bioinformatics requires collaboration across genomics, clinical, and healthcare disciplines. That means the repository needs to expand, from storing clinically relevant data to storing genetic data. The potential of this application is significant when it is possible to extract information from clinical procedures and epidemiological studies. Successful research teams of the future need to master between laboratory experimentation, clinical practice, and the use of algorithms.

References

Akhoon N (2021) Precision medicine: a new paradigm in therapeutics. Int J Prev Med 12:12

Arute F et al (2019) Quantum supremacy using a programmable superconducting processor. Nature 574(7779):505–510

Bayat A (2002) Science, medicine, and the future: bioinformatics. BMJ 324(7344):1018–1022

Bui AA, Taira RK (2009) Medical imaging informatics. Springer Science & Business Media

Catherine JP, Russell MV, Peter CH (2021) The impact of race and socioeconomic factors on paediatric diabetes. EClinicalMedicine 42:101186

Collins FS, Riley WT (2016) NIH's transformative opportunities for the behavioral and social sciences. Sci Transl Med 8(366):366ed14

Collins FS, Guyer MS, Charkravarti A (1997) Variations on a theme: cataloging human DNA sequence variation. Science 278(5343):1580–1581

Delavar A et al (2022) Racial and ethnic disparities in cost-related barriers to medication adherence among patients with glaucoma enrolled in the National Institutes of Health all of us research program. JAMA Ophthalmol 140(4):354–361

Dennis EL, Thompson PM (2014) Functional brain connectivity using fMRI in aging and Alzheimer's disease. Neuropsychol Rev 24(1):49–62

Doyle S et al (2006) Detecting prostatic adenocarcinoma from digitized histology using a multi-scale hierarchical classification approach. Conf Proc IEEE Eng Med Biol Soc 2006:4759–4762

Fernald GH et al (2011) Bioinformatics challenges for personalized medicine. Bioinformatics 27(13):1741–1748

Garten RJ et al (2009) Antigenic and genetic characteristics of swine-origin 2009 a(H1N1) influenza viruses circulating in humans. Science 325(5937):197–201

Geis JR (2007) Medical imaging informatics: how it improves radiology practice today. J Digit Imaging 20(2):99–104

Gujar R, Panwar B, Dhanda SK (2020) Bioinformatics drives discovery in biomedicine. Bioinformation 16(1):13–16

Hafiane A, Bunyak F, Palaniappan K (2008) Level set-based histology image segmentation with region-based comparison. In Microscopic image analysis with applications in biology workshop

Hsu W, Markey MK, Wang MD (2013) Biomedical imaging informatics in the era of precision medicine: progress, challenges, and opportunities. J Am Med Inform Assoc 20(6):1010–1013

Hull LE, Natarajan P (2022) Self-rated family health history knowledge among all of us program participants. Genet Med 24(4):955–961

Hurt AC et al (2011) Increased detection in Australia and Singapore of a novel influenza A(H1N1) 2009 variant with reduced oseltamivir and zanamivir sensitivity due to a S247N neuraminidase mutation. Euro Surveill 16(23)

Jemal A et al (2010) Cancer statistics, 2010. CA Cancer J Clin 60(5):277–300

Kuznetsov V et al (2013) How bioinformatics influences health informatics: usage of biomolecular sequences, expression profiles and automated microscopic image analyses for clinical needs and public health. Health Inf Sci Syst 1:2

Lista S et al (2015) Evolving evidence for the value of neuroimaging methods and biological markers in subjects categorized with subjective cognitive decline. J Alzheimers Dis 48(Suppl 1): S171–S191

Luo J (2013) Applied bioinformatics tools. In Basics of bioinformatics. Lecture Notes of the Graduate Summer School on Bioinformatics of China, p 271–301

Maurer-Stroh S et al (2009) Mapping the sequence mutations of the 2009 H1N1 influenza A virus neuraminidase relative to drug and antibody binding sites. Biol Direct 4:18; discussion 18

Maurer-Stroh S et al (2010) A new common mutation in the hemagglutinin of the 2009 (H1N1) influenza A virus. PLoS Curr 2:RRN1162

Miller JB, Barr WB (2017) The technology crisis in neuropsychology. Arch Clin Neuropsychol 32(5):541–554

Miller JB et al (2018) Biomedical informatics applications for precision management of neurodegenerative diseases. Alzheimers Dement (N Y) 4:357–365

Moore JH, Asselbergs FW, Williams SM (2010) Bioinformatics challenges for genome-wide association studies. Bioinformatics 26(4):445–455

Muhammad A, Rajpoot N (2007) Classification of potential nuclei in prostate histology images using shape manifold learning. In 2007 international conference on machine vision

Naik S et al (2007) Gland segmentation and computerized gleason grading of prostate histology by integrating low-, high-level and domain specific information. In Proceedings of 2nd workshop on microsopic image analysis with applications in biology, Piscataway, NJ

Naik S, et al (2008) Automated gland and nuclei segmentation for grading prostate and breast cancer histopathology. p 284–287

Naithani N et al (2021a) Precision medicine: concept and tools. Med J Armed Forces India 77(3): 249–257

Naithani N et al (2021b) Precision medicine: uses and challenges. Med J Armed Forces India 77(3): 258–265

Nguyen HT et al (2012) Analysis of influenza viruses from patients clinically suspected of infection with an oseltamivir resistant virus during the 2009 pandemic in the United States. Antivir Res 93(3):381–386

Oxtoby NP et al (2017) Data-driven sequence of changes to anatomical brain connectivity in sporadic Alzheimer's disease. Front Neurol 8:580

Pool R, Esnayra J (eds) (2000) Bioinformatics: converting data to knowledge: workshop summary. Washington, DC

Rubinov M, Sporns O (2010) Complex network measures of brain connectivity: uses and interpretations. NeuroImage 52(3):1059–1069

Seow A et al (2004) Trends in cancer incidence in Singapore 1968–2002. Singapore Cancer Registry 2004, report

Sethi P, Theodos K (2009) Translational bioinformatics and healthcare informatics: computational and ethical challenges. Perspect Health Inf Manag 6(Fall):1h
Sinha U et al (2002) A review of medical imaging informatics. Ann N Y Acad Sci 980:168–197
Smith GJ et al (2009) Origins and evolutionary genomics of the 2009 swine-origin H1N1 influenza A epidemic. Nature 459(7250):1122–1125
Stone A (2016) Precision medicine: health care tailored to you. The White House Blog
Talwar P et al (2014) Genomic convergence and network analysis approach to identify candidate genes in Alzheimer's disease. BMC Genomics 15(1):199
Tang A et al (2022) Translational bioinformatics to enable precision medicine for all: elevating equity across molecular, clinical, and digital realms. Yearb Med Inform 31(1):106–115
Teverovskiy M et al (2004) Improved prediction of prostate cancer recurrence based on an automated tissue image analysis system. vol 1. p 257–260
van der Vries E et al (2011) Multidrug resistant 2009 a/H1N1 influenza clinical isolate with a neuraminidase I223R mutation retains its virulence and transmissibility in ferrets. PLoS Pathog 7(9):e1002276
Wang ZG, Zhang L, Zhao WJ (2016) Definition and application of precision medicine. Chin J Traumatol 19(5):249–250

Role of Bioinformatics in Data Mining and Big Data Analysis

14

Santosh Kumar Mishra, Avinash Singh, Krishna Bihari Dubey, Prabir Kumar Paul, and Vijai Singh

Abstract

In the past few decades, tremendous growth has been reported in the biological data due to development in the area of genomics, proteomics, microarray as well as biomedical imaging. These biological data are rapidly increasing but due to the availability of limited tools and techniques, the scientific community is able to generate relevant information from this data to a very limited extent. Due to advancements in the area of information technology, data mining and big data analysis tools are being used for the generation of significant results from biological databases to enrich the bioinformatics knowledge for storing, analyzing, and utilizing these data. With the help of data mining techniques and models, this has been possible to identify novel patterns from large-scale biological data and shifted the focus of the research community towards data-dependent discovery. In this chapter, we tried to give a brief insight into different

S. K. Mishra (✉)
Department of Life Sciences, Sharda School of Basic Sciences and Research, Sharda University, Greater Noida, Uttar Pradesh, India

A. Singh
Department of Biotechnology, Meerut Institute of Engineering & Technology, Meerut, Uttar Pradesh, India

K. B. Dubey
Department of Computer Science and Engineering, ABES Institute of Technology, Ghaziabad, Uttar Pradesh, India

P. K. Paul
Department of Biotechnology, Manav Rachna International Institute of Research and Studies, Faridabad, Haryana, India

V. Singh
Department of Biosciences, School of Science, Indrashil University, Mehsana, Gujarat, India

V. Singh, A. Kumar (eds.), *Advances in Bioinformatics*,
https://doi.org/10.1007/978-981-99-8401-5_14

processes of data exploration of biological data for establishing a bridge between data mining techniques and bioinformatics.

Keywords

Genomics, proteomics, microarray, biological data · Sequences · Bioinformatics · Data mining · Analysis · Biomedical imaging

14.1 Introduction

Bioinformatics plays a crucial role in data mining and big data analysis by providing the tools, techniques, and methodologies to handle and extract valuable information from the vast amount of biological data generated through various high-throughput technologies. Bioinformatics also helps in the organization and management of biological data in large-scale databases. These databases store a vast array of genomic, proteomic, and metabolomics data, making them readily accessible for analysis. Data mining has a wide array of applications in bioinformatics, aiding in the discovery and interpretation of complex biological data. Data mining helps in the discovery and understanding of genetic sequences. Techniques such as clustering are useful in determining patterns within sequences and identifying similarities or differences between different genes or organisms. Data mining is a process used to extract useful information from large datasets. It involves methods at the intersection of artificial intelligence, machine learning, statistics, and database systems. In bioinformatics, data mining is primarily used to uncover hidden patterns and correlations in biological data, such as genomic sequences, protein structures, and medical records (Herland et al. 2014). While data mining provides powerful tools for bioinformatics, it also presents several challenges. The complexity and high dimensionality of biological data, as well as the need for robust and interpretable models, are ongoing issues. However, as machine learning and artificial intelligence technologies continue to advance, the capacity for data mining in bioinformatics will likely improve. The fact is that the growth speed and heterogeneous nature of data make it more challenging to store and handle these biomedical data in comparison to conventional data analysis methods (Campbell et al. 2008). Therefore, there is a need to create better powerful methodologies based on good theoretical knowledge and practical tools for analyzing and exploring meaningful information from complex biological data. In this book chapter we have tried to give a brief insight of the role of data mining in bioinformatics and subsequently its application in the generation of knowledge through the study of large-scale databases.

14.2 Evolution of Large-Scale Databases

The volume of data generated in life science research is rapidly due to extensive work in the area of life science and biomedical research. With the digitization of all areas, and the availability of high-throughput devices due to the reduction of cost huge volume of data is increasing everywhere including bioinformatics studies. It was estimated that the size of a single sequenced human genome is roughly 200 gigabytes (Hashemi et al. 2018). The trend in the increasing data volume is supported by decrease in the computing cost and an increase in the facilities of analytical tools and techniques. This also helps in the emergence of big data technologies. This has also been observed that biologist presently no longer uses traditional laboratories to discover novel biomarkers for different diseases. They rely on genomic data for different research activities. Automated genome sequencing now providing new era in the big data generated in bioinformatics. The intelligent implication of the data may accelerate biological knowledge discovery. Data mining and other related computational approaches attempt to find reliable and useful patterns in large amounts of data (Hashemi et al. 2018).

14.3 Biological Data Mining

In any data mining processes behind the data sets much more important thing requires including large capacity data storage devices as well as advanced analysis tools. During this practice machine learning and data, mining approaches play an essential and necessary role. Effective data mining features ensure to achieve accurate and reliable performance. This has been observed that human analysis and abstraction may not be suitable for large-scale data analysis in many instances. The growth rate of data is much faster than that of conventional and manual analysis technology. Therefore, if we are unable to translate the information in an accurate reliable and user-friendly representation the meaning of the existence of such technology is no mean. So in order to make optimum use of such type of data to help clinical diagnosis as well as determine the clinical impact of drugs on experimental data, the need to provide well-optimized and automatic data analysis tools for the analysis of a large amount of data. With the development in the past few decades in exploring and progressing in the area of bioinformatics, many advanced machine-learning tools are used for data investigation and analysis (Yang et al. 2020). In recent times data mining play a significant role in many biomedical studies, such as biomedical electronics and nervous systems, computational biology, biological biomedical imaging, image processing, and visualization and biomedical modeling, etc.

14.4 Data Mining Applications

Data mining applications have significantly contributed to biological sciences, helping to uncover hidden patterns and correlations within vast amounts of biological data. Data mining techniques are widely used in genomics and proteomics for gene finding, prediction of gene expression, protein function prediction, protein-protein interactions, and understanding genetic pathways. Data mining tools can find patterns in large-scale DNA sequences, and through machine learning, can identify genes and their functions. Data mining tools is also helpful to understand how an individual's genetic makeup affects their response to drugs. It helps in drug discovery and design, predicting drug response, and in the development of personalized medicine. Data mining techniques can also serve in identifying the likelihood of diseases in individuals based on genetic or environmental factors. It can also play a significant role in predicting disease progression as well as personalizing treatment plans. In recent years data mining has proved its usefulness in analyzing biological images like MRIs, X-rays, microscopy, etc. Efficient and interactive data mining is used to analyze and interpret complex neuroimaging and neural signal data. It helps in understanding brain structure, function, and neurological disorders. Cellular metabolites can also be easily identified using complex analytical techniques. It also helps in the better understanding of disease mechanisms and identifying novel biomarkers for disease diagnosis (Lan et al. 2018). The application of data mining in biological sciences is vast and continuously expanding as the amount of biological data is growing. These techniques play a very crucial role in understanding complex biological systems as well as predicting disease outcomes, and discovering new drugs. Researchers proposed that a combination of high-dimensional bioinformatics analysis by using an experimental validation process may be useful to achieve translational neuroscience and related applications which includes biomarker discovery, therapeutic development as well as elucidation of disease mechanisms (O'Connor et al. 2023).

14.5 Data Mining Process

The process of data mining in bioinformatics involves many complex steps, which are similar to the general data mining process. Initially, the problem is defined or identified for particular purposes such as genome sequence analysis or protein structure prediction. In the next steps, necessary data is collected. During this process, we have to gather DNA or Protein sequences from different databases such as GenBank, PDB, NCBI, and EMBL, etc. Since many biological data are incomplete and noisy in nature, therefore, these data must be cleaned and converted into appropriate format before proceeding for data analysis (Branco and Choupina 2021). In the next step, actual data mining applies machine learning algorithms and statistical methods to discover patterns and correlations in the data. This method involves supervised learning techniques (e.g., decision trees, SVMs, neural networks) for classification and prediction of the defined tasks, or unsupervised techniques (e.g., clustering, PCA) for exploration and discovering the appropriate tasks. Once the pattern is identified, they need to be evaluated and interpreted in a biological context. This process involves existing cross-sectional evaluation with existing biological literature, visualization or integration with other pre-existing knowledge. The final step is to validate the findings. This could involve applying the model to a separate test dataset and conducting biological experiments to confirm predictions. Once validated, the data mining model can be deployed in the real world. For instance, a predictive model might be used to solve complex biological problems, i.e., clinical setting to diagnose diseases or predict treatment outcomes. Each of these steps can involve many specific techniques, and the exact process can vary depending on the nature of the problem and the available data (Varshney et al. 2022). But in all cases, successful data mining in bioinformatics requires a strong understanding of both computational knowledge and biological principles.

14.6 Techniques in Data Mining

Due to recent technological advancements in data analysis tools scientists acquire multimodal data from different biological applications. That data may be in the form of images, signals, and sequences. These data are in huge amounts and very complex in nature. In data mining practices pattern recognition is a major challenge in the enormous amount of data, therefore this needs a data-intensive machine learning approach to draw a final conclusion. ANN-based learning system emerges as a well-known system for pattern recognition, which is supported by deep learning also (Mahmud et al. 2021). Recent technological advancements in data acquisition tools allowed life scientists to acquire multimodal data from different biological application domains. Categorized in three broad types (i.e., images, signals, and sequences), these data are huge in amount and complex in nature. Mining such an enormous amount of data for pattern recognition is a big challenge and requires sophisticated data-intensive machine-learning techniques (Mahmud et al. 2021). Data mining plays an important role in different human activities because it extracts unknown useful patterns or knowledge. Due to its unique capabilities, data mining techniques become essential tools in the large number of application domains such, as medical, bioinformatics, and life science study, etc. (Gupta and Chandra 2020).

14.7 Limitations of Biological Data Mining

Biological data mining, or bioinformatics, has become a key part of biological research, particularly in areas like genomics, proteomics, and drug discovery. It involves the application of data mining techniques to biological data to uncover new knowledge. However, several limitations come with biological data mining. Biological data is complex and diverse, spanning from DNA sequences to 3D protein structures, to medical images. It can be challenging to devise methods that can handle such a wide variety of data types effectively. With the advancement of various technologies like next-generation sequencing, the amount of biological data being generated is increasing exponentially. This presents challenges in storing, managing, and analyzing these vast amounts of data. Biological data often contain noise and errors. For instance, DNA sequences may have reading errors, and patient data may have missing or incorrect entries. This makes it challenging to perform accurate data mining. The high dimensionality of biological data, combined with the need to perform complex computations like sequence alignment or structure prediction, means that bioinformatics tasks can be very computationally intensive. It can be difficult to interpret the results of data mining in a meaningful way in the biological context. For instance, a pattern discovered in gene expression data might not have a clear biological interpretation (Li and Ng 2009). Biological data often contains sensitive information. Therefore, privacy and security are important concerns in biological data mining. There is a lack of standard formats for many types of biological data, which makes it difficult to integrate data from different sources. Biological systems are dynamic and constantly changing. Therefore, a model that accurately describes a biological system at one point in time may not be accurate later. The complexity of biological systems and experiments often makes it difficult to replicate findings, which is a key part of the scientific process. Despite these challenges, biological data mining has already led to many significant discoveries, and ongoing research is continuously developing new methods to address these limitations.

14.8 Conclusion

Bioinformatics plays a critical role in harnessing the power of data mining and big data analysis to decode the complexity of biological systems. As we continue to generate biological data at an unprecedented rate, bioinformatics will remain central to extracting meaningful insights from this data deluge, leading to improved understanding of life and disease processes, and accelerating discoveries in biomedical research. In an era defined by big data, bioinformatics stands at the forefront, providing the tools, methodologies, and frameworks needed to make sense of vast amounts of biological data. Through data mining and big data analysis, bioinformatics is unlocking a deeper understanding of biology and paving the way for future breakthroughs in healthcare and medicine. While challenges exist, the potential of bioinformatics in leveraging big data is immense and largely unexplored, offering exciting opportunities for future research and discovery. As technologies, continue to evolve, so too will the role of bioinformatics in data mining and big data analysis, promising an exciting future for this rapidly evolving field.

References

Branco I, Choupina A (2021) Bioinformatics: new tools and applications in life science and personalized medicine. Appl Microbiol Biotechnol 105:937–951

Campbell AJ, Cook JA, Adey G, Cuthbertson BH (2008) Predicting death and readmission after intensive care discharge. Br J Anaesth 100(5):656–662
Gupta MK, Chandra P (2020) A comprehensive survey of data mining. Int J Inf Technol 12(4): 1243–1257
Hashemi A, Vikalo H (2018) Evolutionary self-expressive models for subspace clustering. IEEE Journal of Selected Topics in Signal Processing 12(6):1534–1546
Herland M, Khoshgoftaar TM, Wald R (2014) A review of data mining using big data in health informatics. J Big Data 1(1):1–35
Lan K, Wang DT, Fong S, Liu LS, Wong KK, Dey N (2018) A survey of data mining and deep learning in bioinformatics. J Med Syst 42:1–20
Li XL, Ng SK (eds) (2009) Biological data mining in protein interaction networks. Igi Global
Mahmud M, Kaiser MS, McGinnity TM, Hussain A (2021) Deep learning in mining biological data. Cogn Comput 13:1–33
O'Connor LM, O'Connor BA, Lim SB, Zeng J, Lo CH (2023) Integrative multi-omics and systems bioinformatics in translational neuroscience: a data mining perspective. J Pharm Anal 13:836
Varshney S, Bharti M, Sundram S, Malviya R, Fuloria NK (2022) The role of bioinformatics tools and technologies in clinical trials. In: Bioinformatics tools and big data analytics for patient care. Chapman and Hall/CRC, pp 1–16
Yang A, Zhang W, Wang J, Yang K, Han Y, Zhang L (2020) Review on the application of machine learning algorithms in the sequence data mining of DNA. Front Bioeng Biotechnol 8:1032

Unveiling the Dynamic Role of Bioinformatics in Automation for Efficient and Accurate Data Processing and Interpretation

15

Ghlomareza Abdi, Mukul Jain, Mukul Barwant, Reshma Tendulkar, Mugdha Tendulkar, Mohd Tariq, and Asad Amir

Abstract

The field of bioinformatics has witnessed remarkable advancements in recent years, enabling efficient and accurate processing and interpretation of large-scale biological data. In this article, we delve into the dynamic role of bioinformatics in

The original version of this chapter was revised. A correction to this chapter can be found at https://doi.org/10.1007/978-981-99-8401-5_18

G. Abdi
Department of Biotechnology, Persian Gulf Research Institute, Persian Gulf University, Bushehr, Iran
e-mail: abdi@pgu.ac.ir

M. Jain
Cell and Developmental Biology Lab, Centre of Research for Development, Parul University, Vadodara, Gujarat, India

Department of Life Sciences, Parul Institute of Applied Sciences, Parul University, Vadodara, Gujarat, India

M. Barwant (✉)
Department of Botany, Sanjivani Arts, Commerce and Science College, Kopargaon, Maharashtra 423603, Ahmednagar, Maharashtra, India

R. Tendulkar
Vivekanand Education Society's, College of Pharmacy, Mumbai, Maharashtra, India

M. Tendulkar
K. J. Somaiya Medical College and Research Centre, Mumbai, Maharashtra, India

M. Tariq
Department of Life Sciences, Parul Institute of Applied Sciences, Parul University, Vadodara, Gujarat, India

A. Amir
Department of Biotechnology and Microbiology, Meerut Institute of Engineering and technology, Meerut, Uttar Pradesh, India

V. Singh, A. Kumar (eds.), *Advances in Bioinformatics*,
https://doi.org/10.1007/978-981-99-8401-5_15

automation, specifically focusing on its impact on data processing and interpretation for genomics research. With the rapid advancements in high-throughput technologies, such as next-generation sequencing, the amount of genomic data generated has grown exponentially. Managing and analysing such vast amounts of data manually is impractical, time-consuming, and prone to errors. Here, we highlight the pivotal role of bioinformatics in automating data processing pipelines for genomic analysis. By developing sophisticated algorithms and tools, bioinformatics facilitates efficient data handling, quality control, read alignment, and variant calling. Automation not only accelerates the analysis process but also enhances the reproducibility and reliability of results.

Furthermore, bioinformatics plays a crucial role in automating genome annotation and functional analysis. Through integration of diverse data sources and computational approaches, bioinformatics tools automate the identification of genes, regulatory elements, and functional regions within genomes. This automation enables researchers to swiftly annotate and interpret newly sequenced genomes, as well as compare them with existing genomic knowledge.

Additionally, the field of bioinformatics has made significant strides in automating variant calling and interpretation. Detecting genetic variations accurately is essential for understanding the genetic basis of diseases and traits. Bioinformatics algorithms and pipelines automate the process of variant calling, leveraging reference genomes and statistical models to identify and classify genomic variations. Moreover, automated variant interpretation tools integrate functional annotations, population databases, and disease association studies to prioritize and elucidate the functional impact of identified variants.

In conclusion, bioinformatics plays a dynamic role in automation for efficient and accurate data processing and interpretation in genomics research. The automation of data handling, genome annotation, and variant calling empowers researchers to analyse large-scale genomic datasets with enhanced speed, accuracy, and reproducibility. This article highlights the vital contributions of bioinformatics in advancing genomic analysis and underscores its potential to unravel the complexities of biological systems through automation.

Keywords

Bioinformatics · Automation · Genomics · Transcriptomics

15.1 Introduction

A dynamic ecosystem of bioinformatics tools, large-scale storage, and high-performance computing (HPC) resources make up the e-infrastructure that bioinformaticians must increasingly deal with. Analyses frequently include a number of software tools being used sequentially on input data, and because of the magnitude and complexity of the data involved, these analysis processes are typically carried out on a server or computer cluster (Bux and Leser 2013). Workflow is a frequent term for such a multi-step process. Scientific Workflow Management Systems, which may ease the design and execution of workflows and pipelines in high-performance computing contexts such as local clusters or distributed computing clouds, may be useful for carrying out such analysis effectively (Spjuth et al. 2015). In bioinformatics, a variety of workflow systems are available. Academic HPC resources typically consist of Linux-based compute clusters with batch (queueing) systems for work scheduling. A new technology that provides virtualized environments and the ability to run customized virtual machine images is cloud computing (VMI). The ability to sequence more nucleotides for a given dollar has increased exponentially, and genome sequencing technology has also advanced significantly. But up until a few years ago, DNA sequencing took a little longer to double than computation and storage capacity did to grow. The ecology of genomic informatics benefited greatly from this. The long-term trends enabled the archival databases and the value-added genome distributors to upgrade their capacity quicker than the global sequencing labs could update theirs, so they did not need to be concerned about running out of disc storage space. Because they were constantly a step ahead of the curve, computational biologists did not worry about not having access to powerful enough networks or computing clusters (Stein 2010). It has been studied new opportunities for processes, such as the packaging of full studies or pipelines as VMIs (Schatz et al. 2010). The breadth of scientific inquiry in modern bioinformatics mirrors the breadth of biological research. Sequence analysis (including, for example, sequence alignment, gene discovery, and phylogenetics), determination of the three-dimensional protein structure, visualization, pathway analysis and reconstruction, modelling and simulation of molecular processes, construction of genome maps, statistical analysis of experimental data, development of ontologies, and database development are all active areas of bioinformatics research. Although biological databases have been useful for many genomics applications, accessing these data requires caution. The creation and use of (statistical) algorithms for data processing and interpretation is a significant field of bioinformatics study. These algorithms span a wide range of topics such sequence analysis, visualization (pathway maps, genomic maps, protein structures), statistical analysis of experimental data, and many others. A wide range of topics are covered by these algorithms, including sequence analysis, visualization (pathway maps, genomic maps, protein structure), statistical analysis of experimental data, and modelling and simulation of cellular processes. Most of Algorithms are created to address novel questions and issues brought up by the generation of genome-wide data sets (Van Kampen and Horrevoets 2006).

Diseases including cancer, hepatitis, HIV, and others are spreading quickly and becoming more severe, leading to significant morbidity and mortality. Clinical trials are carried out to determine the safety and effectiveness of pharmaceuticals, whereas clinical research involves the discovery and development of drugs. The identification, validation, and lead optimization of targets are the first steps in the lengthy process of drug discovery. Preclinical trials, extensive clinical trials, and finally post-marketing vigilance for drug safety come after this. Software and bioinformatics technologies are particularly important for both medication development and drug discovery. Data management during clinical trials, the creation of new knowledge about health and disease, and the utilization of clinical data in secondary research are all included (Gill et al. 2016). By using effective statistical algorithms, logical approaches for target selection, validation, and optimization, and computer science applications in biology, bioinformatics might enhance drug discovery. Making databases, predicting protein function, modelling protein structure, identifying the coding regions of nucleic acid sequences, finding suitable drug compounds from a large pool, performing data mining, analysing, and interpreting data faster, and reducing the time and cost of drug discovery are all made possible by computers and software tools (Zerhouni 2006).

A wealth of evidence relating gene activity to disease has been made available by recent advances in genomics. The amount of a protein, as well as its ultimate structure and state of activity, cannot be fully determined from gene sequence data alone, as is now understood to be the case for a number of reasons. Thousands of samples per day throughput for large-scale proteomics for drug discovery and proteome mapping will necessitate a solution for complete automation of the image analysis workflow (Dowsey et al. 2003). Bioinformatics is essential for managing the enormous amounts of data produced by modern, high-throughput techniques as well as for data integration, analysis, and model prediction. The widespread use of bioinformatics in agricultural applications can help with effective crop breeding and the enhancement of plant resistance to diseases. Researchers must clarify the intricate molecular pathways underlying pathogen infection in order to create novel methods for controlling plant diseases. For its use in agriculture, bioinformatics faces both opportunities and challenges in the age of big data. Learning and creating additional bioinformatics tools will enable efficient breeding and plant resistance studies by integrating all currently available bio information resources. The ongoing expansion of the human population is placing enormous strain on food production systems. Many of the world's ecosystems are already overexploited, and it is impossible to meet the rising food demand by increasing the usage of arable land. The advancement of genomics technology has given breeders tremendous technical support, enabling them to consistently develop new varieties that are more tolerant of their environments and produce larger yields, which has improved the seed replacement rate. The technology of whole genome sequencing has made it possible to sequence a growing number of pathogens and amass vast volumes of genetic information. Consequently, to comprehend disease infection processes and pathogenic targets, which are all factors contributing to plant pathology, bioinformatics methods for evaluating pathogen genomes, effectors, and inter-specific interactions have been established (Mu et al. 2022).

15.1.1 Application of Bioinformatics in Clinical Research

The most important link between developments in medical research technologies and better healthcare is the clinical trial. It is a crucial component of medical research that aims to better understand human disease, as well as its prevention, treatment, and promotion of health. The process of conducting a clinical trial for a novel medication candidate is becoming more and more difficult, expensive, and time-

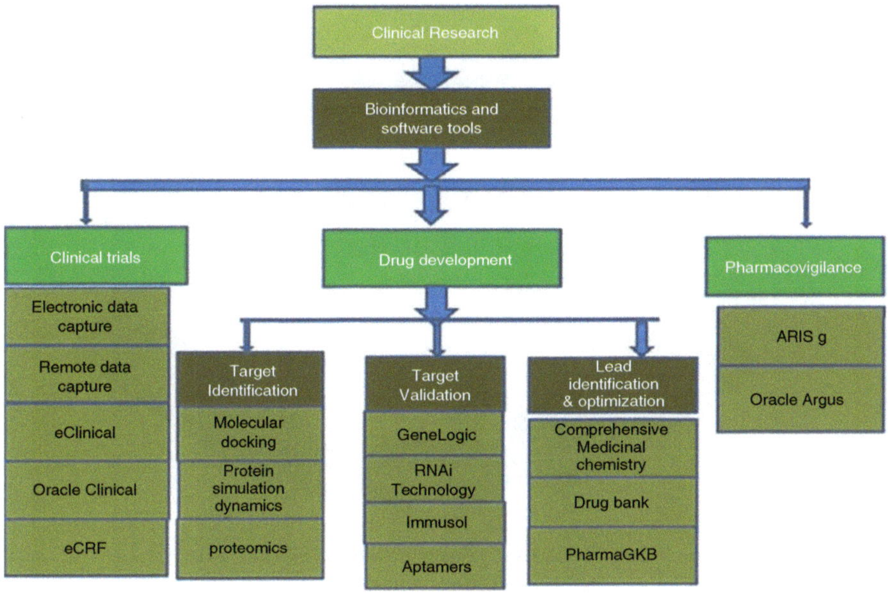

Fig. 15.1 Role of bioinformatics in clinical research (Gill et al. 2016)

consuming. Because of the fierce competition in the pharmaceutical industry, there is a great demand for quick approval of novel drugs. As a result, there is intense pressure on pharmaceutical companies to improve the effectiveness and efficiency of medication discovery and development. The technological initiative is thought to be the only strategy for achieving this objective. The techniques for discovering new drugs and developing them underwent a change with the introduction of electronic clinical trials and computer-aided drug design research (Gill et al. 2016).

A number of clinical trial processes, including target identification, target validation, randomization, data collecting, and data integration, as well as trial management and pharmacovigilance, also become more streamlined, efficient, and manageable. Pharmaceutical businesses and regulators used new technology, which not only increased productivity but also dramatically enhanced data security and the evaluation of clinical data (i.e., turning trial data into information that can be applied). The potential benefits of bioinformatics in clinical research include developing and utilizing a large data strategy for clinical trials, utilizing new techniques to provide patient-centric trial design, bringing evolution to existing processes and systems with new techniques, assisting with case studies from existing data sources for advanced trials, making data sharing simpler, and more Figs. 15.1, 15.2 and 15.3 (Gill et al. 2016).

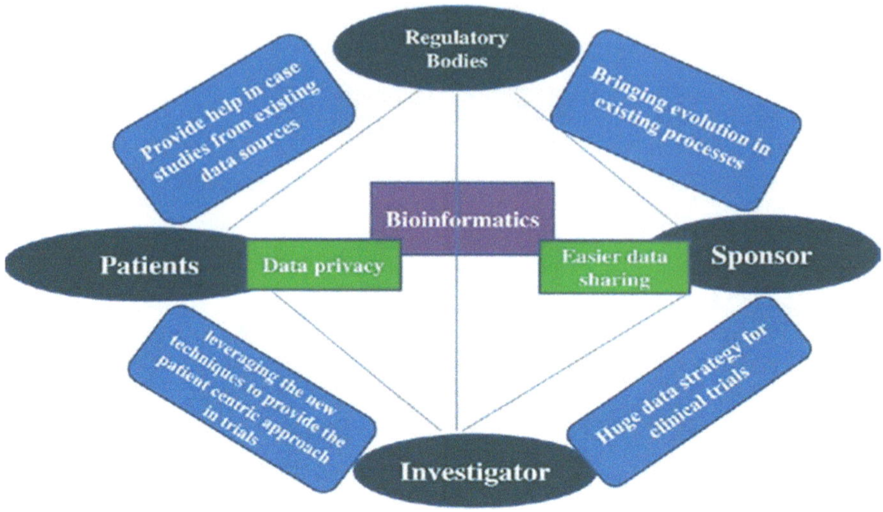

Fig. 15.2 Application of bioinformatics in clinical research (Gill et al. 2016)

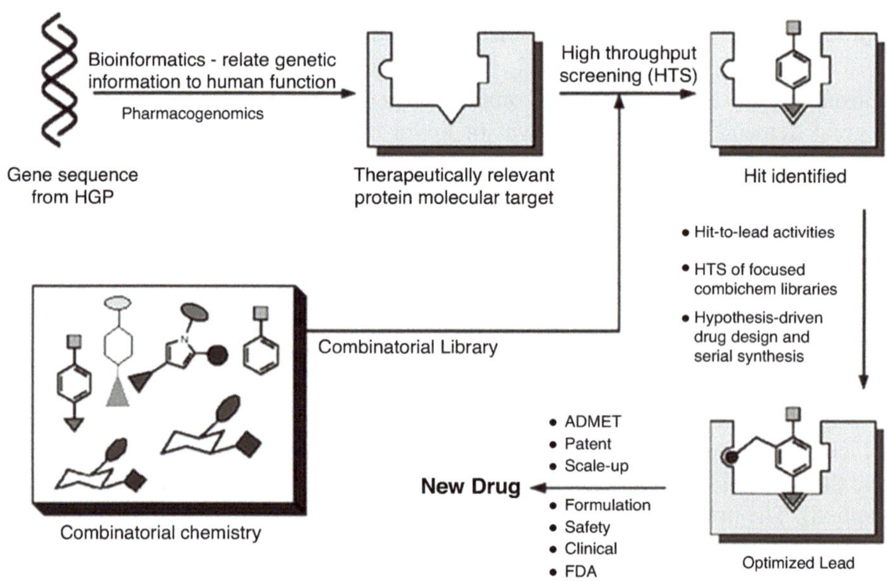

Fig. 15.3 Role of bioinformatics in drug discovery process (Katara 2013)

15.1.2 Bioinformatics Role of Drug Development

To screen vast compound catalogues, many pharmaceutical companies have built automated high throughput screening facilities. Genome-wide analysis is now possible for a variety of scientific domains because to advances in bioinformatics. The

Human Genome Project's tetrabytes of data can be examined by researchers thanks to bioinformatics technologies (Clark and Pickett 2000). Finding out whether and how a specific molecule is directly involved in a disease process with the aid of gene sequence databases, gene expression databases, protein sequence databases, and related analysis tools leads to the discovery of novel and improved medication targets. The time and money required to generate effective pharmacological agents could be decreased with a successful and dependable drug design procedure. In order to identify and exclude candidate compounds that are unlikely to survive the further stages of discovery and development, computational approaches are employed to predict drug-likeness. Approaches based on genetic algorithms and neural networks could forecast how potent a drug would be. The first crucial stage in this process is the capacity to identify new therapeutic targets for additional investigation. According to reports, practically all medications now on the market have as many as 483 pharmacological targets (45% receptors, 28% enzymes, 5% ion channels, and 2% nuclear receptors) (DiMasi et al. 2003; Gill et al. 2016).

The conventional approaches to drug discovery and development are significantly changing as a result of genomics, proteomics, and metabolomics. High-throughput sequencing is extensively used today to identify new therapeutic targets. Nearly every step of drug discovery, drug evaluation, and drug development now involves the use of bioinformatics. This expanding significance is due to the use of bioinformatics tools to anticipate, evaluate, and help interpret clinical and preclinical findings, as well as the role that bioinformatics plays in handling massive volumes of data Both bioinformatics and cheminformatics depend heavily on data and databases (Altschul et al. 1997). Most data searches would be ineffective without a lot of readily available electronic data, and most types of predictive or analytical software could never be created or tested. The quality of biological or chemical data is just as significant as its quantity. Data on gene and protein sequences are increasingly essential to nearly every element of pharmacological research. For instance, standard pathogen sequencing includes Today, it is possible to identify potential protein therapeutic targets or pathogenicity in viruses, parasites, or bacteria impressively quickly and reasonably cheaply (Wishart 2005). Drug metabolism and drug interaction databases are a new class of databases that are likely considerably more useful to pharmaceutical researchers than general metabolism and pathway databases, which are becoming more and more significant in drug development and assessment (Comess and Schurdak 2004). These databases aim to connect the genomic/proteomic data being obtained about the pertinent genes or proteins with the drug compounds themselves and concentrate much more intently on known medicines or drug metabolites. Already, there are a number of commercial drug metabolism or drug interaction databases such as those provided by MDL Information Systems, the University of Washington (Seattle), and Lhasa Ltd., an organization located at the University of Leeds that is not for profit. The MDL METABOLITE system and MDL TOXICITY are maintained and sold by MDL Database (Wishart 2005). Drug development is a difficult, risky, time-consuming, and sometimes very lucrative process. The methodical process through which new candidate medications are found is known as drug discovery. Drug development has traditionally been a challenge for pharmaceutical corporations, who use well-established pharmacology and chemistry-based methods. Drug development is a labour-intensive, expensive, and complex process (Iskar et al. 2012).

Pharmacogenomics and bioinformatics both offer significant assistance in overcoming the cost and time constraints in various ways. Drug-related databases and software are widely available thanks to bioinformatics, and they can be utilized for a variety of processes connected to the process of creating and developing new drugs. In a similar vein, pharmacogenomics gives genome-level data on the varied medication response, which is crucial for pharmaceutical companies to build novel drugs, in addition to orphan drugs (Katara 2013). Moreover, bioinformatics offers methods and algorithms for predicting new drug targets as well as for storing and managing data on existing drug targets. There is hardly any requirement to prove a direct link between a putative target and the disease of concern once "possible" therapeutic targets have been identified. The process of developing new medications is justified by the creation of such a significant relationship. Target validation is a step in this process that bioinformatics is heavily utilizing observe in Fig. 15.3 (Katara 2013).

15.1.3 Role of Bioinformatics in Antibacterial Potential

The development of vaccines and antibacterial medications has advanced significantly since the 1940s, saving many lives. The recent development in organ transplants, intensive chemotherapy, invasive surgeries, liberal and indiscriminate use of antibiotics, and epidemiologic virulence—the spread of resistant strains due to insufficient precautions in the hospitals—has produced a human host with impaired immune systems, which are affected both by antibiotic-resistant strains and other microbes (Casadevall and Pirofski 2000). Despite the existence of powerful vaccines and antibiotics against classical pathogens. In addition to weakening the immune system, the use of antibiotics promotes the emergence of opportunistic infections, which the immune system would normally suppress (Swartz 1994). Some antibiotics disrupt the situation for resource and nutrient competition required to restrict the growth of opportunistic and drug-resistant bacterial strains under the usual conditions by negatively controlling the growth of wild-type bacteria. Simple microbial infection clearance may not always eliminate its long-term clinical effects, and can still result in disease due to immunological damage (Ochman and Moran 2001). For instance, reactive arthritis and rheumatic heart disease may develop as a result of certain bacterial infections in the gastrointestinal tract and streptococcal pharyngitis, respectively (Swartz 1994). There are two ways to combat the bacterial infection: either utilize biostatic antibacterial medications that inhibit the growth of the pathogens or stimulate the immune system by immunizing against the invasive infections. The first strategy relates to the creation of vaccines, whereas the second strategy concerns the creation of antibiotics. Better and faster-acting medications and vaccines are becoming a reality because to advancements in computational and biological techniques. Combinational computational chemistry is one of the recent advances in antibacterial development. It allows for the variation of a 3D structure of an antibacterial compound by computationally modelling the 3D structure using energy-minimization techniques and other molecular modelling techniques to find a better compound that docks to a gene involved in. Automated genomic data extraction and analysis have seen a revolutionary breakthrough over the past 10 years, made possible by a massive increase in computing power. We now have a chance to overcome our inability to combat the threat posed by bacterial resistant strains thanks to the ability to save the genomic and proteomic results in databases and automatically access and evaluate the data at the gene, genome, and proteome level (Wang and Kollman 2001). A small number of potentially pathogenic genes can be examined in wet labs using experimental approaches after being rapidly and cost-effectively trimmed and assessed in silico for different possibilities Genome sequencing, automated preservation and retrieval of genomic and proteomic data, comparative genomics, and proteomics are all key components of the bioinformatics field that are helping to determine the whole function of the genome. With the availability of complete genome sequences, bioinformatics and biochemical analysis both have an integrated and complementary role to play: bioinformatics by reducing the number of potential outcomes and speculating on functionality, and biochemical analysis to validate the speculative results, improve the efficacy, and investigate the solubility, permeability, and diffusion required for drug uptake. Finding similar genes and proteins using bioinformatics methods for comparative study is insufficient because even little structural differences in these proteins might affect how they function and how they bind to different substances (Hagman and Shafer 1995). The scope of the bioinformatics research on vaccinations and antibiotics. Understanding hereditary disorders like cancer and deadly viral infections like HIV is another area where bioinformatics is useful. However, the use of bioinformatics to treat viral and genetic disorders (Bansal 2008).

Understanding the genomic machinery is essential for rational medication design since various infections use different mechanisms and gene sets. Numerous elements, such as gene functionality at the domain level, are necessary to comprehend the pathogenicity (Jeffery 2003). Analysis of the conserved and non-conserved structural features of receptors involved in host-pathogen contact and adhesion. The identification of genes in the microorganism, determining the function of the gene, putting genes together to reconstruct metabolic and regulatory pathways, comparing pathways to identify essential pathways and pathways specific to pathogenic strains, and figuring out what proteins or substances interact with the control region of the genes and operons to inhibit transcription are all steps in understanding the genome function of the pathogenic strains at the systemic level (Zhou et al. 2004). Comparative genomics has several benefits, including the automatic reconstruction of metabolic pathways and the identification of plasmid genes thought to be involved for pathogenicity (Shokhen et al. 2006). The discovery of genes implicated in widespread signalling pathways is another benefit. Comparative investigation of genomes with similar evolutionary histories has revealed that numerous genes are absent from key pathways in pathogenic strains. In recent years, databases of genes, proteins, and protein domains within genes have also been made available thanks to bioinformatics research. With the help of these databases and pattern-based search methods, the labelling of the genes and proteins in recently sequenced genomes has multiplied. Among the databases are those for genomic sequences (Goto et al. 1998). Drug discovery can also be aided by bioinformatics research by rebuilding regulatory and metabolic pathways and analysing the rate at which they react. The integration of wet-lab biochemical procedures, comparative genomics, and proteomics—computational analysis of gene array data—is necessary for this research much as it is for genome sequencing (Bansal 2001) The wet lab serves as the foundation for defining the pathway since it offers information on known reactions, reaction rates, the activities of the original enzymes, substrate information, and known metabolic and signalling pathways. Reconstructing metabolic pathways and identifying gene clusters implicated in signalling pathways have both been accomplished using comparative genomics and cluster analysis of microarray data (Benson et al. 2005). The next step is to compare the pathways of two bacteria in order to find crucial pathways and specific pathways found in various microbes. However, the binding data accessible from the wet labs places a cap on the databases of protein-protein interactions and protein-DNA interactions. By aligning and comparing the regulatory areas before the orthologous genes in evolutionary-close genomes, numerous bioinformatics tools have recently been created to conjecture the binding sites. The binding locations have been well estimated by these alignments' conserved areas. The use of molecular modelling techniques in bioinformatics research is another crucial component. These techniques can: make it easier for bacteria to absorb antibiotics through the lipid layer by creating new pores or making better use of already-existing ones; identify the proteins that can prevent operons, a group of co-regulated genes, from forming channels; and/or facilitate the uptake of antibiotics through the lipid layer (Bansal 2008).

15.2 Bioinformatics Tools for Data Analysis and Automation

15.2.1 Introduction to Bioinformatics Tools

In recent years, the field of bioinformatics has seen remarkable advancements in automation, revolutionizing the way biological data is processed, analysed, and interpreted. Automation has become an indispensable tool in bioinformatics, allowing researchers to handle vast amounts of data and perform complex analyses with unprecedented efficiency and accuracy. This chapter explores the pivotal role of bioinformatics in automation, highlighting its applications in data handling, analysis, and interpretation, and discussing the challenges and future prospects of this rapidly evolving field.

Bioinformatics tools are software applications and algorithms specifically designed to process, analyse, and interpret biological data (Table 15.1). These tools play a crucial role in extracting valuable insights from vast amounts of genomic, proteomic, and other biological data, enabling researchers to unravel the complexities of biological systems and make meaningful discoveries.

Table 15.1 Comparison of bioinformatics tools for data analysis and automation

Sequencing platform	Read length	Sequence yield per run	Run time	Error Rate (%)	Instrument expenses (USD)
First generation sequencing					
ABI Sanger	75 bp	1.2–1.4 Gb	14 day	0.30	690,000
Second generation sequencing					
Ion torrent PGM	200 bp	20–50 Mb on 314 chip	2 h	1.71	80,000
Illumina MiSeq	300 bp	1.5–2 Gb	27 h	0.80	125,000
Genexus system	400 bp	19.2–24 Gb per chip	30 h for one chip	<1.0	288,000
Illumina HiSeq 2000	150 bp	600 Gb	11 days	0.26	750,000
Third generation sequencing					
Oxford Nanopore	>5000 bp	2 Gb	48 h	12.0	1000
Pac bio RS	1300 to >10,000 bp	100 Mb	2 h	12.6	750,000

Bioinformatics tools can be broadly categorized into several areas:

15.2.1.1 Sequence Analysis Tools

These tools focus on analysing DNA, RNA, and protein sequences. They include sequence alignment tools like BLAST and ClustalW, which compare a query sequence against a database of known sequences to identify similarities and evolutionary relationships. Genome assemblers, such as Velvet and SPAdes, are used to reconstruct complete genomes from fragmented DNA sequencing data. Gene prediction tools like GeneMark and AUGUSTUS help identify protein-coding genes within genomic sequences (Mount 2014).

15.2.1.2 Structural Bioinformatics Tools

Structural bioinformatics tools are employed to study the three-dimensional structures of biological macromolecules, particularly proteins. Protein structure prediction tools like I-TASSER and Phyre2 use computational algorithms to predict protein structures based on sequence information and known protein structures. Protein-ligand docking tools like AutoDock and Vina simulate the binding of small molecules (ligands) to protein targets, aiding in drug discovery and understanding molecular interactions. Molecular visualization tools such as PyMOL and Chimera provide platforms to visualize and analyse protein structures (Buffalo 2015).

15.2.1.3 Genomics and Transcriptomics Tools

These tools focus on the analysis of large-scale genomic and transcriptomic data. Differential gene expression analysis tools like DESeq2 and edgeR help identify genes that are differentially expressed between different biological conditions. Genome browsers like the UCSC Genome Browser and Ensembl provide interactive platforms to explore genome sequences, annotations, and various genomic datasets (Jones and Pevzner 2004). Variant calling tools like GATK and SAMtools detect genetic variants from DNA sequencing data, including single nucleotide polymorphisms (SNPs) and structural variants.

15.2.1.4 Systems Biology and Network Analysis Tools

Systems biology tools aim to understand biological systems as a whole by considering the interactions among genes, proteins, and other molecular components. Pathway analysis tools like KEGG and Reactome integrate various data sources and provide functional annotations, aiding in the identification and analysis of biological pathways. Network visualization and analysis tools like Cytoscape enable the study of gene regulatory networks, protein-protein interaction networks, and metabolic networks (Mount 2014).

Flux balance analysis tools employ mathematical modelling to simulate and analyse metabolic networks, predicting metabolic fluxes and optimizing cellular phenotypes.

15.2.1.5 Data Integration and Analysis Platforms

These platforms integrate multiple bioinformatics tools, databases, and analysis pipelines to provide comprehensive and user-friendly interfaces for researchers. Examples include Galaxy, an open-source platform for creating reproducible workflows by integrating diverse bioinformatics tools, and Bioconductor, a collection of R packages and tools for the analysis of high-throughput genomic data (Attwood et al. 1999).

15.2.2 Data Analysis and Automation in Bioinformatics

Data analysis is a fundamental aspect of bioinformatics, and automation has played a transformative role in streamlining and accelerating this process. This section explores the applications of automation in various areas of data analysis within bioinformatics.

15.2.2.1 Genome Assembly and Annotation

Genome assembly refers to the process of reconstructing complete genomes from fragmented DNA sequences. Automation tools, such as genome assemblers, have significantly improved the efficiency and accuracy of this process. Similarly, genome annotation, which involves identifying functional elements within a genome, can be automated using computational algorithms that analyse sequence features and compare them against existing databases (Sachdeva and Kumar 2014).

15.2.2.2 Variant Calling and Analysis

Identifying genetic variations or mutations within genomes is crucial for understanding disease susceptibility and personalized medicine. Automation tools, such as variant callers, enable the detection of single nucleotide polymorphisms (SNPs), insertions, deletions, and other genomic variations. These tools can analyse large datasets and compare them to reference genomes, making variant analysis faster and more reliable.

15.2.2.3 Comparative Genomics

Comparative genomics involves comparing and analysing the genomes of different species to identify conserved regions, evolutionary relationships, and functional elements. Automation tools, such as sequence alignment algorithms and phylogenetic analysis pipelines, allow for efficient and comprehensive comparative genomics studies. These tools automate the process of sequence alignment, tree building, and evolutionary analysis, facilitating large-scale comparative genomics studies (Schadt et al. 2010).

15.2.2.4 Structural Bioinformatics

Structural bioinformatics focuses on analysing the three-dimensional structures of biological macromolecules, such as proteins and nucleic acids. Automation has greatly impacted structural bioinformatics, enabling the prediction of protein structures, protein-ligand docking, and drug design.

15.2.2.5 Protein Structure Prediction

Automated methods for protein structure prediction, such as homology modelling and ab initio methods, utilize computational algorithms to predict the 3D structure of proteins based on their amino acid sequences. These methods have accelerated the process of protein structure determination, which is crucial for understanding protein function and drug discovery (Fig. 15.4).

15.2.2.6 Protein-Ligand Docking

Automation tools in protein-ligand docking facilitate the prediction of how a small molecule (ligand) interacts with a protein target. These tools use algorithms to predict the binding affinity and orientation of ligands within protein binding sites. Automation has enabled high-throughput virtual screening of potential drug candidates, significantly expediting the drug discovery process.

15.2.2.7 Drug Design and Discovery

Automation plays a vital role in computer-aided drug design (CADD) by automating various steps, including virtual screening, lead optimization, and toxicity prediction. By employing algorithms and machine learning approaches, automation tools can efficiently analyse large chemical libraries and predict the properties and activities of potential drug candidates.

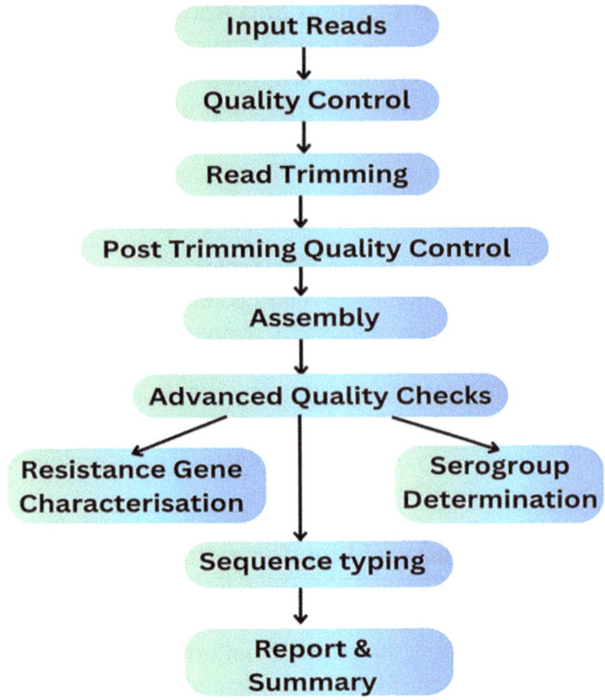

Fig. 15.4 Workflow of bioinformatics in automation process

15.2.2.8 Systems Biology and Network Analysis

Systems biology aims to understand biological systems as a whole, considering interactions among genes, proteins, and other molecular components. Automation tools have facilitated the analysis and modelling of complex biological networks.

15.2.2.9 Biological Pathway Reconstruction

Automated methods for reconstructing biological pathways integrate diverse data sources, such as gene expression data, protein-protein interactions, and literature mining, to infer the relationships and interactions among genes and proteins. These methods help in understanding the underlying mechanisms of diseases and identifying potential therapeutic targets (Jenney and Petritis 2010).

15.2.2.10 Gene Regulatory Network Inference

Automation tools enable the inference of gene regulatory networks from gene expression data. By employing computational algorithms and statistical models, these tools can identify regulatory interactions among genes and unravel the complex regulatory networks that govern cellular processes.

15.2.2.11 Metabolic Modelling and Simulation

Automation in metabolic modelling allows the construction and simulation of metabolic networks. These models can predict the metabolic behaviour of organisms under different conditions, facilitating the study of cellular metabolism and the identification of metabolic engineering targets (Jenney and Petritis 2010).

Automation has transformed data analysis in bioinformatics, making it more efficient and enabling researchers to extract valuable insights from biological data. Automation tools in sequence analysis, structural bioinformatics, systems biology, and network analysis have revolutionized various aspects of bioinformatics research and paved the way for advancements in personalized medicine, drug discovery, and understanding complex biological systems.

15.2.3 Popular Bioinformatics Tools for Automation

15.2.3.1 Galaxy

Galaxy is an open-source, web-based platform that enables the creation and execution of reproducible bioinformatics workflows. It provides a user-friendly interface for designing and running data analysis pipelines. Galaxy integrates a vast collection of bioinformatics tools and resources, allowing researchers to automate complex analysis tasks without the need for extensive programming skills. Workflows in Galaxy can be easily shared and reused, promoting collaboration and reproducibility (Afgan et al. 2018).

15.2.3.2 Snakemake

Snakemake is a workflow management system that facilitates the creation of scalable and reproducible bioinformatics pipelines. It uses a Python-based domain-specific language to define rules and dependencies between tasks. Snakemake automatically manages the execution of tasks based on input/output files, ensuring efficient and reliable pipeline execution. It supports parallel and distributed computing, making it suitable for large-scale data analysis. With its intuitive syntax and flexibility, Snakemake enables researchers to automate complex bioinformatics workflows with ease (Köster and Rahmann 2012).

15.2.3.3 Nextflow

Nextflow is a bioinformatics workflow management system designed for building and executing data analysis pipelines. It allows researchers to define workflows using a domain-specific language that is highly portable and supports multiple computing environments, including local machines, clusters, and cloud platforms. Nextflow simplifies the integration of diverse bioinformatics tools and resources by providing a unified framework. It offers features like parallelization, fault tolerance, and process isolation, enabling efficient and scalable automation of data analysis workflows (Di Tommaso et al. 2017).

15.2.3.4 Bioconductor

Bioconductor is a collection of open-source R packages specifically designed for the analysis and comprehension of high-throughput genomic data. It provides a comprehensive suite of tools and workflows for genomics, transcriptomics, proteomics, and other biological data types. Bioconductor packages cover a wide range of analysis tasks, including data pre-processing, statistical analysis, visualization, and interpretation. With its extensive set of functions and resources, Bioconductor facilitates the automation of bioinformatics data analysis in the R programming environment (Huber et al. 2015).

15.2.3.5 Bpipe

Bpipe is a lightweight pipeline manager that enables the automation of bioinformatics workflows. It allows researchers to define and execute command-line tools and processes in a scalable and reproducible manner. Bpipe provides a simple yet powerful scripting language for describing pipelines, making it easy to incorporate existing tools into workflows. It supports parallelization, input/output management, and error handling, ensuring reliable and efficient pipeline execution. Bpipe's simplicity and flexibility make it a popular choice for automating various bioinformatics analysis tasks (Sadedin et al. 2012).

These popular bioinformatics tools for automation provide researchers with the means to design, execute, and manage complex data analysis workflows. They offer features and functionalities that enhance reproducibility, scalability, and collaboration in bioinformatics research, ultimately facilitating efficient and reliable data analysis.

15.2.4 Integration of Automation Tools in Data Analysis Workflows

The integration of automation tools into data analysis workflows in bioinformatics is crucial for streamlining and accelerating the analysis process. Here's an explanation of the integration process:

15.2.4.1 Workflow Design

Automation tools are integrated into data analysis workflows right from the design stage. Researchers define the overall structure and steps of the workflow, including data pre-processing, analysis, and result generation. This design involves selecting appropriate automation tools that can handle specific tasks within the workflow (Sakharkar and Sakharkar 2007).

15.2.4.2 Tool Selection

Integration begins by identifying the automation tools that are best suited for each step of the data analysis workflow. These tools can range from workflow management systems (e.g., Galaxy, Snakemake, Nextflow) to specialized bioinformatics software packages (e.g., alignment tools, sequence analysis tools, statistical analysis tools). The choice of tools depends on the specific analysis requirements and the capabilities of the tools to automate those tasks.

15.2.4.3 Parameterization and Configuration

Automation tools allow for the parameterization and configuration of analysis steps. Researchers can set up the desired parameters and customize the tool's behaviour to align with their specific analysis requirements. This flexibility ensures that the workflow can be adapted and fine-tuned as needed, without manual intervention at each step.

15.2.4.4 Data Integration and Transformation

Automation tools facilitate the integration and transformation of diverse data types within the analysis workflow. They enable seamless data retrieval from various sources, such as databases or external repositories, and support the transformation of data into compatible formats for downstream analysis. This integration ensures that different data sources and formats are harmonized and effectively utilized during the analysis (Vayssière and Licznar 2010).

15.2.4.5 Error Handling and Reporting

Automation tools provide mechanisms for error handling and reporting within the workflow. They can detect errors or failures in task execution, provide notifications, and allow for automatic recovery or re-execution of failed tasks. Additionally, automation tools can generate comprehensive reports summarizing the results and analysis steps for documentation and reproducibility purposes (Sachdeva and Kumar 2014).

By integrating automation tools into data analysis workflows, researchers can streamline the entire analysis process, reduce manual intervention, enhance reproducibility, and improve overall efficiency. These tools enable researchers to focus on data interpretation and scientific discovery, rather than spending excessive time on repetitive and time-consuming tasks (Cock and Van Der Lelij 2020). An improvized depiction of automation is summarized in Fig. 15.5.

15.3 Automation in Genome Sequencing and Analysis

The quest for genome sequencing began with breakthroughs introduced in sequencing technology by Friedrick Sanger in 1977. The technological advancements in the field of sequencing took a giant leap in the next few decades, involving the discovery of next-generation sequencing, robust development in the field of bioinformatics etc.

15.3.1 Automation in Genome Sequencing and Analysis

The quest for genome sequencing began with breakthrough introduced in sequencing technology by Friedrick Sanger in 1977. The technological advancements in the field of sequencing took a giant leap in the next few decades involving the discovery of next-generation sequencing, robust development in the field of bioinformatics et cetera (Table 15.2). The whole scientific perspective towards genome sequencing

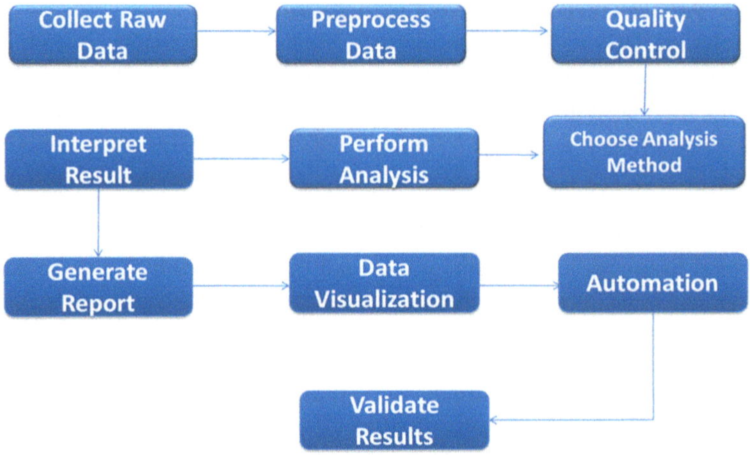

Fig. 15.5 Data analysis and automation in bioinformatics

Table 15.2 Examples of automated genome sequencing platforms

Platform	Technology	Model	Length of read	Throughput (per day)	Company
454	Pyrosequencing	GS FLX^{+}	500–700 bp	700 Mb	454 life sciences (Roche)
SOLiD™	Sequencing by ligation	5500 xl	75 bp	30 Gb	Life Technologies (ABI)
Automated Sanger sequencing	Capillary electrophoresis, BigDye®-terminator chemistry	3730 xl	Upto 900 bp	<3 Mb	Applied Biosystems
Ion torrent	Hydrogen ion semiconductor	Ion 316 Chip	100 bp	100 Mb	Life Technologies (ABI)
Illumina	Clonal single molecule array	HiSeq2000	50–150 bp	Upto 55 Gb	Illumina, Inc.
Complete genomics	DNA nanoball array, ligation-based sequencing	–	70 bp	8.8 Gb	Complete Genomics
HeliScope™	Imaging single nucleotide incorporation	Single molecule sequencer	35 bp	1 Gb	Helicos
PacBio	SMRT™ technology	PacBio RS	>1000 bp	500 Mb	Pacific Biosciences

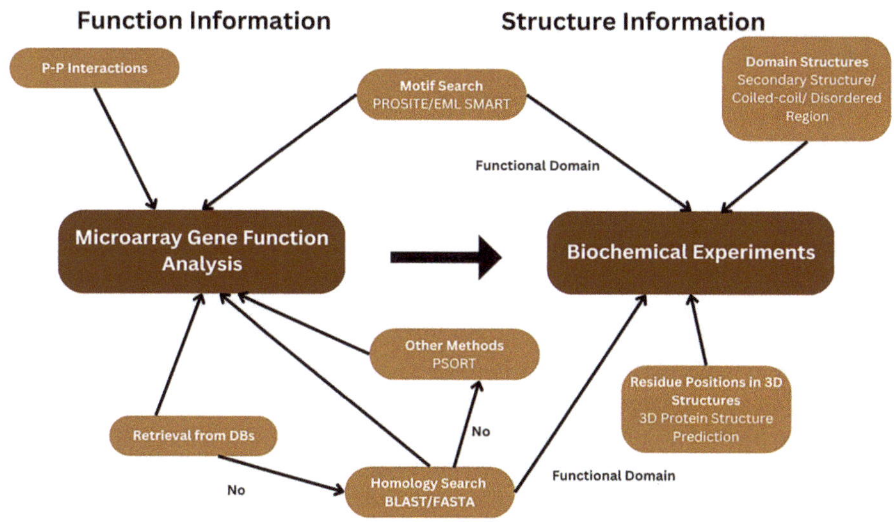

Fig. 15.6 Integration of automation tools in genomic data analysis

was revolutionized with the inception of the Human Genome Project (HGP) in 1990. The manual procedures adopted for genome sequence analysis were the chemical degradation method developed by (Maxam and Gilbert 1977) and enzymatic method developed by Sanger et al. (1977). Both of these methods belonged to the classical DNA sequencing technology. These methods involve a set of radioactive DNA of common origin which terminates at a specific unknown nucleotide sequence. The fundamental difference between both of these methods is the procedures followed to obtain the DNA fragments. Each of these methods possesses their own characteristic pros and cons which ultimately create the need for minimizing the errors encountered during the process.

This gave rise to introducing automation in these sequencing procedures, as the contemporary methods were manual in both interpretative and experimental aspects. Moreover, automation aimed at accelerating the process to boost efficiency and improve acquiring of genomic data (Martin and Davies 1986). Seiko Instruments developed a microprocessor-led robot to aid the Japanese Human Genome Sequencing Program (Wada et al. 1983). This developed robot was programmed to automate the chemical degradation reactions (Fig. 15.6).

To achieve automation in enzymatic methods, reagent-manipulating automated system was designed by University of Manchester, Institute of Science and Technology for European Sequence Automation Project (Martin and Davies 1986). To assess from a wider perspective, sufficient research has been conducted focussed on the automation of the postreaction procedures involved in sequencing. For instance, electrophoretic segregation of the genome fragments and evaluation of the separated fragments are some of the successful automation projects conducted (Lee et al. 1987). This section aims to thoroughly explain the automation techniques employed in the sequencing process and the various challenges encountered during achieving our goals.

15.3.2 Introduction to Genome Sequencing and Analysis

The concept of genome sequencing has gained much importance after the discovery of the first-generation sequencing technologies in 1977. This field then attained skyrocketed attention when the United States initiated the Human Genome Project (HGP) in 1990 followed by its conclusion in 2003. HGP gave science enormous data about the human genome and most importantly converted the blurry misconceptions about DNA to direct experimentally derived answers. This revolutionized the field of Genetics and Bioinformatics (Egeland et al. 1987).

Genome sequencing is the process of basically "decoding" the entire genomic sequence. It involves acquiring information about the entire strand of DNA and ascertaining the sequence of nucleotides on the strand (St. George-Hyslop et al. 1987). We can employ the collected data for identification, exploration of genomic similarities and differences, and much more. It plays a pivotal role in the diagnosis of ailments, distinguishing mutations from normal gene sequences and pharmacogenomics.

Bioinformatics presents a term 'sequence analysis' which includes subjecting the genomic sequence to undergo various analytical tools in order to extract as much information as possible from the sequence (Horowitz et al. 1986). The rapid development of high-throughput technologies has made the idea of maximizing the data extraction from the genomic sequence a reality. High-throughput techniques also make it possible to add novel genomic sequences to the databases with great ease (Estivill et al. 1987). The sequence analysis in genetics covers wide areas of research:

- Finding similarities between the genomic sequences.
- Exploration of features like placement of introns and exons, regulatory elements, active sites et cetera.
- Finding numerous variations like point mutations, single nucelotide polymorphism (SNPs) et cetera.
- Finding evolutionary relationships and extrapolating the findings in the population to generalize the results, if any.
- Predicting the molecular structure solely from the genomic sequence.

Genomic sequencing finds innumerable applications in science and technology. It influences a wide array of fields which have helped improve the quality of human life in the past few decades. Incessant efforts for improving the overall efficiency of the sequencing process have been made.

15.3.3 Challenges in Genome Sequencing and Analysis

It is extremely important to consider the numerous challenges posed with the advent of novel sequencing technologies. It is pivotal to understand the mechanism of these challenges and the way in which these can be tackled (Gnirke et al. 2009). Handling these challenges so as to increase the efficiency of the entire genome sequencing process. To mention a few, the core obstacles encountered in these processes are contamination of the concerned sample from various sources, variable run quality, library chimaeras, and sample mix-ups.

15.3.3.1 Contamination of the Sample

One of the most complicated challenges faced during the sequencing procedures is the contamination of the sample. Next-generation sequencing has allowed us two strategies to deal with this issue. The bacterial cloning step was known to be a primary source of contamination as observed in the capillary-based sequencing (Okou et al. 2007). NGS can be directly preformed on the libraries constructed, skipping the bacterial cloning stage. This ultimately significantly nullifies the possibility of the sample getting contaminated.

Moreover, each and every read obtained as a result of NGS inspects a single DNA molecule. This individual interrogation of the DNA molecule helps to identify and eliminate contaminated reads from each molecule under inspection. This allows us to narrow down our search for any contamination in each molecule rather than the whole sequenced genome, which might make the whole task tedious (Ng et al. 2009). The progress in the field of bioinformatics also allows us to map the reads in the contaminated genome sequencing databases hence allowing us to swiftly screen the libraries and eliminate the contaminated reads. These approaches greatly aid us to increase the efficiency of the sequencing process thereby dealing effectively with the challenges encountered.

15.3.3.2 Variable Run-Quality

Due to the rapid revolution observed in the sequencing instruments from technology labs to production floors, it is inevitable to maintain unswerving run quality. This is greatly influenced by the skill of the concerned technician and the amount and quality on the starting material, DNA as well as the reagents employed in the sequencing. Considering the enormous costs of NGS sequencing processes, it is obvious to avoid variabilities in experiments to curb the expenses (Yeager et al. 2008). Most genome centres tackle this challenge by employing streamlined workflows and automated liquid handling techniques. Regular quality control checks, like microfluidics fragment size selection, quantification of DNA by Picogreen et cetera deals with the problems as soon as they arise.

15.3.3.3 Library Chimaeras

Statistically, almost 5% of the paired-ends long-insert libraries possess chimeric ends. These libraries can have consequential implications for SV prediction algorithms and de novo assembly. These primarily rely on mate pairing knowledge.

This assembly of chimeric ends creates a plethora of challenges as these ends generate false pathways for assemblies (Li et al. 2009).

One alternative to tackle this can be use in short-end paired reads instead of long-end for assemblies.

15.3.3.4 Sample Mix-Ups

Human error has always been an inevitable challenge of almost every process in science and technology. Different errors like mislabelling, switching, and contamination of samples are some of the human errors observed. Many genome centres have adopted novel approaches to identify and tackle these issues without compromising with the efficiency of the process. To solve the problem of mislabelled data, genome centres resort to high-density SNP array data (Langmead et al. 2009). These provide us with millions of genotypes across genomes accurately and precisely. Additionally, it furnishes us with reference points for estimation of diploid coverage and composes an accurate individualistic DNA profile of the sample under consideration.

15.3.3.5 Tumour-Normal Switches

NGS of the cancer genomic sequences aims at comparing the tumour and normal samples from the same patient. In these process, the accurate identification of the sample for characterizing the somatic changes so as to compare the tumour genome with the normal genome is critical. Lamentably, as large amount of sample shares common genetic origin it renders SNP arrays insufficient in solving this problem. The possible solution employed for this issue is the CNV detection algorithms to the procured data for evaluating the sample switches.

15.3.4 Role of Automation in Genome Sequencing

As mentioned before, both the enzymatic and chemical methods are manual processes at their core and as time passed, there was an ever increasing need for automating the entire sequencing procedure for boosting the efficacy. With a wider perspective of reducing costs and delimiting variations in data gave rise to the union of automation with the genome sequencing procedures. Automated DNA sequencing follows the principle of the chain-termination method proposed by Sanger. The chain termination method involves amplifying the DNA fragment that needs to be sequenced by DNA polymerases. Followed by incorporation of altered nucleotides specifically, dideoxynucleotides (ddNTPs). In a nutshell, this method relies on the random integration of chain-terminating ddNTPs via DNA polymerases during the course of DNA replication.

The striking difference between the original Sanger method and automated method is that each ddNTPs are marked with a novel fluorescent marker. Automating the procedure makes the procurement of the enormous amount of data plausible (Koboldt et al. 2009). In order to achieve smooth automated sequencing of lambda and cosmids phage clones, the following subcloning of the DNA becomes necessary. Numerous strategies have been employed by genome centres to achieve the desired results. Classical perspective to achieve this goal entails creation of a precise clone restriction map. For instance, novel primer walking approach, transposon-facilitated sequencing, shot-gun approach et cetera have been employed.

15.3.5 Automated Platforms for Genome Sequencing

Automation in genome sequencing plays an intricate role of intertwining engineering, chemistry, and molecular genetics, all into one technology. This automation technique when intertwined with novel physical approaches made it possible to ascertain the long-range link among the cloned fragments of the genome. The discovery of fluorescent DNA sequencers made it possible to give rise to standard genomic sequences for model organisms and for reference human genomes (Simpson et al. 2009). Various automated sequencing technologies have been developed that allow us to skyrocket the relative speed of the collection of genomic data.

Many sequencing platforms have been developed for the automated sequencing of the large amount of genomic data:

15.3.5.1 Roche/454 Sequencing

It is the first commercialized NGS platform available. It employs pyrosequencing which is based on the sequencing by synthesis principle. Each time a nucleotide is attached to the DNA molecule; a pyrophosphate molecule is observed to be released. It ultimately allows a cascade of interlinked enzymatic reactions which radiates light that is further detected for ascertaining the added nucleotide.

15.3.5.2 Ion Torrent/Proton Sequencing

It is based on the semiconductor technology; also based on sequencing by synthesis perspective. Whenever a nucleotide is integrated to a DNA molecule, hydrogen ions are released which are detected. This whole sequencing process takes place on a metal-oxide semiconductor chip.

15.3.5.3 ABI/SOLiD Sequencing

It is based on the principle of sequencing by ligation thereby exploiting the mismatch sensitivity of the enzyme DNA ligase to determine the sequence of the added nucleotide. Upon matching of the sequence of the fluorescent tagged probes with the sequences of the concerned DNA fragment, ligation occurs and the fluorescent signal can be used to determine the sequence of the nucleotide.

15.3.5.4 Illumina Sequencing

It works on the principle of sequencing by synthesis. Detecting the fluorescent signals produced upon addition of nucleotide to the DNA fragment, generates images of the each step when the nucleotide is added. High-quality data is procured when these images are analysed.

15.3.6 Bioinformatics Automation in Genome Analysis

It is unescapable to analyse the high quality data obtained from the genome sequencing processes. The field of sequencing gave rise to a new field of science namely, bioinformatics which solely revolved around the storage, analysing and monitoring of large amount of data. It mainly comprises of statistical tools to manage complex and enormous data, it also ensures easy data retrieval systems for future perusal. Bioinformatics strives to justify interconnection between the data obtained, identify mutations, and most importantly to store all the data into easily accessible systems. This third-generation technology employed for automating the role of bioinformatics can also construct long-reads for ascertaining transcript in forms and overlapping reads, if any. The automated tools for the purpose of analysing the data obtained are:

15.3.6.1 Sequence Alignment Tools

This genome-processing stage consists of data-analysing and quality control. The popular tool known as regional Hashing-based Alignment Tool (rHAT) is used for processing of the SMRT data. rHAT can be employed for only long-reads. On the contrary, the tool marginAlign employs Oxford Nanopore for long-read alignment. In a nutshell, marginAlign helps to create superior quality alignments of the sequences (Chi et al. 2009). It enables the users to identify the single-nucleotide variants precisely with the help of its built-in software called marginCaller. It also figures out the unresolved region of the repetitive sequences in the reads.

15.3.6.2 Base Calling and Polishing Tools

The software NanoCall uses Oxford Nanopore for the purpose of base calling. The pros of NanoCall include its double-strand pore scaling functions better than the single-strand ones. Another tool employed for base calling is Albacore, a command-line base caller for ultra-long reads. It is extremely efficient in its data retention and it directly base calls the FASTQC file obtained. It is the sole base caller that resolves the base calling problems encountered in sequencing of homopolymers. Some of the sequence polishing tools include Racon and Nanopolish.

15.3.6.3 Halotype Assembly Tools

Halotype assemble comes into the picture due to the computational errors faced while regenerating the halotypes. Halotypes are basically the two parental copies of a diploid genome. Some of the tools used for this purpose are—HapCol, WhatsHap and HapCut2.

15.3.6.4 Error Correction Tools

The error correction stage is inevitable as it is an important step in haplotype interference, sequence assembly, and single nucleotide variant calling. Error correction approaches belongs to two main categories, de novo and hybrid techniques. A few hybrid detectors are, proovread, PacBioToCA, LORDEC, Jabba, LSC et cetera. The de novo detectors are LORMA and PacBioToCA.

15.4 Bioinformatics Automation in Protein Structure Prediction

Predicting the structure of the protein is a pivotal step for filling the gap between the sequence and structure of the genomes (Fischer 2006). Protein structure prediction by employing bioinformatics involves:

- Ascertaining and characterization of domains,
- Prediction of the secondary structure,
- Sequence similarity searches,
- Prediction of the accessibility of the solvent,
- Multiple sequence alignment,
- Generation of three-dimensional models to atomic precision,
- Automated recognition of protein fold, and
- Model validation.

Various methods have been discussed in the following sections explaining the role of automated bioinformatics tools in the protein structure prediction.

15.4.1 Protein Structure Prediction: An Overview

Protein structure prediction implies creating a three-dimensional structure from the amino acid sequence. In short, it is the process of predicting the secondary and tertiary structures of the protein from the primary structure. It fills the void between sequence and structure of the protein (Fig. 15.7). It primarily involves ascertaining the whole structure of the protein from the amino acid sequence (Wu et al. 2006). It is of paramount importance in the genome sequencing procedure to determine the protein structure, as its structure heavily influences its function. The fundamental core of protein structure prediction lies in the recognition of an apt structural target from which the required three-dimensional information can be pooled to ascertain the sequence. On the basis of the process adopted, there are three pathways through which prediction can be done.

- The first approach involves using standardized scientific techniques.
- In case of the structure of the protein remains to be illusory, the adoption of nontrivial techniques is adopted.
- The third approach suggests if we are unable to extrapolate reliable results from the nontrivial techniques, it can be categorized as virtually inconceivable to accomplish.

15.4.2 Challenges in Protein Structure Prediction

In the recent decades, the effort of structure genomics for achieving the goal of protein structure prediction is monumental. The main objective of structure

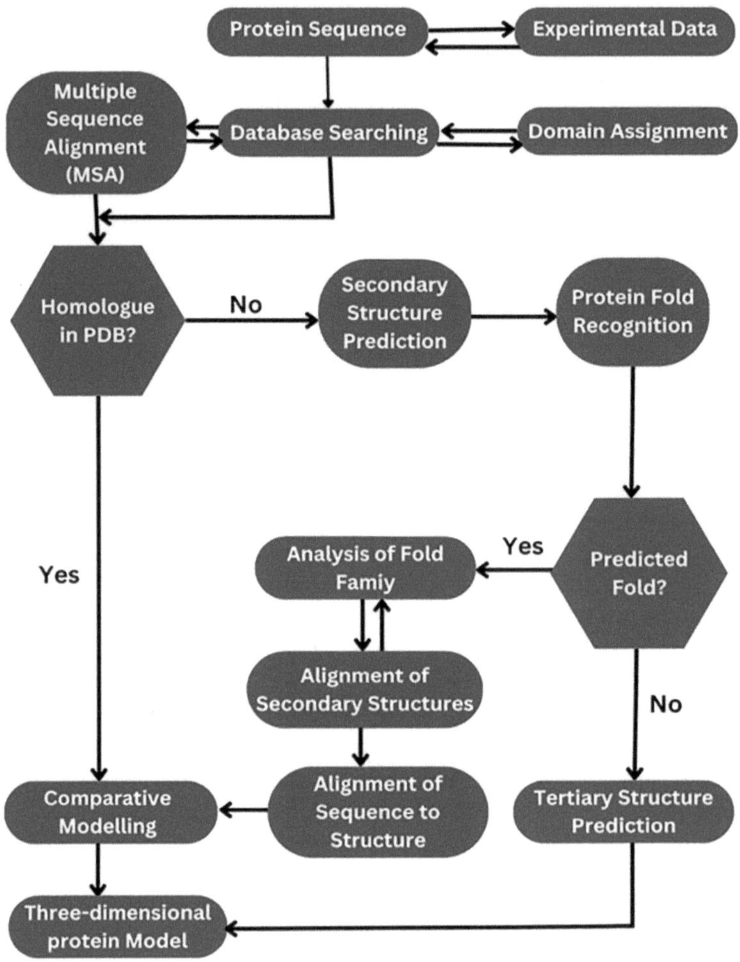

Fig. 15.7 Bioinformatics automation in protein structure prediction

genomics is to generate 3D models for refined integration of computationally derived structure prediction and experimentally derived structure prediction. The factors which heavily influence the successful completion of the prediction procedure are meticulous modelling algorithms and structural prediction of selected proteins experimentally (Katoh et al. 2005).

The key obstacles in the unhindered completion of the task involves ambiguity regarding the sequences comprising of similar structures in PDB file format, the obscurity experiences during selection of apt templates, refining the template structure similar to the native et cetera. Moreover, the other roadblocks in this journey are creation of precise topology files from scratch for those sequences which lack pertinent templates.

15.4.3 Automation Techniques in Protein Structure Prediction

Extrapolating enormous information about the desired protein by just knowing its primary structure is important to ponder on while aiming the completion of genomic sequencing (Table 15.3). In the last few years, the explosion of protein structures in popular databases like PDB etcetera has made its structure prediction a field of research. Due to the rapid inflow of structure without accurate analysation, there is deep void between experimentally observed protein structures and the ones which have not yet been deduced. To deal with such an overwhelming amount data and to perform such tedious tasks on them, poses innumerable challenges when decided to tackle them manually (O'Sullivan et al. 2004). This creates the need to handle these issues by automating the whole procedure with the aid of bioinformatics which helps us evaluate and analyse the data procured. In the postgenomic era, proteins are looked upon as 'drug targets' and also the avalanche of proteins in the sequencing programs highlights the gravity for understanding the structure of proteins. Experimental structure prediction techniques are horrifyingly labour-intensive, tedious, time-consuming and unviable due to its extortionate expenses. The only solution to this rising problem is automation the whole process using computational bioinformatics tools.

Table 15.3 Automation techniques in protein structure prediction

Sr. No	Software	Type
1	PSIPRED	Secondary structure of protein
2	PredictProtein	Secondary structure of protein and others
3	SABLE	Secondary structure of protein and solvent accessibility
4	SAM-T02	Secondary structure of protein
5	PONDR	Disordered region
6	PORTER	Secondary structure of protein
7	COILS	Coiled-coil region
8	GlobPlot	Disordered region
9	TMHMM	Transmembrane domain
10	HHPred	Three-dimensional structure, homology modelling
11	SWISS-MODEL	Three-dimensional structure, homology modelling
12	FUGUE	Three-dimensional structure, threading
13	HMMTOP	Transmembrane domain
14	MODELLER	Three-dimensional structure, homology modelling
15	Phyre	Three-dimensional structure, threading
16	Robetta	Three-dimensional structure, ab initio
17	SPARKS	Three-dimensional structure, threading

15.4.4 Role of Bioinformatics in Automating Protein Structure Prediction

Bioinformatics has put forwards two important approaches—evolutionary or phylogenetic interrelations and physical interactions, through which we can predict the protein structure. The physical interaction approach is quite challenging to follow due to the various practical difficulties faced while conceiving it. The initial stage in phylogenetic analysis approach is to compare the protein to be predicted with the deduced structure of an evolutionarily related protein. Further, the shared evolutionary similarities are then assessed by phylogenetic analysis using bioinformatics tools.

- Biological Sequence Databases
  The publicly accessible repositories comprise of data mining and analysis tools. The data extrapolated from the genome sequencing projects is stored individualistically. Some of the databases known are—European Molecular Biology Laboratory Nucleotide Sequence Database (EMBL), GenBank, DNA DataBank of Japan (DDBJ) and many more. There are certain specialized genomic databases comprising of sequenced genomes of numerous living organisms (Markowitz et al. 2006). To name a few, ENSEMBL, JCI, Entrez Genome, FlyBase et cetera.
  The universal repository consisting of three-dimensional protein and nucleic acid structures is the Protein Data Bank (PDB).
- Multiple Sequence Alignment (MSA)
  Multiple Sequence Alignment is a very important step in phylogenetic reconstruction. Most of the bioinformatics tools follow the 'progressive sequencing' method for achieving MSA. This procedure consists of creation of a 'guide tree' that dictates the alignment of the amino acids in the sequence. Some of the bioinformatics tools employed for this purpose are—ProbCons, CLUSTALW, MAFFT 5.3, T-Coffee, MUSCLE and many more.

15.4.5 Examples of Bioinformatics Tools for Protein Structure Prediction

The advancements in technology for predicting the protein structure has progressed in two main pathways—through phylogenetic links or physical interactions. The pathway dealing with physical interactivity largely relies on intertwining our knowledge of molecular forces with either kinetic or thermodynamic simulation of proteins. Although the concept behind this approach seems intriguing, the actualization of the methodology creates tons of issues while conceiving the idea. Practical factors like protein stability, generation of precise models et cetera are some of the unavoidable challenges faced with this approach.

The phylogenetic or evolutionary approach in which necessary knowledge is extrapolated from the bioinformatics analysis of the protein history, pair-based phylogenetic correlations, homology to solved structures and many more (Benson et al. 2009). This approach has been quite popular in the recent decades with its great compatibility with the protein structures deposited in PDB.

Some of the important tools employed for this purpose are mentioned in Table 15.3.

15.5 Application of Bioinformatics in Clinical Research

The most important link between developments in medical research technologies and better healthcare is the clinical trial. It is a crucial component of medical research that aims to better understand human disease, as well as its prevention, treatment, and promotion of health. The process of conducting a clinical trial for a novel medication candidate is becoming more and more difficult, expensive, and time-consuming. Because of the fierce competition in the pharmaceutical industry, there is a great demand for quick approval of novel drugs. As a result, there is intense pressure on pharmaceutical companies to improve the effectiveness and efficiency of medication discovery and development. The technological initiative is thought to be the only strategy for achieving this objective. The techniques for discovering new drugs and developing them underwent a change with the introduction of electronic clinical trials and computer-aided drug design research (Gill et al. 2016).

A number of clinical trial processes, including target identification, target validation, randomization, data collecting, and data integration, as well as trial management and pharmacovigilance, also become more streamlined, efficient, and manageable. Pharmaceutical businesses and regulators used new technology, which not only increased productivity but also dramatically enhanced data security and the evaluation of clinical data (i.e., turning trial data into information that can be applied). The potential benefits of bioinformatics in clinical research include developing and utilizing a large data strategy for clinical trials, utilizing new techniques to provide patient-centric trial design, bringing evolution to existing processes and systems with new techniques, assisting with case studies from existing data sources for advanced trials, making data sharing simpler (Gill et al. 2016).

15.6 Bioinformatics Role of Drug Development

To screen vast compound catalogues, many pharmaceutical companies have built automated high throughput screening facilities. Genome-wide analysis is now possible for a variety of scientific domains because of advances in bioinformatics. The Human Genome Project's terabytes of data can be examined by researchers thanks to bioinformatics technologies (Clark and Pickett 2000). Finding out whether and how a specific molecule is directly involved in a disease process with the aid of gene sequence databases, gene expression databases, protein sequence databases, and related analysis tools leads to the discovery of novel and improved medication targets. The time and money required to generate effective pharmacological agents could be decreased with a successful and dependable drug design procedure. In order to identify and exclude candidate compounds that are unlikely to survive the further stages of discovery and development, computational approaches are employed to predict drug-likeness. Approaches based on genetic algorithms and neural networks could forecast how potent a drug would be. The first crucial stage in this process is the capacity to identify new therapeutic targets for additional investigation. According to reports, practically all medications now on the market have as many as 483 pharmacological targets (45% receptors, 28% enzymes, 5% ion channels, and 2% nuclear receptors) (DiMasi et al. 2003; Gill et al. 2016).

The conventional approaches to drug discovery and development are significantly changing as a result of genomics, proteomics, and metabolomics. High-throughput sequencing is extensively used today to identify new therapeutic targets. Nearly every step of drug discovery, drug evaluation, and drug development now involves the use of bioinformatics. This expanding significance is due to the use of bioinformatics tools to anticipate, evaluate, and help interpret clinical and preclinical findings, as well as the role that bioinformatics plays in handling massive volumes of data Both bioinformatics and cheminformatics depend heavily on data and databases (Altschul et al. 1997). Most data searches would be ineffective without a lot of readily available electronic data, and most types of predictive or analytical software could never be created or tested. The quality of biological or chemical data is just as significant as its quantity. Data on gene and protein sequences are increasingly essential to nearly every element of pharmacological research. For instance, standard pathogen sequencing includes Today, it is possible to identify potential protein therapeutic targets or pathogenicity in viruses, parasites, or bacteria impressively quickly and reasonably cheaply (Wishart 2005). Drug metabolism and drug interaction databases are a new class of databases that are likely considerably more useful to pharmaceutical researchers than general metabolism and pathway databases, which are becoming more and more significant in drug development and assessment (Comess and Schurdak 2004). These databases aim to connect the genomic/proteomic data being obtained about the pertinent genes or proteins with the drug compounds themselves and concentrate much more intently on known medicines or drug metabolites. There are already a number of commercial drug metabolism or drug interaction databases, such as those provided by MDL Information Systems and the University of Washington. (Seattle) and Lhasa Ltd., an organization located at the University of Leeds that is not for profit. The MDL METABOLITE system and MDL TOXICITY are maintained and sold by MDL. Database (Wishart 2005). Drug development is a difficult, risky, time-consuming, and sometimes very lucrative process. The methodical process through which new candidate medications are found is known as drug discovery. Drug development has traditionally been a challenge for pharmaceutical corporations, who use well-established pharmacology and chemistry-based methods. Drug development is a labour-intensive, expensive, and complex process (Iskar et al. 2012).

Pharmacogenomics and bioinformatics both offer significant assistance in overcoming the cost and time constraints in various ways. Drug-related databases and software are widely available thanks to bioinformatics, and they can be utilized for a variety of processes connected to the process of creating and developing new drugs. In a similar vein, pharmacogenomics gives genome-level data on the varied medication response, which is crucial for pharmaceutical companies to build novel drugs, in addition to orphan drugs (Katara 2013). Moreover, bioinformatics offers methods and algorithms for predicting new drug targets as well as for storing and managing data on existing drug targets. There is hardly any requirement to prove a direct link between a putative target and the disease of concern once "possible" therapeutic targets have been identified. The process of developing new medications is justified by the creation of such a significant relationship. Target validation is a step in this process that bioinformatics is heavily utilizing observe in Fig. 15.3 (Katara 2013).

15.7 Role of Bioinformatics in Antibacterial Potential

The development of vaccines and antibacterial medications has advanced significantly since the 1940s, saving many lives. The recent development in organ transplants, intensive chemotherapy, invasive surgeries, liberal and indiscriminate use of antibiotics, and epidemiologic virulence—the spread of resistant strains due to insufficient precautions in the hospitals—has produced a human host with impaired immune systems, which are affected both by antibiotic-resistant strains and other microbes (Casadevall and Pirofski 2000). Despite the existence of powerful vaccines and antibiotics against classical pathogens. In addition to weakening the immune system, the use of antibiotics promotes the emergence of opportunistic infections, which the immune system would normally suppress (Swartz 1994). Some antibiotics disrupt the situation for resource and nutrient competition required to restrict the growth of opportunistic and drug-resistant bacterial strains under the usual conditions by negatively controlling the growth of wild-type bacteria. Simple microbial infection clearance may not always eliminate its long-term clinical effects, and can still result in disease due to immunological damage (Ochman and Moran 2001). For instance, reactive arthritis and rheumatic heart disease may develop as a result of certain bacterial infections in the gastrointestinal tract and streptococcal pharyngitis, respectively (Swartz 1994). There are two ways to combat the bacterial infection: either utilize biostatic antibacterial medications that inhibit the growth of the pathogens or stimulate the immune system by immunizing against the invasive infections. The first strategy relates to the creation of vaccines, whereas the second strategy concerns the creation of antibiotics. Better and faster-acting medications and vaccines are becoming a reality because of advancements in computational and biological techniques. Combinational computational chemistry is one of the recent advances in antibacterial development. It allows for the variation of a 3D structure of an antibacterial compound by computationally modelling the 3D structure using energy-minimization techniques and other molecular modelling techniques to find a better compound that docks to a gene involved in. Automated genomic data extraction and analysis have seen a revolutionary breakthrough over the past 10 years, made possible by a massive increase in computing power. We now have a chance to overcome our inability to combat the threat posed by bacterial resistant strains thanks to the ability to save the genomic and proteomic results in databases and automatically access and evaluate the data at the gene, genome, and proteome level (Wang and Kollman 2001). A small number of potentially pathogenic genes can be examined in wet labs using experimental approaches after being rapidly and cost-effectively trimmed and assessed in silico for different possibilities Genome sequencing, automated preservation and retrieval of genomic and proteomic data, comparative genomics, and proteomics are all key components of the bioinformatics field that are helping to determine the whole function of the genome. With the availability of complete genome sequences, bioinformatics and biochemical analysis both have an integrated and complementary role to play: bioinformatics by reducing the number of potential outcomes and speculating on functionality, and biochemical analysis to validate the speculative results, improve the efficacy, and investigate the solubility, permeability, and diffusion required for drug uptake. Finding similar genes and proteins using bioinformatics methods for comparative study is insufficient because even little structural differences in these proteins might affect how they function and how they bind to different substances (Hagman and Shafer 1995). The scope of the bioinformatics research on vaccinations and antibiotics. Understanding hereditary disorders like cancer and deadly viral infections like HIV is another area where bioinformatics is useful. However, the use of bioinformatics to treat viral and genetic disorders (Bansal 2008).

Understanding the genomic machinery is essential for rational medication design since various infections use different mechanisms and gene sets. Numerous elements, such as gene functionality at the domain level, are necessary to comprehend the pathogenicity (Jeffery 2003). Analysis of the conserved and non-conserved structural features of receptors involved in host-pathogen contact and adhesion. The identification of genes in the microorganism, determining the function of the gene, putting genes together to reconstruct metabolic and regulatory pathways, comparing pathways to identify essential pathways and pathways specific to pathogenic strains, and figuring out what proteins or substances interact with the control region of the genes and operons to inhibit transcription are all steps in understanding the genome function of the pathogenic strains at the systemic level (Zhou et al. 2004). Comparative genomics has several benefits, including the automatic reconstruction of metabolic pathways and the identification of plasmid genes thought to be involved for pathogenicity (Shokhen et al. 2006). The discovery of genes implicated in widespread signalling pathways is another benefit. Comparative investigation of genomes with similar evolutionary histories has revealed that numerous genes are absent from key pathways in pathogenic strains. In recent years, databases of genes, proteins, and protein domains within genes have also been made available thanks to bioinformatics research. With the help of these databases and pattern-based search methods, the labelling of the genes and proteins in recently sequenced genomes has multiplied. Among the databases are those for genomic sequences (Goto et al. 1998). Drug discovery can also be aided by bioinformatics research by rebuilding regulatory and metabolic pathways and analysing the rate at which they react. The integration of wet-lab biochemical procedures, comparative genomics, and proteomics—computational analysis of gene array data—is necessary for this research much as it is for genome sequencing (Bansal 2001) The wet lab serves as the foundation for defining the pathway since it offers information on known reactions, reaction rates, the activities of the original enzymes, substrate information, and known metabolic and signalling pathways. Reconstructing metabolic pathways and identifying gene clusters implicated in signalling pathways have both been accomplished using comparative genomics and cluster analysis of microarray data (Benson et al. 2005). The next step is to compare the pathways of two bacteria in order to find crucial pathways and specific pathways found in various microbes although being. However, the binding data accessible from the wet labs places a cap on the databases of protein-protein interactions and protein-DNA interactions. By aligning and comparing the regulatory areas before the orthologous genes in evolutionary-close genomes, numerous bioinformatics tools have recently been created to conjecture the binding sites. The binding locations have been well estimated by these alignments' conserved areas. The use of molecular modelling techniques in bioinformatics research is another crucial component. These techniques can: make it easier for bacteria to absorb antibiotics through the lipid layer by creating new pores or making better use of already-existing ones; identify the proteins that can prevent operons, a group of co-regulated genes, from forming channels and/or facilitate the uptake of antibiotics through the lipid layer (Bansal 2008).

15.8 Robotics and Automation in Biological Experiments

The basic questions that arise in our minds are, is the integration of robotics and automation necessary for increasing the efficiency of the sequencing procedures, and what difference does it make to employ robots to perform the same task a human would do?

The overall lower costs of automation tools as compared to the contemporary methods when precisely coupled with the essence of artificial intelligence and data-extracting algorithms; has the power to completely transform the existing paradigm. Incorporation of automated tools and robotics in the procedures significantly enhances the reproducibility of the experiments performed, lowers the expenses, offers enormous speed, and saves our precious time (Pei et al. 2008). It enables the genomics to focus on the overall design and novelty of the procedure rather than worrying about the daunting procedures to be followed. It tries to integrate the view of shifting the perspective of carrying out single experiments manually by considering each variable at a time to conducting experiments considering multiple variables simultaneously. These are some of the factors that boost the efficiency and highlight the need for automating the whole procedure.

15.8.1 Robotics and Automation in Biological Research

While considering all the factors that tend to decrease the productivity of the sequencing procedures, many genome centres employ liquid-handling automation to transform the entire workflow into a more efficient one. The inclusion of the field of robotics in genetics gave rise to the state-of-the-art technology known as do-it-yourself (DIY) robotic liquid handlers. Automated robotic instruments comprise of both workstations and handheld devices. Some of the popular instruments used to deal with the mundane repetitive tasks of sample preparation are the automated pipettes and syringes. The robotic workflow becomes completely independent once the concerned experiment is commenced. These instruments are bound to work continuously and tirelessly to provide with close to ideal efficiency factors provided the calibration errors are nullified (Pruitt et al. 2007). These robotic devices have also mastered the act of multi-tasking which significantly curbs the expenses and saves an enormous amount of time.

Bio robotics is another field of integrating the knowledge of robotics with different disciplines of science particularly biology to achieve the idealistic goals. Efforts have also been made to completely automate the surgeries undertaken by the hospitals considering the variables of the operation room environment. These advancements transfigure the whole contemporary perspective of science and genomics which guarantees the unravelling of a lot of information.

15.8.2 Applications of Robotics and Automation in Laboratories

Science is constantly engaged in the quest for boosting its productivity by incorporating novel technologies (Hunter et al. 2009). Robotics and automation prove to be an excellent combination in transfiguring the whole paradigm of dealing with laboratory tasks.

15.8.2.1 Pharmaceutical Applications

The crucial application of robotics in pharmaceutical research is the determination of the structures of concerned molecules. The task of sample preparation involved in procedures like HPLC-MS and NMR can be carried out using robotic arm. The task of accomplishing structural protein analysis can be done by combining X-ray crystallography and NMR automatically.

15.8.2.2 Verification of Reproducibility

The scientists have proposed a semi-automated procedures for evaluating reproducibility. The role played by robotics in this field is reproducing the experiments mentioned in the research papers and verifying the reliability of the objective the experiments are aiming to prove. This was done by the robot scientist 'Eve' for reproducing the experiments mentioned in non-semantic expression of genes in oncology research articles.

15.8.2.3 Biological Laboratory Robotics

The problem of contamination has always been a menace in achieving the desired results in any scientific process. Numerous companies have designed robots that are capable of interfacing to volumetric pipettes. Robotic instruments like plate readers which are specifically designed to detect and monitor the biological/chemical activities taking place in the plates under consideration.

15.8.2.4 Pathogen Diagnostic Testing

During the pandemic era, robots were designed to analyse the swabs procured from the possible COVID-19 patients. These automated robotic liquid handling systems were designed for lateral flow assays.

15.8.3 Bioinformatics Integration in Robotic Experiments

The current scenario of ever-evolving sphere of genome sequencing techniques has aided us to educate ourselves with ample of information but also generated the challenge of handling enormous amount of data and the issues that arise allied with it namely, analysing, monitoring and evaluating the data. The most popular bioinformatics tools which aid us in introspecting the data obtained are differential expression analysis and RNA-Seq. These analyses play a crucial role in identifying the genomic transformations in the organisms (Bagos et al. 2004). Most genomics struggle with the complexity of the software interfaces to work with due to its development in the UNIX environment. Consequently, it has led to the design of a novel web server IDEAMEX (Integrative Differential Expression Analysis for Multiple EXperiments). The user-friendly interface of IDEAMEX enables the genomics to select the factors to be compared individually instead of doing an leave-one-out comparison which is obviously tedious. The IDEAMES workflow comprises of three basic stages:

Stage 1: Data Analysis

Quality control checks of preliminary level on the data distributed to each sample based on numerous types of graphs.

Stage 2: Differential Expression

Conducts differential expression analyses either with or without considering the batch effect errors with the help of various bioconductor packages like limma-Voom, edgeR, NOISeq, DESeq2 and many more. It also generates reports based on the procured data.

Stage 3: Result Integration

The results obtained from the above processes are then presented using different graphical representations, for instance—Venn diagrams, heatmaps, text lists, and correlograms.

The integration of two highly developed fields and incorporating the benefits of their union in the field of genomics is surely a revolutionary concept. This will create innumerable research opportunities in the future.

15.8.4 Benefits and Challenges of Robotics and Automation in Biology

As every technology has its own pros and cons, robotics and automation are no exception to this generalization. As no technology is bound to be perfectly ideal in terms of productivity or efficiency (Garrow et al. 2005). The associated benefits of this revolutionary union of fields are described below-

- **Increased Productivity**
  - Rise in the rate of processing diagnostic samples
  - Significant reduction of expenses
  - Enhanced laboratory workflow
- **Reproducibility and Quality**
  - Improved incubation—enhanced bacterial growth
  - Improved inoculation—enhanced yield of isolated colonies
- **Reduction in Time to obtain results**
  - Reduction in hospitalization time
  - Decrement in risks associated with nosocomical infections
  - Enhanced treatment approach

The associated challenges with this field can be discussed as follows-

- Crash of the Automated Softwares
  - Good support and maintenance of the instruments and softwares required
  - Expensive maintenance budget
- Dismissal of the staff
  - Staff escaping from the job, as they are no longer needed due to the automation of each laboratory project.
- Lack of laboratory adaptation to automation
  - Not achieving the expectations of increased productivity
  - Misusing the instruments

We encounter numerous challenges every time we try to upgrade and expand our horizon of knowledge. It is important to ascertain our approach towards handling these challenges and progressing in the field at the same time.

15.9 Conclusion

There is dynamic role of bioinformatics in automation, showcasing its significant contribution to efficient and accurate data processing and interpretation. The field of bioinformatics has evolved rapidly, driven by advancements in computational technology and the ever-increasing volume of biological data generated. By harnessing automation techniques, researchers and scientists have been able to streamline data processing pipelines, overcome data analysis challenges, and extract valuable insights from complex biological datasets. Through the integration of various computational algorithms, machine learning, and artificial intelligence techniques, bioinformatics has revolutionized the way biological data is handled. Automation has facilitated the development of sophisticated tools and pipelines that can process vast amounts of genomic, proteomic, and metabolomic data with enhanced speed, precision, and reliability. This has greatly expedited the research process, enabling scientists to extract meaningful information and make data-driven decisions more efficiently.

Moreover, the role of automation in bioinformatics extends beyond data processing to data interpretation. By integrating different data sources, utilizing advanced statistical methods, and leveraging machine learning algorithms, automated bioinformatics tools can uncover hidden patterns, identify biomarkers, predict molecular interactions, and even facilitate the discovery of novel therapeutic targets. This ability to effectively interpret complex biological data has immense implications in diverse fields, including medicine, agriculture, biotechnology, and environmental sciences. However, it is important to acknowledge that automation in bioinformatics is not without its challenges. Ensuring the accuracy and reliability of automated processes, addressing data quality issues, and handling the ethical implications of automated decision-making are among the key areas that require ongoing attention and research. Additionally, continued collaboration between bioinformaticians, biologists, and computer scientists is crucial to harness the full potential of automation and develop robust, user-friendly tools that can be readily adopted by the scientific community.

In summary, it emphasizes the indispensable role of bioinformatics in automation for efficient and accurate data processing and interpretation. The advancements in automation techniques have empowered researchers to handle large-scale biological datasets, extract meaningful insights, and drive scientific discoveries. As bioinformatics continues to evolve, integrating automation will remain essential in unravelling the complexities of biological systems and accelerating progress in various domains, ultimately leading to significant advancements in human health, agriculture, and our understanding of life itself.

References

Afgan E, Baker D, Batut B et al (2018) Galaxy: a comprehensive approach for supporting accessible, reproducible, and transparent computational research in the life sciences. Genome Biol 19(1):151

Altschul SF, Madden TL, Schäffer AA, Zhang J, Zhang Z, Miller W, Lipman DJ (1997) Gapped BLAST and PSI-BLAST: a new generation of protein database search programs. Nucleic Acids Res 25(17):3389–3402

Teresa K. Attwood, David J. Parry-Smith, Attwood, T. K., & Parry-Smith, D. J. (1999). Introduction to bioinformatics. Pearson Education

Bagos PG, Liakopoulos TD, Spyropoulos IC, Hamodrakas SJ (2004) PRED-TMBB: a web server for predicting the topology of betabarrel outer membrane proteins. Nucleic Acids Res 32: W400–W404

Bansal AK (2001) Integrating co-regulated gene-groups and pair-wise genome comparisons to automate reconstruction of microbial pathways. In: Proceedings 2nd annual IEEE international symposium on bioinformatics and bioengineering (BIBE 2001). IEEE, pp 209–216

Bansal AK (2008) Role of bioinformatics in the development of new antibacterial therapy. Expert Rev Anti-Infect Ther 6(1):51–65

Benson DA, Karsch-Mizrachi I, Lipman DJ, Ostell J, Wheeler DL (2005) GenBank. Nucleic Acids Res 33(suppl_1):D34–D38

Benson DA, Karsch-Mizrachi I, Lipman DJ, Ostell J, Sayers EW (2009) GenBank. Nucleic Acids Res 37:D26–D31

Buffalo V (2015) Bioinformatics data skills: reproducible and robust research with open source tools. O'Reilly Media, Inc.

Bux M, Leser U (2013) Parallelization in scientific workflow management systems. arXiv preprint arXiv:1303.7195

Casadevall A, Pirofski LA (2000) Host-pathogen interactions: basic concepts of microbial commensalism, colonization, infection, and disease. Infect Immun 68(12):6511–6518

Chi SW, Zang JB, Mele A, Darnell RB (2009) Argonaute HITS-CLIP decodes microRNA-mRNA interaction maps. Nature 460(7254):479–486. https://doi.org/10.1038/nature08170

Clark DE, Pickett SD (2000) Computational methods for the prediction of 'drug-likeness'. Drug Discov Today 5(2):49–58

Cock PJA, Van Der Lelij P (2020) Bioinformatics data skills: reproducible and robust research with open source tools. O'Reilly Media

Comess KM, Schurdak ME (2004) Affinity-based screening techniques for enhancing lead discovery. Curr Opin Drug Discov Devel 7(4):411–416

Di Tommaso P, Chatzou M, Floden EW (2017) Nextflow enables reproducible computational workflows. Nat Biotechnol 35(4):316–319

DiMasi JA, Hansen RW, Grabowski HG (2003) The price of innovation: new estimates of drug development costs. J Health Econ 22(2):151–185

Dowsey AW, Dunn MJ, Yang GZ (2003) The role of bioinformatics in two-dimensional gel electrophoresis. Proteomics 3(8):1567–1596

Egeland JA, Gerhard DS, Pauls DL, Sussex JN, Kidd KK, Allen CR, Hostetter AM, Housman DE (1987) Bipolar affective disorders linked to DNA markers on chromosome 11. Nature 325:783–787

Estivill X, Farrall M, Scambler PJ, Bell GM, Hawley KMF, Lench NJ, Gillian PB, Kruyer HC, Frederick PA, Stanier P, Watson EK, Williamson R, Wainwright BJ (1987) A candidate for the cystic fibrosis locus isolated by selection for methylation-free islands. Nature 326:840–845

Fischer D (2006) Servers for protein structure prediction. Curr Opin Struct Biol 16:178–182

Garrow AG, Agnew A, Westhead DR (2005) TMB-hunt: a web server to screen sequence sets for transmembrane beta-barrel proteins. Nucleic Acids Res 33:W188–W192

Gill SK, Christopher AF, Gupta V, Bansal P (2016) Emerging role of bioinformatics tools and software in evolution of clinical research. Perspect Clin Res 7(3):115

Gnirke A, Melnikov A, Maguire J, Rogov P, LeProust EM, Brockman W, Fennell T, Giannoukos G, Fisher S, Russ C, Gabriel S, Jaffe DB, Lander ES, Nusbaum C (2009) Solution hybrid selection with ultra-long oligonucleotides for massively parallel targeted sequencing. Nat Biotechnol 27(2):182–189. https://doi.org/10.1038/nbt.1523

Goto S, Nishioka T, Kanehisa M (1998) LIGAND: chemical database for enzyme reactions. Bioinformatics (Oxford, England) 14(7):591–599

Hagman KE, Shafer WM (1995) Transcriptional control of the mtr efflux system of Neisseria gonorrhoeae. J Bacteriol 177(14):4162–4165

Horowitz R, Kempner ES, Bisher ME, Podolsky RJ (1986) A physiological role for titin and nebulin in skeletal muscle. Nature 323:160–164

Huber W, Carey VJ, Gentleman R et al (2015) Orchestrating high-throughput genomic analysis with Bioconductor. Nat Methods 12(2):115–121

Hunter S, Apweiler R, Attwood TK et al (2009) InterPro: the integrative protein signature database. Nucleic Acids Res 37:D211–D215
Iskar M, Zeller G, Zhao XM, van Noort V, Bork P (2012) Drug discovery in the age of systems biology: the rise of computational approaches for data integration. Curr Opin Biotechnol 23(4): 609–616
Jeffery CJ (2003) Moonlighting proteins: old proteins learning new tricks. Trends Genet 19(8): 415–417
Jenney A, Petritis K (2010) Automation in genomics and proteomics—an engineering case study. Biotechnol J 5(1):20–30
Jones NC, Pevzner PA (2004) An introduction to bioinformatics algorithms. MIT press
Katara P (2013) Role of bioinformatics and pharmacogenomics in drug discovery and development process. Netw Model Anal Health Inform Bioinforma 2:225–230
Katoh K, Kuma K, Toh H, Miyata T (2005) MAFFT version 5: improvement in accuracy of multiple sequence alignment. Nucleic Acids Res 33:511–518
Koboldt DC, Chen K, Wylie T, Larson DE, McLellan MD, Mardis ER, Weinstock GM, Wilson RK, Ding L (2009) VarScan: variant detection in massively parallel sequencing of individual and pooled samples. Bioinformatics 25(17):2283–2285. https://doi.org/10.1093/bioinformatics/btp373
Köster J, Rahmann S (2012) Snakemake-A scalable bioinformatics workflow engine. Bioinformatics 28(19):2520–2522
Langmead B, Schatz MC, Lin J, Pop M, Salzberg SL (2009) Searching for SNPs with cloud computing. Genome Biol 10(11):R134. https://doi.org/10.1186/gb-2009-10-11-r134
Lee W-H, Bookstein R, Hong F, Young L-J, Shew J-Y, Lee EY-HP (1987) Human retinoblastoma susceptibility gene: cloning, identification, and sequence. Science 235:1394–1399
Li H, Handsaker B, Wysoker A, Fennell T, Ruan J, Homer N, Marth G, Abecasis G, Durbin R, 1000 Genome Project Data Processing Subgroup (2009) The sequence alignment/map format and SAMtools. Bioinformatics 25(16):2078–2079. https://doi.org/10.1093/bioinformatics/btp352
Markowitz VM, Korzeniewski F, Palaniappan K, Szeto E, Werner G, Padki A, Zhao X, Dubchak I, Hugenholtz P, Anderson I, Lykidis A, Mavromatis K, Ivanova N, Kyrpides NC (2006) The integrated microbial genomes (IMG) system. Nucleic Acids Res 34:D344–D348. https://doi.org/10.1093/nar/gkj024
Martin WJ, Davies RW (1986) Automated DNA sequencing: progress and prospects. BioITechnology 4:890–895
Maxam AM, Gilbert W (1977) A new method for sequencing DNA. Proc Natl Acad Sci U S A 74: 560–564
Mount DW (2014) Bioinformatics: sequence and genome analysis, 2nd edn. Cold Spring Harbor Laboratory Press
Mu H, Wang B, Yuan F (2022) Bioinformatics in plant breeding and research on disease resistance. Plan Theory 11(22):3118
Ng SB, Turner EH, Robertson PD, Flygare SD, Bigham AW, Lee C, Shaffer T, Wong M, Bhattacharjee A, Eichler EE, Bamshad M, Nickerson DA, Shendure J (2009) Targeted capture and massively parallel sequencing of 12 human exomes. Nature 461(7261):272–276. https://doi.org/10.1038/nature08250
O'Sullivan O, Suhre K, Abergel C, Higgins DG, Notredame C (2004) 3DCoffee: combining protein sequences and structures within multiple sequence alignments. J Mol Biol 340:385–395
Ochman H, Moran NA (2001) Genes lost and genes found: evolution of bacterial pathogenesis and symbiosis. Science 292(5519):1096–1099
Okou DT, Steinberg KM, Middle C, Cutler DJ, Albert TJ, Zwick ME (2007) Microarray-based genomic selection for high-throughput resequencing. Nat Methods 4(11):907–909. https://doi.org/10.1038/nmeth1109
Pei J, Kim BH, Grishin NV (2008) PROMALS3D: a tool for multiple protein sequence and structure alignments. Nucleic Acids Res 36:2295–2300

Pruitt KD, Tatusova T, Maglott DR (2007) NCBI reference sequences (RefSeq): a curated non-redundant sequence database of genomes, transcripts, and proteins. Nucleic Acids Res 35:D61–D65

Sachdeva G, Kumar K (2014) Automation of bioinformatics tools: a critical review. Mol Biol Rep 41(10):6477–6486

Sadedin SP, Dashnow H, James PA et al (2012) Bpipe: a tool for running and managing bioinformatics pipelines. Bioinformatics 28(11):1525–1526

Sakharkar MK, Sakharkar KR (2007) Automation in bioinformatics: the role of workflow management systems. Drug Discov Today 12(15–16):684–691

Sanger F, Nicklen S, Coulson AR (1977) DNA sequencing with chain-terminating inhibitors. Proc Natl Acad Sci U S A 74:5463–5467

Schadt EE, Linderman MD, Sorenson J et al (2010) Automation in high-throughput genomics. Cold Spring Harb Protoc 2010(10):pdb.top95

Schatz MC, Langmead B, Salzberg SL (2010) Cloud computing and the DNA data race. Nat Biotechnol 28(7):691–693

Shokhen M, Khazanov N, Albeck A (2006) Enzyme isoselective inhibitors: application to drug design. ChemMedChem 1(6):639–643

Simpson JT, Wong K, Jackman SD, Schein JE, Jones SJ, Birol I (2009) ABySS: a parallel assembler for short read sequence data. Genome Res 19(6):1117–1123. https://doi.org/10.1101/gr.089532.108

Spjuth O, Bongcam-Rudloff E, Hernández GC, Forer L, Giovacchini M, Guimera RV et al (2015) Experiences with workflows for automating data-intensive bioinformatics. Biol Direct 10(1): 1–12

St. George-Hyslop PH, Tanzi RE, Polinsky RJ, Haines JL, Nee L, Watkins PC, Myers RH, Feldman RB, Pollen D, Drachman D, Growdon J, Bruni A, Foncin J-F, Salmon D, Frommelt P, Amaducci L, Sorbi S, Piacentini S, Stewart GD, Hobbs WJ, Conneally PM, Gusella JF (1987) The genetic defect causing familial Alzheimer's disease maps on chromosome 21. Science 235:885–889

Stein LD (2010) The case for cloud computing in genome informatics. Genome Biol 11:1–7

Swartz MN (1994) Hospital-acquired infections: diseases with increasingly limited therapies. Proc Natl Acad Sci 91(7):2420–2427

Van Kampen AHC, Horrevoets AJG (2006) The role of bioinformatics in genomic medicine. In Cardiovascular research: new technologies, methods, and applications. p 103–119

Vayssière JL, Licznar P (2010) The role of workflow Management Systems in Bioinformatics. Bioinformatics 26(6):844–851

Wada A, Yamamoto M, Soeda E (1983) Automatic DNA sequencer: computer-programmed microchemical manipulator for the Maxam Gilbert sequencing method. Rev Sci Instrum 54: 1569–1572

Wang W, Kollman PA (2001) Computational study of protein specificity: the molecular basis of HIV-1 protease drug resistance. Proc Natl Acad Sci 98(26):14937–14942

Wishart DS (2005) Bioinformatics in drug development and assessment. Drug Metab Rev 37(2): 279–310

Wu CH, Apweiler R, Bairoch A, Natale DA, Barker WC, Boeckmann B, Ferro S, Gasteiger E, Huang H, Lopez R, Magrane M, Martin MJ, Mazumder R, O'Donovan C, Redaschi N, Suzek B (2006) The universal protein resource (UniProt): an expanding universe of protein information. Nucleic Acids Res 34:D187–D191. https://doi.org/10.1093/nar/gkj161

Yeager M, Xiao N, Hayes RB, Bouffard P, Desany B, Burdett L, Orr N, Matthews C, Qi L, Crenshaw A, Markovic Z, Fredrikson KM, Jacobs KB, Amundadottir L, Jarvie TP, Hunter DJ, Hoover R, Thomas G, Harkins TT, Chanock SJ (2008) Comprehensive resequence analysis of a 136 kb region of human chromosome 8q24 associated with prostate and colon cancers. Hum Genet 124(2):161–170. https://doi.org/10.1007/s00439-008-0535-3

Zerhouni EA (2006) Clinical research at a crossroads: the NIH roadmap. J Investig Med 54(4): 171–173

Zhou J, Thompson DK, Xu Y, Tiedje JM (2004) Microbial functional genomics. Wiley-Liss, Hoboken, NJ, pp 141–147

Artificial Intelligence and Machine Learning in Bioinformatics 16

Shabroz Alam, Juveriya Israr, and Ajay Kumar

Abstract

Artificial intelligence (AI) and machine learning (ML) have emerged over the past decade as the cutting-edge technologies most expected to revolutionize the research and development sector. This is fueled in part by game-changing developments in computer technology and the concomitant evaporation of barriers to collecting massive amounts of data. Meanwhile, the cost of researching, testing, manufacturing, and distributing new pharmaceuticals has risen. In light of these challenges, the pharmaceutical industry is interested in AI/ML methods because to their automation, predictability, and the ensuing anticipated boost in efficiency. The use of ML techniques in the pharmaceutical industry has matured during the past 15 years. Clinical trial design, management, and analysis are the most recent drug development process steps to benefit from AI and ML. As we move toward a world in which AI/ML is increasingly integrated into R&D, it is essential to sort through the corresponding jargon and hype. Equally crucial is the understanding that the scientific method is still relevant for drawing conclusions from evidence. By doing so, we can better evaluate the potential benefits of AI/ML in the pharmaceutical industry and make well-informed decisions on their best application. The purpose of this paper is to clarify certain fundamental ideas, provide some examples of their

S. Alam
Department of Biotechnology, Era University, Lucknow, Uttar Pradesh, India

J. Israr
Department of Biotechnology, Era University, Lucknow, Uttar Pradesh, India

Institute of Biosciences and Technology, Shri Ramswaroop Memorial University, Lucknow- Deva Road, Barabanki, Uttar Pradesh, India

A. Kumar (✉)
Department of Biotechnology, Rama University, Kanpur, Uttar Pradesh, India

V. Singh, A. Kumar (eds.), *Advances in Bioinformatics*,
https://doi.org/10.1007/978-981-99-8401-5_16

application, and then provide some helpful perspective on how to best apply AI/ML techniques to research and development.

Keywords

Artificial intelligence · Machine learning · Deep learning · Drug discovery

16.1 Introduction

Machine Learning, a subset of Artificial Intelligence, develops algorithms and models to help robots learn and behave like people. Knowledge, comprehension, and competence are the focus of the study, teaching, and experience that make up the area of machine learning, which integrates computer science and statistics. Assimilation of new information leads to a dynamic shift in behavior (Alpaydin 2020).

Machine learning, a bioinformatics field, transforms computing systems to do complex AI-like processes. The above bioinformatics activities include pattern recognition, disease diagnostics, computational planning, robotic control systems, and predictive modeling. The "alterations" may include system enhancements or new system building (Chetty et al. 2022).

In recent years, medical oncology has gained a remarkable understanding of cancer biology and pathogenesis. Bioinformatics has improved our ability to study and model complex biological processes thanks to next-generation sequencing technologies, particularly single-cell RNA sequencing. This includes incredibly deep and exact research and characterization of complicated issues like cancer heterogeneity, resistance mechanisms, and illness causation. In addition, collaborative efforts and extensive projects in bio specimen collection and bioinformatics, such as The Cancer Genome Atlas (TCGA), have helped consolidate, organize, and examine an unprecedented volume of patient data. This has led to the identification of novel therapeutic targets and the examination of established targets in previously unexplored illness contexts (Alpaydin 2020).

Despite the growth of cancer biology, drug discovery still faces several hurdles. Despite high-throughput screening technology, development timetables and expenses are long and expensive. Bringing a pharmaceutical molecule to market takes years, usually a decade. This complex procedure requires enormous R&D and financial investments of over $2.8 billion. Suboptimal pharmacokinetics, toxicity, and clinical efficacy can cause candidate medication failure in the drug development pipeline (Gupta et al. 2021).

In bioinformatics, using pre-existing medications to treat new diseases is a promising way to overcome the challenges of drug development for novel chemicals. To enter the market, approved pharmaceuticals have passed rigorous clinical trials, including preclinical studies, human testing, and careful evaluation. Therefore, these medications have a well-known safety profile. Bioinformatics can greatly benefit from discovering a new clinical indication for an approved medicine. This fascinating idea allows the medicine to re-enter Phase II clinical trials. This strategy reduces research and development risks and time and money expenses (Vamathevan et al. 2019).

The extensive use of computational algorithms spanning a variety of methodologies and approaches has advanced medication repurposing research in recent years. The structural biology of therapeutic protein targets can be fully explored using molecular modeling. It also enables high-throughput virtual screenings, which identify interesting drug candidates with therapeutic potential. Bioinformatics has advanced rapidly because of advances in machine learning and artificial intelligence, particularly in deep learning (Nosi et al. 2021). These cutting-edge technologies have transformed our understanding of drug-target interactions and the complex link between drug physicochemical features and phenotypic changes. These methods also help find new cancer targets in the vast cancer data repositories accumulated via many joint efforts. Due to the extensive use of high-throughput and multi-omics drug profiling experiments, chemical and bioactivity data is growing, making bioinformatics crucial to cancer treatment discovery. Additionally, the increased accessibility of these publicly available dataset collections considerably improves computational techniques (Min et al. 2017). These methods can be used for more than only experimental and biological data. Bioinformatics benefits from clinical dataset integration, notably electronic health records. This in-depth chapter discusses state-of-the-art computational techniques for oncology drug repurposing. Machine learning and deep neural networks are highlighted.

16.2 Artificial Intelligence

Machine learning is a subfield of artificial intelligence (AI). Academic interest in machine learning from data dates back to the earliest days of artificial intelligence. They tried to solve it using a wide range of symbolic techniques, including "neural networks" (primarily perceptron's and related models, which were later shown to be statistical generalized linear re-imaginings). Automated medical diagnosis, in particular, made extensive use of probabilistic reasoning (Sarle Warren 1994).

However, a divide between AI and machine learning was produced by an increased focus on the logical, knowledge-based approach. Issues with data collection and representation, both theoretical and practical, afflicted probabilistic systems. By 1980, expert systems had supplanted statistics as the dominant approach to artificial intelligence. While research into symbolic/knowledge-based learning and its offshoot, inductive logic programming, continued inside AI, work along a more statistical line of inquiry moved out of AI and into pattern recognition and information retrieval. Around the same time, artificial intelligence and computer science ceased their investigation into neural networks. Hopfield, Rumelhart, and Hinton, who had previously worked in artificial intelligence and computer science, went on to develop this line of thought as "connectionism" in their new fields of study. In the mid-1980s, when they rediscovered backpropagation, they saw their greatest success (Stuart and Peter 2003).

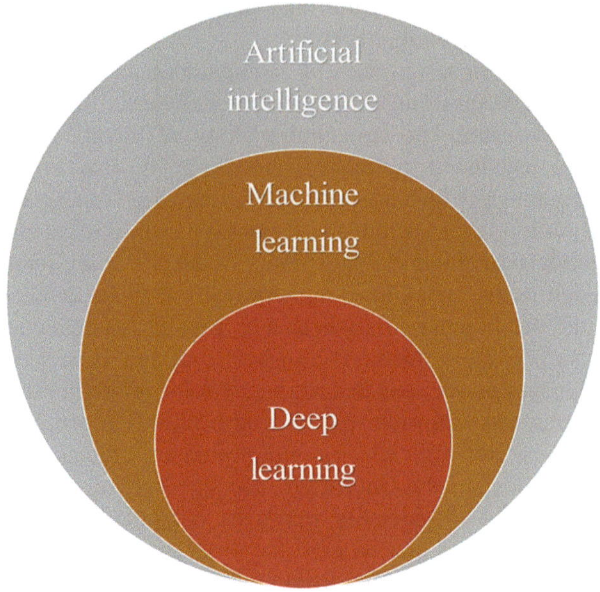

Fig. 16.1 The subfield of artificial intelligence that is known as machine learning

In the 1990s, machine learning (ML) began to flourish as a distinct discipline. The field shifted its focus from developing artificial intelligence to solving real-world issues. It abandoned the symbolic methodologies it had received from AI in favor of statistical, fuzzy logic, and probability theory-based procedures and models in Fig. 16.1 (Langley 2011).

16.3 Importance of Machine Learning

In the realm of bioinformatics, certain computational challenges elude precise definition, save for the provision of illustrative instances. These instances may consist of well-defined input/output pairs, while the connection between what is put in and what comes out remains elusive to articulate succinctly. The objective is to enable machines to dynamically adapt their internal configuration, allowing them to generate accurate outputs for a vast array of sample inputs. This process aims to effectively restrict their input/output mechanism, thereby approximating the underlying relationship inherent in the provided examples.

In the vast expanse of data, lies the potential for unearthing concealed connections and intricate correlations. Machine learning techniques, commonly employed in the field of bioinformatics, have proven to be highly effective in extracting intricate relationships from complex datasets, a process commonly referred to as data mining (Ngiam and Khor 2019).

The individual in question possesses a keen interest in the field of bioinformatics, a discipline that combines the phenomenon of human designers frequently encountering challenges in achieving optimal performance of machines within their designated environments is a well-documented observation. In reality, the comprehensive understanding of all aspects of the working environment may not be fully ascertainable during the initial design phase. When it comes to bioinformatics, machine learning applications are becoming increasingly popular. It has demonstrated its potential for enhancing the performance and optimization of existing machine designs (Mohsen et al. 2021).

The user has provided a brief statement. In the area of bioinformatics, the vast expanse of knowledge pertaining to specific tasks often exceeds the capacity for direct human encoding, the potential for machines to acquire knowledge incrementally holds great promise in surpassing the limitations of human documentation. These intelligent systems have the capacity to assimilate a wealth of information that may surpass the extent to which humans are inclined to transcribe (Erickson 2021).

The individual in question has a keen interest in the field of bioinformatics. They possess a deep understanding Environmental conditions undergo dynamic transformations throughout the course of temporal progression. The development of adaptable machines capable of dynamically responding to environmental changes holds great potential in mitigating the necessity for recurrent redesign efforts.

Humans are perpetually unearthing novel insights pertaining to various tasks. The user's text will be transformed to incorporate bioinformatics terminology and vocabulary. The ever-evolving landscape of global affairs presents a perpetual influx of novel occurrences. The ongoing endeavor to reengineer artificial intelligence (AI) systems in accordance with emerging insights presents inherent challenges. However, leveraging the potential of machine learning techniques holds promise in effectively monitoring and assimilating a substantial portion of this evolving knowledge landscape (Munjal et al. 2023).

16.4 Types of Machine Learning

Machine learning, a subfield of bioinformatics, encompasses a wide range of computational techniques that enable the analysis and interpretation of complex biological data. While classification is indeed a fundamental aspect of machine learning, it is important to recognize that this field extends far beyond this single task. By leveraging advanced algorithms and statistical models, machine learning enables researchers to uncover in the field of bioinformatics, a diverse range of problem classes can be identified (Fig. 16.2). These problem classes serve as the foundation for addressing various biological and computational challenges.

1. **Classification learning**: an essential task in bioinformatics, where the goal is to acquire the ability to accurately assign instances to predetermined classes. This process involves the utilization of various computational algorithms and statistical techniques to train models that can effectively distinguish between different classes based on specific features or attributes. Classification learning's ability to harness the power of machine learning is essential in many bioinformatics

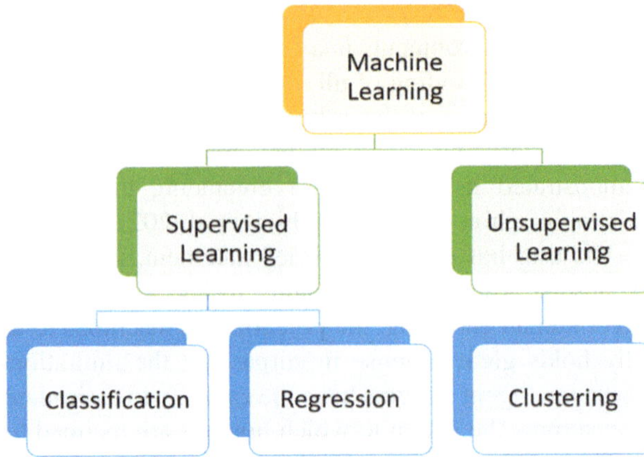

Fig. 16.2 Types of machine learning

applications, including those for analyzing gene expression, predicting protein function, and diagnosing disease (Medin and Schaffer 1978).

2. Through in the realm of bioinformatics, **association learning** is a fundamental concept that involves the acquisition of knowledge regarding the intricate relationships that exist between various attributes. Through meticulous analysis and exploration, researchers strive to uncover and comprehend the intricate connections and dependencies that may exist within biological datasets. By employing sophisticated algorithms and statistical techniques, association learning enables the identification of significant associations and patterns.
3. Thereby shedding light on **Clustering:** Uncovering cohesive groups of instances that exhibit similar characteristics (Karim et al. 2021).
4. In the realm of bioinformatics, one fascinating area of study involves the task of **numeric prediction**. Rather than focusing on classifying data into distinct categories, this branch of research delves into the realm of forecasting numeric quantities. By employing sophisticated algorithms and machine learning techniques, scientists and researchers strive to develop models that can accurately predict numerical values associated with various biological phenomena. This research has far-reaching ramifications in areas like genetics, proteomics, and drug development, and holds tremendous promise for enhancing our knowledge of complex biological systems through the utilization of vast datasets and cutting-edge computational methodologies.

16.5 Supervised and Unsupervised Learning

Supervised learning, a fundamental concept in bioinformatics, refers to the learning process wherein training instances are meticulously annotated with the correct outcomes. This meticulous labelling enables the system to receive valuable feedback, facilitating an understanding of the progress made in the learning journey. In the area of unsupervised learning, the objective becomes more challenging as it necessitates the absence of predetermined categorizations (Goudbeek et al. 2009).

16.5.1 Supervised Learning

In bioinformatics, supervised learning is often used in categorization issues. The main goal is to teach computational systems categorization systems. Training neural networks and decision trees relies heavily on supervised learning, a common bioinformatics technique. Both computational approaches use predetermined classification data substantially. Classification helps neural networks measure inaccuracy and fine-tune parameters to reduce discrepancies. In decision trees, classifications help identify attributes with the most informational value, solving complex classification problems (Le et al. 2020).

Supervised learning is essential in bioinformatics. These methods can be used to create prediction models that can identify patterns and relationships in input and output data. These models can learn from the dataset and accurately anticipate unseen variables by carefully analyzing the available data. This method has great potential in bioinformatics applications, helping researchers understand complicated biological events and living organisms (Chen and Gao 2016).

16.5.2 Unsupervised Learning

Unsupervised learning is a difficult bioinformatics activity that trains computers to learn and accomplish tasks without explicit instructions. The goal is to let machines learn and do tasks without human involvement. In computational biology and bioinformatics, unsupervised learning has two ways. The initial technique instructs the agent via rewards rather than explicit categorizations to indicate achievement. Clustering is a popular unsupervised learning paradigm and a second bioinformatics computational method. In computational biology, this learning paradigm seeks to identify patterns and resemblances in the training dataset rather than optimize a utility function. The clusters identified are expected to match an intuitive categorization. Demographic clustering can divide people into two groups: affluent and impoverished (Goudbeek et al. 2009).

Unsupervised learning in bioinformatics groups and interprets data based on input data. This approach explores underlying data patterns and structures without labels or annotations. Unsupervised learning algorithms use algorithms and statistics to get insights from unannotated datasets. This method is essential for clustering analysis, dimensionality analysis, and other bioinformatics applications (Chen and Gao 2016).

16.5.3 Semi Supervised Learning

In the realm of bioinformatics, semi-supervised learning is a computational approach that combines features of supervised and unsupervised learning. The dataset under examination is a hybrid of unannotated and annotated data, including a wide variety of sources. The fundamental objective of this research is to create a computational technique that can reliably predict output values for inputs that are either poorly described or for which no outputs are available. There is a little amount of labeled data and a huge amount of unlabeled data in the given database. In addition to the well-established paradigms of supervised and unsupervised learning, the field of bioinformatics encompasses a diverse array of learning algorithms, including reinforcement learning, among others. Both supervised and unsupervised learning methods have gained significant popularity and are extensively utilized in various domains, including computational biology and pattern recognition. These approaches play a crucial role in real-world applications, facilitating advancements in diverse fields (Yan and Wang 2022).

16.5.4 Reinforcement Learning

The field of reinforcement learning is a subset of machine learning that seeks to create intelligent decision-making algorithms and models via trial and error. The algorithms employed in this context are specifically designed to identify an optimal policy that effectively maps various states of the world to corresponding actions. The selection of actions is determined from a set of available options that an agent is expected to undertake based on the prevailing states, with the ultimate objective of optimizing a measure of cumulative reward over an extended period. Bioinformatics has revolutionized the field of machine learning by introducing a novel approach that sets it apart from traditional methods. One of its key differentiating factors lies in its ability to leverage biological data to drive predictive models and uncover hidden patterns. This distinctive characteristic has propelled bioinformatics to the forefront of cutting-edge research, enabling scientists to tackle complex problems in diverse domains such as genomics, proteomics, and drug discovery. By harnessing the power of biological information, bioinformatics has opened up new avenues for understanding and manipulating biological systems, paving the way for ground breaking advancements in the field of the absence of input-output pairs within a database characterizes this system, which is primarily designed to optimize online performance (Weltz et al. 2022; Liu et al. 2021).

16.5.5 Optimization

Optimization, a fundamental concept in bioinformatics, plays a crucial role in the field's pursuit of identifying the most optimal solution within a vast array of potential solutions. In the realm of bioinformatics, the pursuit of knowledge through data analysis is akin to a quest for the most suitable model that accurately captures the intricacies of the data. Consequently, the utilization of optimization techniques becomes an integral component in the process of constructing these models. In the past decade, there has been a significant proliferation of both exact and heuristic optimization algorithms across various domains.

16.5.6 Machine Learning and Statistics

In the field of bioinformatics, statistical analysis plays a crucial role in hypothesis testing, allowing researchers to assess the significance of their findings. Conversely, machine learning approaches in bioinformatics focus on the development of algorithms that facilitate the process of generalization by exploring and evaluating various hypotheses. By leveraging computational power, machine learning techniques aid in the discovery of patterns and relationships within complex biological datasets, enabling researchers to make informed predictions and decisions. Statistics is a multifaceted discipline that extends beyond the realm of hypothesis testing. In the realm of bioinformatics, it plays a crucial role in analyzing and interpreting complex biological data. Moreover, it is worth noting that numerous machine learning methodologies exist that do not rely on traditional search algorithms. These techniques leverage sophisticated computational models to uncover patterns and make predictions, thereby enhancing our understanding of biological systems. Machine learning algorithms commonly employ statistical tests during the construction of rules or trees, as well as for the purpose of rectifying models that exhibit "overfitting" tendencies. Overfitting occurs when models excessively rely on specific examples utilized during their creation, leading to a lack of generalizability. Statistical tests play a crucial role in the realm of bioinformatics by serving as a means to validate and evaluate machine learning models and algorithms. These tests enable researchers to assess the performance and reliability of such computational tools, ensuring their efficacy in addressing complex biological problems. Through rigorous statistical analysis, bioinformaticians can confidently determine the accuracy, precision, and generalizability of machine learning approaches, thereby facilitating their integration into various biological research domains (Venkatesh et al. 2020).

16.6 Selecting the Right Algorithm

In the field of bioinformatics, the task of algorithm selection can be a daunting endeavor. With a multitude of both supervised and unsupervised machine learning algorithms at one's disposal, each algorithm exhibits a unique methodology for acquiring knowledge. In the field of bioinformatics, it is widely acknowledged that the absence of a universally optimal approach or a one-size-fits-all solution is a prevailing reality. The process of identifying the optimal algorithm involves a combination of empirical exploration and meticulous analysis. Even seasoned bioinformaticians acknowledge that the efficacy of an algorithm cannot be ascertained a priori, necessitating iterative experimentation. In the field of bioinformatics, it is widely acknowledged that models exhibiting a high degree of flexibility possess the inherent risk of succumbing to overfitting. This phenomenon occurs when such models, in their quest to capture intricate patterns and nuances within the data, inadvertently incorporate even the minutest variations that may potentially be attributed to mere noise. In the field of bioinformatics, it is widely acknowledged that the interpretability of models is inversely proportional to their complexity. Consequently, simpler models tend to offer a more straightforward understanding of the underlying biological phenomena. However, it is important to note that this simplicity often comes at the cost of reduced accuracy. The selection of an appropriate algorithm necessitates a careful consideration of various factors, wherein the trade-offs between different advantages come into play. These considerations encompass crucial aspects such as the computational efficiency, precision, and intricacy of the model at hand. The iterative process of experimentation and algorithmic exploration lies at the heart of machine learning, wherein the pursuit of optimal solutions necessitates the continuous evaluation and refinement of various approaches.

16.6.1 Machine Algorithms in Omics Field

In the ever-expanding state of bioinformatics, the imperative to remain at the forefront is twofold: to seamlessly assimilate burgeoning data and to continuously advance algorithmic methodologies. In the field of bioinformatics, the integration of machine learning (ML) algorithms has become indispensable for conducting predictive analytics and unravelling the intricate biological mechanisms inherent in the human body. The adoption of machine learning techniques has improved some difficult areas of bioinformatics. Genomics, proteomics, microarrays, systems biology, evolutionary biology, and text mining are all examples of these disciplines (Li et al. 2022; Perakakis et al. 2018).

16.6.2 Genomics

The burgeoning demand for the advancement of machine learning algorithms designed to autonomously identify the precise genomic coordinates of protein-coding genes within a provided DNA sequence has become increasingly evident. The issue at hand pertains to the field of computational biology, specifically gene prediction. Machine learning techniques have been effectively employed in the realm of bioinformatics to address the intricate task of multiple sequence alignment. This intricate process entails the alignment of numerous DNA or amino acid sequences, with the aim of identifying regions of similarity that may signify a common evolutionary lineage. Bioinformatics is a powerful tool that finds utility not only in the identification and visualization of genome rearrangements, but also in a myriad of other applications (Libbrecht and Noble 2015; Esposito et al. 2019).

16.6.3 Proteomics

A novel bioinformatics method classifies amino acids in a protein sequence into their structural classes using machine learning methods. Helix, sheet, and coil structural motifs can be accurately identified using this novel method. This ground breaking technology revolutionizes protein analysis by using machine learning to reveal the complex link between amino acid content and protein structure. For secondary structure prediction in bioinformatics, Deep CNF is the latest method. This advanced method uses artificial neural networks, a machine learning model, to achieve 84% accuracy. Theoretical studies estimate that three-state protein secondary structure occurs around 88–90%. Machine learning has solved complex proteomics problems. These include protein side-chain prediction, loop modeling, and contact map estimate (Mou et al. 2022; Kelchtermans et al. 2014).

16.6.4 Microarrays

One of the primary challenges encountered in the area of bioinformatics revolves around the discernment of gene expression patterns through the analysis of gathered data. Moreover, owing to the vast multitude of genes encompassed in the microarray dataset, a substantial volume of extraneous data is present, thereby exacerbating the intricacy of the expressed gene identification task. Machine learning, a cutting-edge field at the intersection of computer science and biology, offers a promising avenue to address this challenge. Leveraging a diverse range of classification techniques, machine learning algorithms can be harnessed to effectively carry out the task of identification in question. In the realm of bioinformatics, a plethora of methodologies has emerged as prominent tools for data analysis and pattern recognition. Radial basis function networks, deep learning methods, Bayesian classification, decision trees, and random forest models are popular. These methods, renowned for their versatility and efficacy, have proven instrumental in unravelling complex biological phenomena and extracting meaningful insights from vast datasets. By leveraging the power of these computational approaches, researchers in the field of bioinformatics are able to navigate the intricacies of biological systems and make significant strides towards advancing our understanding of life's fundamental processes (Ekins and Chu 1999; Pirooznia et al. 2008).

16.6.5 Systems Biology

Machine learning has made computational modeling complex biological system interactions easier. This is notably the case in the context of metabolic pathways, signal transduction pathways, and genetic networks. Probabilistic graphical models, a popular bioinformatics computational framework, can reveal complex variable interactions. These methods use machine learning to untangle genomic networks' complicated structure. Probabilistic graphical models have become a standard tool for modeling genetic networks, enabling extensive studies of biological systems' mechanisms. Complex systems biology issues have also been addressed by machine learning in the bioinformatics community. Locating binding sites for transcription factors is crucial for controlling gene expression. The intricate patterns of these binding sites can be revealed by using machine learning methods in conjunction with Markov chain optimization. Natural selection-based genetic algorithms have found widespread usage in simulating biological regulation and control networks. These methods employ machine learning to recreate the interactions between genetic elements, illuminating the complex dynamics of biological systems (Muggleton 2005).

Machine learning in systems biology is one of several bioinformatics applications. Machine learning methods are used to predict enzyme function based on molecular characteristics. Machine learning is also used to analyze high-throughput microarray data, allowing researchers to gain insights from massive genetic data. Genome-wide association studies use machine learning methods to reveal complex genetic marker-disease susceptibility correlations. Last but not least, machine learning helps identify and characterize proteins based on their structural and functional properties. These applications demonstrate how machine learning improves our understanding of complicated biological processes (Liu et al. 2013).

16.6.6 Text Mining

The utilization of machine learning in the field of bioinformatics has paved the way for efficient knowledge extraction methodologies. By employing modern methods like natural language processing, valuable insights can be extracted from vast repositories of human-generated reports stored within databases. The utilization of this methodology has been extensively employed in the pursuit of discovering innovative pharmaceutical targets. This endeavor necessitates the meticulous scrutiny of data repositories encompassing biological databases and scholarly publications. Protein databases frequently lack comprehensive annotations that encompass the entirety of available knowledge for each protein. Consequently, it becomes necessary to extract supplementary information from the vast pool of biomedical literature. The application of machine learning techniques has revolutionized the field of bioinformatics by enabling automated annotation of gene and protein functions, prediction of subcellular localization of proteins, analysis of DNA-expression arrays, exploration of large-scale protein interaction networks, and investigation of molecular interactions. Text mining has emerged as a valuable tool in the realm of bioinformatics, with diverse applications including the identification and graphical representation of unique DNA regions, provided an ample amount of reference data is available (Mohsen et al. 2021).

16.7 Commonly Used Machine Learning Algorithms in Bioinformatics

In the field of bioinformatics, some of the most commonly used learning algorithms are Support Vector Machines, Linear Regression, Logistic Regression, Naive Bayes, Linear Discriminant Analysis, Decision Trees, K-Nearest Neighbor Algorithm, and Neural Networks (especially Multilayer Perception).

16.7.1 Decision Tree Classifier

Decision tree classifiers are extensively employed in the field of bioinformatics due to their numerous advantageous features. These classifiers are highly favored for their simplicity, efficiency, and effectiveness in analyzing complex biological data. Moreover, their ability to provide visually intuitive graphical representations further enhances their utility in bioinformatics research and analysis. The decision tree model is constructed using a recursive top-down approach, a widely employed methodology in bioinformatics. This approach facilitates the creation of a model that is both comprehensible and verifiable, making it highly suitable for analysis and interpretation. In the field of bioinformatics, a decision tree is a widely used computational model for classification and regression analysis. It consists of nodes that represent various features or attributes, with the topmost node referred to as the root. The remaining nodes within the tree structure are known as internal nodes, which aid in the decision-making process by evaluating different criteria and branching out accordingly. The construction of the tree follows a recursive approach, starting from the root node and considering each feature individually. Each node in the tree represents an input parameter, allowing for a systematic evaluation of the data. The sample is partitioned through the iterative process of posing recursive inquiries. The terminal node, also known as the leaf node, serves as the final prediction node in the bioinformatics analysis (Charbuty and Abdulazeez 2021; Navada et al. 2011).

16.7.2 Naïve Bayes Classifier

In bioinformatics, classification tasks are often handled using the Naive Bayes classifier, a machine learning method. It functions on the premise that the parameters employed in classification are not reliant on one another, which is to say that it operates on the assumption of feature independence. Since this assumption simplifies the computation of probabilities and reduces the computational cost of the algorithm, it permits efficient and successful categorization. The Naive Bayes classifier is useful in a wide variety of bioinformatics applications due to its ability to reliably categorize data points based on their feature values by exploiting the independence assumption. A common probabilistic machine learning approach in bioinformatics is the Naive Bayes classifier. Assuming that the features are conditionally independent given the class label, Bayes' theorem provides a method for classifying data. Equation 16.1 is a mathematical representation of the classifier that captures its core functional principles.

$$\mathbf{P}\,(\mathbf{C1}|\mathbf{P1},\mathbf{P2}) = \mathbf{P}(\mathbf{P1}|\mathbf{C1})\ \mathbf{P}(\mathbf{P2}|\mathbf{C2}).\mathbf{P}(\mathbf{C1})/\mathbf{P}(\mathbf{P1})\ \mathbf{P}(\mathbf{P2}) \tag{16.1}$$

The probability that the input will fall into class C1 can be calculated using Eq. (16.1), using the parameters P1 and P2.

The conditional probability of observing event C1 given events P1 and P2 can be expressed as the expression (16.1) represents the conditional probability of event C1 given events P1 and P2, divided by the joint probability of events. Equation (16.1) provides the probabilistic assessment of the input's membership in class C1, utilizing the parameters P1 and P2. The probability of obtaining class C1, given parameters P1 and P2, can be expressed as the ratio between the product of the probabilities of P1 occurring with class C1 and P2 occurring with class C2, and the product of the probabilities of P1 and P2 occurring. The utilization of the Bayes formula is evident in this context (Berrar 2018; Saritas and Yasar 2019).

16.7.3 Support Vector Machines

One of the most popular classification approaches in modern bioinformatics is practiced by this person, who is considered an authority in the field. It has risen to the top as a favorite amongst industry professionals thanks to its solid computational base and outstanding accuracy in a wide range of practical applications. Classification of data points is made possible in bioinformatics with the use of Support Vector Machines (SVMs), which work by projecting them into a higher dimensional space. By using this transformation, we may generate a hyperplane that cleanly demarcates between several types of situations. SVMs reliably identify new instances by finding the hyperplane that minimizes the distance to the nearest data points of each class. Building two extra parallel hyperplanes, one on each side of the initial hyperplane, is what is meant by the proposed method. Finding the hyperplane that optimizes the gap between two parallel hyperplanes is the goal of the support vector machine (SVM) method. It is hypothesized that increasing the distance between these hyperplanes will improve the classifier's ability to forecast. Large portions of this domain's division appear to be controlled by two tests that are almost coincident with parallel hyperplanes. Support vectors is a term that is frequently used to describe these cases in the field of bioinformatics. Because of the difficulty in correctly categorizing them, these samples are notoriously difficult to study in the field of bioinformatics. In bioinformatics, it might be difficult to accurately and completely separate training points into their respective classes. These incorrectly classified locations cannot be located too far from the partition zone's outermost boundary. Since support vector machines (SVMs) are so effective at classifying data and addressing a wide range of computational problems, they have become increasingly prominent in the field of bioinformatics. However, they have been criticized for not being sufficiently expressive and understandable in terms of the mathematics they employ (Meyer and Wien 2001; Burbidge et al. 2001).

16.8 Commonly Used Unsupervised Machine Learning Algorithms

16.8.1 Partitional Clustering

This family of clustering algorithms uses a strategy in which each sample is placed into a unique cluster, creating a division in the data set. The user must decide ahead of time how many groups should be created in the dataset before applying a partitional clustering technique. Despite the availability of a number of heuristic approaches in bioinformatics, determining the appropriate cluster size remains a persistent problem. In bioinformatics, the k-implies computation is a standard, go-to method for partitional cluster analysis. This computational method seeks to reduce the sum of squares for each cluster of tests by grouping them into K distinct clusters. At its core, the algorithm relies on the transformational interplay of two fundamental and expedient processes in the realm of bioinformatics. Before the initiation of the sequential progression of these two distinct phases, a preliminary assessment involving a series of examinations is conducted on K initial clusters. During the initial phase, the provided examples are assigned to specific groups based on their proximity to the centroid, typically determined by the Euclidean distance. During the subsequent iteration, the recalibration of group centroids is performed as part of the algorithmic process. The culmination of the dual phases is terminated upon the cessation of protest development, as an alternative assemblage shall diminish the aggregate count of internal blocks. The author explores various computational approaches in the field of bioinformatics, with a specific focus on optimizing the efficiency of K implies calculation. The study aims to enhance processing times, thereby improving the overall performance of high jumper sity. The main limitation of this approach lies in its inability to consistently produce identical results across different runs, as the final configuration of clusters is contingent upon the initial random assignment of points to K initial clusters. In the context of bioinformatics, fluffy and probabilistic clustering methods are employed to analyze and classify biological data sets. These methods allow for a more nuanced approach to clustering, as they do not enforce strict membership of examples to a single cluster. Instead, they consider the likelihood or probability of an example belonging to each cluster, allowing for a more flexible and probabilistic assignment. This approach recognizes the inherent complexity and uncertainty in biological data, enabling a more comprehensive understanding of the underlying patterns and relationships within the data set. Through the utilization of these bioinformatics methodologies, each data point possesses a distinct degree of membership within the various clusters. Driven by the principle of reducing intracluster variation, the aforementioned composition showcases captivating methodologies in the realm of fluffy and probabilistic clustering. The domain remains ripe with untapped prospects for further dissemination endeavors (Celebi 2014; Sonagara and Badheka 2014).

16.8.2 Hierarchical Clustering

The idea of clustering presented here is widely employed in the field of bioinformatics. The output of hierarchical clustering algorithms is a dendrogram, or stable and progressive tree structure, in which the lowest level represents individual samples and the highest level represents a cluster containing all elements. Agglomerative approaches typically used in bioinformatics start at the root of the tree and work their way up. Though also used in this context, disruptive algorithms tend to cluster around the optimal starting point. Agglomerative methods are used to construct dendrograms in bioinformatics by combining clusters based on individual occurrences. Difficult techniques typically don't have a lot of ties between them because of their inefficiency. The expert can strategically cut the dendrogram at a particular level to partition a segment into a desired number of disjoint groups due to its simplicity and intuitiveness. Hierarchical clustering in bioinformatics has been made easier by the ability to choose which clusters to consider. In bioinformatics, a difference grid controls the complex agglomerative combining process. This procedure of merging bunches uses the difference grid to guide each step. The difference grid helps this sophisticated bioinformatics technique run smoothly by separating these sets. Scientific literature offers many clustering analysis separation metrics. Several bioinformatics clustering analysis methods are well-known. Single-linkage measures the distance between two groups' closest people. Complete linkage, which defines distance between two groups as the maximum distance between any two points inside each group, is another popular metric. However, Ward's progressive clustering technique merges the two groups with the lowest increase in the total within-group sum of squares at each algorithm stage. Commonly used centroid distance measures the distance between cluster centroids. Bioinformatics clustering techniques also use the median distance and group average linkage, which calculate the average dissimilarity between all pairs of individuals, one from each group (Nunez-Iglesias et al. 2013; Contreras and Murtagh 2015).

16.9 Open Source Machine Learning Software Tools

16.9.1 Weka 3: Machine Learning Software in Java

Weka uses advanced machine learning methods to solve complicated data mining problems. The bioinformatics toolset includes data preparation, predictive modeling, pattern identification, data grouping, knowledge finding, and data representation.

Open-source The Weka software is available for use under the GNU Public License. A popular bioinformatics application, it offers machine learning algorithms and data mining methods. Weka is famous among bioinformatics researchers and practitioners because to its user-friendly interface and vast capability. Weka's adaptable and customized platform lets users study and interpret complicated biological data, advancing bioinformatics research (Bouckaert et al. 2010).

A carefully designed set of free online courses in machine learning and data mining uses the powerful Weka software suite as the main teaching tool. The classes' multimedia content is available on YouTube.

Popular open-source machine learning program Weka supports deep learning. This feature lets Weka customers employ neural networks and other deep learning algorithms. Integrating deep learning (Frank et al. 2010).

16.9.2 The R Project for Statistical Computing

The R Core Team and the Foundation for Statistical Computing advocate for the use of R, a high-level programming language for statistical computing and graphical representation. Legends in the fields of bioinformatics and computational biology include Ross Ihaka and Robert Gentleman. They are famous for their ground breaking contributions to R, a high-level language and software environment for data processing and statistical modeling in bioinformatics. By revolutionizing data analysis and the development of statistical software, Ihaka and Gentleman have pushed bioinformatics forward. The fields of bioinformatics, data mining, and statistics all benefit from this potent resource. R, a sophisticated programming language and software environment, has many extension packages with reusable code and extensive documentation. Bioinformaticians and researchers use these tools to rapidly analyze and interpret complicated biological data. These extensions enable data manipulation, statistical analysis, visualization, and machine learning. R uses bioinformatics community expertise (Persson Hoden et al. 2021).

User polls and scholarly literature database analysis show that R, a popular programming language, dominates data mining. R, a bioinformatics programming language, ranks 16th in the TIOBE index as of April 2023. It dropped somewhat from 8th in August 2020. Bioinformaticians like R for its versatility and wide selection of biological data analysis tools, as well as its statistical computation and graphical capabilities (Ripley 2001).

R, developed by the GNU Project, is open-source and free under the GNU General Public License. The software framework uses C, FORTRAN, and R, with partial self-hosting. Many bioinformatics operating systems offer precompiled executables. These expert-crafted executables are essential for biological data computational analyses and simulations. By harness R, a strong and adaptable programming language, has a command line interface (CLI) for easy software interaction. This CLI lets users perform R scripts and instructions from the terminal, making data analysis, statistical modeling, and visualization easy and efficient. The bioinformatics community values third-party GUIs like RStudio, an IDE, and Jupyter, a notebook interface (Tierney 2012).

16.9.3 Bioconductor

Bioconductor is an esteemed and revolutionary software project that operates under the principles of freedom, openness, and collaborative development. It is specifically designed to facilitate the intricate analysis and comprehensive understanding of genomic data derived from wet lab experiments in the field of molecular biology. Bioconductor, a prominent bioinformatics platform, is predominantly built upon the robust statistical capabilities of the R programming language. However, it also encompasses valuable contributions from various other programming languages, augmenting its versatility and functionality. The software exhibits a biannual release pattern, synchronizing with the semi-annual updates of the R programming language. In the realm of bioinformatics, a dynamic ecosystem exists where two distinct versions coexist harmoniously. The first is the release version, meticulously aligned with the currently unleashed iteration of the esteemed R programming language. The second is the development version, intricately intertwined with the ongoing evolution of R, as it progresses towards its forthcoming manifestation. The majority of users will discover that the release version is well-suited to fulfil their requirements in the realm of bioinformatics. Furthermore, a plethora of genome annotation packages exists, primarily designed for various microarray applications, although not exclusively limited to such (Gentleman et al. 2004; Reimers and Carey 2006).

16.9.4 RapidMiner

RapidMiner, an innovative bioinformatics tool, uses a client/server design for data analysis and processing. Users can access RapidMiner's sophisticated features and capabilities through a server infrastructure housed on-premises or in public or private clouds. This flexible deployment option lets academics and scientists easily use RapidMiner's broad set of tools and resources for bioinformatics study (Kotu and Deshpande 2014).

RapidMiner is state-of-the-art bioinformatics software that provides an extensive suite of data mining and machine learning techniques. Data loading and transformation (ETL), data pre-treatment and visualization, predictive analytics and statistical modeling, comprehensive review, and rapid deployment are just some of the areas in which it shines. Using bioinformatics, scientists are able to gain new insights with the help of RapidMiner. RapidMiner, a popular data mining and machine learning package, uses Java. One of the most sophisticated bioinformatics tools, RapidMiner, has a simple graphical interface for designing and running complex analytical workflows. RapidMiner "Processes" are collections of "Operators" that perform computational tasks. Bioinformatics operators are carefully built to do a certain duty in the complex process. Each operator's result feeds the next, accelerating workflow. External software applications or APIs can call the engine. The command line interface supports individual function execution. The comprehensive bioinformatics program RapidMiner includes a variety of learning techniques, models, and algorithms for data analysis and interpretation. It integrates well with R and Python, allowing users to add own scripts. RapidMiner, a comprehensive data science platform, can integrate several plugins from the RapidMiner Marketplace to expand functionality. The RapidMiner Marketplace allows developers to carefully create and share powerful data analysis algorithms with the dynamic and collaborative data enthusiast community.

The RapidMiner Studio Free Edition bioinformatics software helps computational biologists analyze and interpret data. Following open-source development principles, this edition is licensed under AGPL. One logical processor may handle up to 10,000 data rows, making bioinformatics data manipulation and exploration efficient (Hofmann and Klinkenberg 2016).

16.9.5 Orange

Bioinformatics-specific Orange is cutting-edge, modular software. Using data visualization, machine learning, data mining, and analysis, Orange aids researchers and scientists in gaining insights from large biological datasets. Users may quickly and effectively integrate several data sources and algorithms into complex processes and pipelines because to its straightforward visual programming interface. Through the analysis of molecular networks, the prediction of protein 3D structures, and the identification of genetic relationships, Orange contributes to the unraveling of life's secrets.

"Orange components" are like widgets in the world of bioinformatics. Data visualization, subset selection, preprocessing, experimental evaluation of learning methods, and predictive modeling are all examples of what fall under the umbrella of bioinformatics.

In bioinformatics, "visual programming" refers to the use of an interface for the connection of pre-existing or user-created widgets in order to design workflows. Python experts can use Orange as a library to modify data and interface components (Demšar et al. 2013).

16.10 Applications of Machine Learning in Bioinformatics

16.10.1 Facilitating Gene Editing Experiments

Gene editing, a revolutionary bioinformatics approach, involves complex genomic changes. Specific DNA segments are deleted, inserted, and replaced during these alterations. Gene editing allows scientists to comprehend and manipulate life's fundamental building elements in new ways. Bioinformatics analysis relies on CRISPR, a highly effective approach. The search for optimal DNA sequence selection for manipulation in bioinformatics continues, with space for improvement. However, the promising field of machine learning (ML) aids this effort. Scientists can optimize gene editing studies and reliably predict their results using machine learning in bioinformatics. The team used machine learning methods to find the best amino acid residue combinations for Cas9 binding to target DNA. Due to the massive number of genetic differences, a large-scale experiment would have been impracticable. By using machine learning-driven engineering, screening was greatly simplified, reducing it by 95% (Krohannon et al. 2022).

16.10.2 Identifying Protein Structure

Proteomics, a bioinformatics area, studies proteins' complicated nature, interactions, composition, and vital role in the body's complex machinery. Bioinformatics analyzes and interprets large biological databases, which demand a lot of processing power. Bioinformatics jobs are computationally complex and require advanced algorithms and high-performance computing to handle and analyze data. Innovative technologies like machine learning are crucial in bioinformatics. A major bioinformatics success is the use of convolutional neural networks (CNNs) to classify protein amino acids into sheet, helix, and coil categories. Neural networks have achieved 84% accuracy, reaching the theoretical top bounds of 88–90%.

Machine learning (ML) has been used in proteomics, a topic that combines biology and computer science. Protein model score, essential for protein structure prediction, is one use. Researchers use ML algorithms to improve protein structure prediction, improving protein function and drug development. ML in proteomics has helped resolve the intricate link between protein structure and function, advancing bioinformatics. Fayetteville State University bioinformatics researchers used machine learning. ML was used to improve protein model scoring accuracy. The protein models were grouped and analyzed using a machine learning method. This approach determined the most important features for evaluating models in each group. The data feature vectors were used to improve machine learning algorithms during training, with each group trained separately.

16.10.3 Spotting Genes Associated with Diseases

Bioinformatics researchers increasingly use machine learning to uncover disease-related genes. The process uses RNA sequencing and gene expression microarray analysis. In cancer research, gene identification helps locate cancer-causing genes and classify tumors molecularly. Cancer prediction and classification were evaluated using decision tree, support vector machine, and neural network bioinformatics at the University of Washington. RNA sequencing data from The Cancer Genome Atlas project showed that linear support vector machine identified cancer best with 95.8% accuracy. Using gene expression data using ML, another study categorized breast cancer types. This team used Cancer Genome Atlas data. Researchers categorized breast cancer samples into triple negative and non-triple negative. Support vector machine classifiers excelled again (Athreya et al. 2018). Penn researchers employed machine learning to uncover CAD drug targets in non-cancerous illnesses. The researchers uncovered CAD-related SNPs using

ML-powered Tree-based Pipeline Optimization Tool. They detected 28 relevant SNPs in UK Biobank genomic data. This study confirmed that the top SNPs on this list were connected to CAD in the literature (Liu et al. 2022).

16.10.4 Traversing the Knowledge Base in Search of Meaningful Patterns

Researchers are trying to gain insights from genomic databases that double every 2.5 years thanks to advanced sequencing technologies. Biomedical articles and studies can be analyzed using machine learning to find genes and proteins and their functions. It can also annotate protein databases and provide literature information. A group of researchers used bioinformatics and machine learning in literature mining to score protein models. Multiple protein-protein docking models are usually produced and scored based on structural constraints. The team utilized ML techniques to search PubMed papers on protein-protein interactions for residues to establish model score constraints. To ensure the limitations are meaningful, scientists tested machine learning techniques to examine all residues for relevance.

This study found that computationally expensive neural networks and less resource-intensive support vector machines performed similarly (Zhou et al. 2022).

16.10.5 Repurposing Drugs

In the area of bioinformatics, researchers adeptly leverage the strategy of drug repurposing, also known as reprofiling, to explore novel applications for existing pharmaceutical agents. The utilization of artificial intelligence (AI) methodologies by bioinformatics researchers enables the comprehensive analysis of vast datasets from Binding DB and DrugBank. Drug repurposing, also known as drug repositioning, encompasses a multifaceted strategy that involves the exploration of existing drugs for novel therapeutic applications. This innovative field of research employs three primary approaches to identify potential drug candidates for repurposing (Pushpakom et al. 2019). These approaches include:

Target-based approach field of drug-target interaction encompasses the investigation of the direct binding between drugs and their target proteins.

Drug-drug interaction studies elucidate the intricate interplay between pharmaceutical agents, shedding light on the multifaceted mechanisms by which these compounds interact within biological systems.

The exploration of intracellular protein surfaces for hotspots and allosteric regions is a fundamental aspect of protein-protein interaction searches in the field of bioinformatics.

Researchers from China University of Petroleum and Shandong University employed a cutting-edge deep neural network methodology to analyze and extract valuable insights from the extensive DrugBank database. The primary focus of their research revolved around investigating the drug-target interactions involving mitochondrial fusion protein 2 (MFN2), a protein that has been implicated as a potential etiological factor in Alzheimer's disease. A recent investigation has successfully identified a collection of 15 distinct medicinal compounds exhibiting promising binding potential. Subsequent investigations have revealed that the protein 11 exhibits the capability to engage in docking interactions with the mitochondrial fusion protein MFN2. The quintet exhibits a range of medium-to-strong binding affinities (Wang et al. 2021).

16.11 Conclusion

The integration of Artificial Intelligence (AI) and Machine Learning (ML) methodologies has exhibited remarkable promise within the realm of bioinformatics. AI, an expansive domain encompassing machine learning (ML), empowers systems to acquire knowledge from data and subsequently generate predictions or make informed decisions. Bioinformatics, a burgeoning field at the intersection of biology and computer science, has witnessed the utilization of cutting-edge artificial intelligence (AI) algorithms to meticulously scrutinize vast and intricate datasets. These datasets encompass a wide array of genetic variations, harboring invaluable information that can be harnessed to unravel patterns and glean profound insights. By leveraging the power of AI, bioinformaticians strive to unlock novel avenues for drug discovery and treatment development, thus revolutionizing the landscape of modern medicine. In conclusion, the integration of artificial intelligence (AI) and machine learning (ML) methodologies has emerged as indispensable assets within the realm of bioinformatics. These cutting-edge technologies empower scientific investigators to scrutinize vast and intricate datasets, thereby facilitating the identification of intricate patterns and invaluable insights that would otherwise prove arduous or unattainable through conventional approaches. The burgeoning field of bioinformatics is witnessing a remarkable surge in the utilization of Artificial Intelligence (AI) and Machine Learning (ML) methodologies. This trend is anticipated to persist in the foreseeable future, driven by the scientific community's pursuit of novel therapeutic interventions and pharmaceutical advancements targeting diverse ailments and medical conditions.

References

Alpaydin E (2020) Introduction to machine learning, 4th ed. p 1–3, 13–18

Athreya AP, Gaglio AJ, Cairns J, Kalari KR, Weinshilboum RM, Wang L, Kalbarczyk ZT, Iyer RK (2018) Machine learning helps identify new drug mechanisms in triple-negative breast cancer. IEEE Trans Nanobioscience 17(3):251–259. https://doi.org/10.1109/TNB.2018.2851997

Berrar D (2018) Bayes' theorem and naive Bayes classifier. In: Encyclopedia of bioinformatics and computational biology: ABC of bioinformatics. Elsevier, pp 403–412

Bouckaert RR, Frank E, Hall MA, Holmes G, Pfahringer B, Reutemann P, Witten IH (2010) WEKA—experiences with a Java open-source project. J Mach Learn Res 11:2533–2541

Burbidge R, Trotter M, Buxton B, Holden S (2001) Drug design by machine learning: support vector machines for pharmaceutical data analysis. Comput Chem 26(1):5–14
Celebi ME (ed) (2014) Partitional clustering algorithms. Springer
Charbuty B, Abdulazeez A (2021) Classification based on decision tree algorithm for machine learning. J Appl Sci Technol Trends 2(01):20–28
Chen XW, Gao JX (2016) Big data bioinformatics. Methods 111:1–2. https://doi.org/10.1016/j.ymeth
Chetty M, Hallinan J, Ruz GA, Wipat A (2022) Computational intelligence and machine learning in bioinformatics and computational biology. Biosystems 222:104792. https://doi.org/10.1016/j.biosystems.2022.104792
Contreras P, Murtagh F (2015) Hierarchical clustering. In: Handbook of cluster analysis, pp 103–123
Demšar J, Curk T, Erjavec A, Gorup Č, Hočevar T, Milutinovič M et al (2013) Orange: data mining toolbox in python. J Mach Learn Res 14(1):2349–2353
Ekins R, Chu FW (1999) Microarrays: their origins and applications. Trends Biotechnol 17(6): 217–218
Erickson BJ (2021) Basic artificial intelligence techniques: machine learning and deep learning. Radiol Clin N Am 59(6):933–940. https://doi.org/10.1016/j.rcl.2021.06.004
Esposito S, Carputo D, Cardi T, Tripodi P (2019) Applications and trends of machine learning in genomics and phenomics for next-generation breeding. Plan Theory 9(1):34
Frank E, Hall M, Holmes G, Kirkby R, Pfahringer B, Witten IH, Trigg L (2010) Weka-A machine learning workbench for data mining. In: Data mining and knowledge discovery handbook, pp 1269–1277
Gentleman RC, Carey VJ, Bates DM, Bolstad B, Dettling M, Dudoit S et al (2004) Bioconductor: open software development for computational biology and bioinformatics. Genome Biol 5(10): 1–16
Goudbeek M, Swingley D, Smits R (2009) Supervised and unsupervised learning of multidimensional acoustic categories. J Exp Psychol Hum Percept Perform 35(6):1913–1933. https://doi.org/10.1037/a0015781
Gupta R, Srivastava D, Sahu M, Tiwari S, Ambasta RK, Kumar P (2021) Artificial intelligence to deep learning: machine intelligence approach for drug discovery. Mol Divers 25(3):1315–1360. https://doi.org/10.1007/s11030-021-10217-3
Hofmann M, Klinkenberg R (eds) (2016) RapidMiner: data mining use cases and business analytics applications. CRC Press
Karim MR, Beyan O, Zappa A, Costa IG, Rebholz-Schuhmann D, Cochez M, Decker S (2021) Deep learning-based clustering approaches for bioinformatics. Brief Bioinform 22(1):393–415. https://doi.org/10.1093/bib/bbz170
Kelchtermans P, Bittremieux W, De Grave K, Degroeve S, Ramon J, Laukens K et al (2014) Machine learning applications in proteomics research: how the past can boost the future. Proteomics 14(4–5):353–366
Kotu V, Deshpande B (2014) Predictive analytics and data mining: concepts and practice with rapidminer. Morgan Kaufmann
Krohannon A, Srivastava M, Rauch S, Srivastava R, Dickinson BC, Janga SC, Sowary CA (2022) CRISPR-Cas13 guide RNA predictor for transcript depletion. BMC Genomics 23(1):172. https://doi.org/10.1186/s12864-022-08366-2
Langley P (2011) The changing science of machine learning. Mach Learn 82(3):275–279. https://doi.org/10.1007/s10994-011-5242-y
Le NQK, Do DT, Hung TNK, Lam LHT, Huynh TT, Nguyen NTK (2020) A computational framework based on ensemble deep neural networks for essential genes identification. Int J Mol Sci 21:9070. https://doi.org/10.3390/ijms21239070
Li R, Li L, Xu Y, Yang J (2022) Machine learning meets omics: applications and perspectives. Brief Bioinform 23(1):bbab460

Libbrecht MW, Noble WS (2015) Machine learning applications in genetics and genomics. Nat Rev Genet 16(6):321–332

Liu C, Che D, Liu X, Song Y (2013) Applications of machine learning in genomics and systems biology. Comput Math Methods Med 2013:587492

Liu Y, Qiao N, Altinel Y (2021) Reinforcement learning in Neurocritical and neurosurgical care: principles and possible applications. Comput Math Methods Med 6657119:1. https://doi.org/10.1155/2021/6657119

Liu L, Zhai W, Wang F, Yu L, Zhou F, Xiang Y, Huang S, Zheng C, Yuan Z, He Y, Yu Z, Ji J (2022) Using machine learning to identify gene interaction networks associated with breast cancer. BMC Cancer 22(1):1070. https://doi.org/10.1186/s12885-022-10170-w

Medin DL, Schaffer MM (1978) Context theory of classification learning. Psychol Rev 85(3): 207–238. https://doi.org/10.1037/0033-295X.85.3.207

Meyer D, Wien FT (2001) Support vector machines. R News 1(3):23–26

Min S, Lee B, Yoon S (2017) Deep learning in bioinformatics. Brief Bioinform 18:851–869. https://doi.org/10.1093/bib/bbw068

Mohsen Y-N, Earl H, Dan T, John S, Milad E (2021) Application of machine learning algorithms in plant breeding: predicting yield from hyperspectral reflectance in soybean? Front Plant Sci 11: 624273. https://doi.org/10.3389/fpls.2020.624273

Mou M, Pan Z, Lu M, Sun H, Wang Y, Luo Y, Zhu F (2022) Application of machine learning in spatial proteomics. J Chem Inf Model 62(23):5875–5895

Muggleton SH (2005) Machine learning for systems biology. In: International conference on inductive logic programming. Springer Berlin Heidelberg, Berlin, Heidelberg, pp 416–423

Munjal NK, Fleischman AD, Coller RJ (2023) Machine learning, predicting future hospitalizations, and the importance of perception. Hosp Pediatr 13(5):e114–e116. https://doi.org/10.1542/hpeds.2023-007224

Navada A, Ansari AN, Patil S, Sonkamble BA (2011) Overview of use of decision tree algorithms in machine learning. In IEEE control and system graduate research colloquium. IEEE. p 37–42

Ngiam KY, Khor IW (2019) Big data and machine learning algorithms for health-care delivery. Lancet Oncol 20(5):e262–e273. https://doi.org/10.1016/S1470-2045(19)30149-4. Erratum in: Lancet Oncol. 20(6):293

Nosi V, Luca A, Milan M, Arigoni M, Benvenuti S, Cacchiarelli D, Cesana M, Riccardo S, Di Filippo L, Cordero F et al (2021) MET exon 14 skipping: a case study for the detection of genetic variants in cancer driver genes by deep learning. Int J Mol Sci 22:4217. https://doi.org/10.3390/ijms22084217

Nunez-Iglesias J, Kennedy R, Parag T, Shi J, Chklovskii DB (2013) Machine learning of hierarchical clustering to segment 2D and 3D images. PLoS One 8(8):e71715

Perakakis N, Yazdani A, Karniadakis GE, Mantzoros C (2018) Omics, big data and machine learning as tools to propel understanding of biological mechanisms and to discover novel diagnostics and therapeutics. Metabolism 87:A1–A9

Persson Hoden K, Hu X, Martinez G, Dixelius C (2021) Smart PARE: an R package for efficient identification of true mRNA cleavage sites. Int J Mol Sci 22:4267. https://doi.org/10.3390/ijms22084267

Pirooznia M, Yang JY, Yang MQ, Deng Y (2008) A comparative study of different machine learning methods on microarray gene expression data. BMC Genomics 9:1–13

Pushpakom S, Iorio F, Eyers PA, Escott KJ, Hopper S, Wells A, Doig A, Guilliams T, Latimer J, McNamee C, Norris A, Sanseau P, Cavalla D, Pirmohamed M (2019) Drug repurposing: progress, challenges and recommendations. Nat Rev Drug Discov 18(1):41–58. https://doi.org/10.1038/nrd.2018.168

Reimers M, Carey VJ (2006) Bioconductor: an open source framework for bioinformatics and computational biology. Methods Enzymol 411:119–134

Ripley BD (2001) The R project in statistical computing. MSOR Connections. The Newsletter of the LTSN Maths, Stats & OR Network 1(1):23–25

Saritas MM, Yasar A (2019) Performance analysis of ANN and Naive Bayes classification algorithm for data classification. Int J Intell Syst Appl Eng 7(2):88–91

Sarle Warren S (1994) Neural networks and statistical models. In SUGI 19: proceedings of the nineteenth annual SAS users group international conference. SAS Institute, p 1538–1550. ISBN 9781555446116. OCLC 35546178

Sonagara D, Badheka S (2014) Comparison of basic clustering algorithms. Int J Comput Sci Mob Comput 3(10):58–61

Stuart R, Peter N (2003) Artificial intelligence: a modern approach, 2nd edn. Prentice Hall. ISBN 978-0137903955

Tierney L (2012) The R statistical computing environment. In: Statistical challenges in modern astronomy V. Springer New York, New York, NY, pp 435–447

Vamathevan J, Clark D, Czodrowski P, Dunham I, Ferran E, Lee G, Li B, Madabhushi A, Shah P, Spitzer M, Zhao S (2019) Applications of machine learning in drug discovery and development. Nat Rev Drug Discov 18(6):463–477. https://doi.org/10.1038/s41573-019-0024-5

Venkatesh KK, Strauss RA, Grotegut CA, Heine RP, Chescheir NC, Stringer JSA, Stamilio DM, Menard KM, Jelovsek JE (2020) Machine learning and statistical models to predict postpartum hemorrhage. Obstet Gynecol 135(4):935–944. https://doi.org/10.1097/AOG.0000000000003759

Wang S, Liu D, Ding M, Du Z, Zhong Y, Song T, Zhu J, Zhao R (2021) SE-onion net: a convolution neural network for protein-ligand binding affinity prediction. Front Genet 11:607824. https://doi.org/10.3389/fgene.2020.607824

Weltz J, Volfovsky A, Laber EB (2022) Reinforcement learning methods in public health. Clin Ther 44(1):139–154. https://doi.org/10.1016/j.clinthera.2021.11.002

Yan J, Wang X (2022) Unsupervised and semi-supervised learning: the next frontier in machine learning for plant systems biology. Plant J 111(6):1527–1538. https://doi.org/10.1111/tpj.15905. Epub 2022 Jul 27

Zhou Y, Shi W, Zhao D, Xiao S, Wang K, Wang J (2022) Identification of immune-associated genes in diagnosing aortic valve calcification with metabolic syndrome by integrated bioinformatics analysis and machine learning. Front Immunol 13:937886. https://doi.org/10.3389/fimmu.2022.937886

Bioinformatics in Preventive Medicine and Epidemiology

17

Linh Thao Tran, Hue Vu Thi, and Dinh-Toi Chu

Abstract

Bioinformatics is a promising science for the future. Bioinformatics tools help analyze complex computer-based biological data. Currently, scientists have applied bioinformatics in many different fields. Notably, its application in preventive medicine and epidemiology is significantly important. Bioinformatics has been thought to support preventive medicine in screening and early detection. From there, it helps countries develop effective prevention and prognosis strategies. Bioinformatics also supports epidemiology in identifying genes involved in disease, detecting and predicting disease outbreaks, and seeking targeted therapies. Grasping contemporary bioinformatics applications in preventative medicine and epidemiology is essential, particularly during epidemics. This chapter has summarized the bioinformatics applications in preventive medicine and epidemiology to illustrate its capabilities and prospects.

Keywords

Bioinformatics · Preventive medicine · Epidemiology · Diseases · Public health

L. T. Tran
Center for Biomedicine and Community Health, International School, Vietnam National University, Hanoi, Vietnam

H. V. Thi · D.-T. Chu (✉)
Center for Biomedicine and Community Health, International School, Vietnam National University, Hanoi, Vietnam

Faculty of Applied Sciences, International School, Vietnam National University, Hanoi, Vietnam
e-mail: toicd@vnu.edu.vn

V. Singh, A. Kumar (eds.), *Advances in Bioinformatics*,
https://doi.org/10.1007/978-981-99-8401-5_17

17.1 Introduction

In the era of modern biology and related sciences, the dependence of bioinformatics related fields is increasing. Bioinformatics is a computer-oriented field pertaining to biological information (Bhardwaj et al. 2021). Bioinformatics is the research and application of computational approaches and methods used for collecting, utilizing, storing, arranging, and analyzing biological, medical, behavioral or other health-related information according to the U.S. National Institutes of Health (NIH), (Bioinformatics 2023). Bioinformatics first emerged not from DNA analysis but from protein analysis (Gauthier et al. 2019). Margaret Dayhoff was an American chemical physicist who pioneered the application of computational methods to the field of biochemistry. In 1958, she and Robert A. Ledley, a physicist, combined to design a composite to determine the primary structure of a protein, and the composite was also the first bioinformatics software published in 1962 (Gauthier et al. 2019). By 1979, Fredrick Sanger's research group had released the first software specifically for the analysis of reading DNA sequences. Thanks to the advent of desktop computers designed for scientific and engineering applications, bioinformatics software written in Perl has emerged since 1980. Ten years later in 1990, with the appearance of web, information sites were established such as Pubmed (1997) and Human genome (1999) (Gauthier et al. 2019). After the completion of the Human Genome project in 2003, the bioinformatics department was established by biologists, computer scientists, and statistical scientists (Oyelade et al. 2015).

Health is paid special attention in any countries of the world. A country having sustainable growth or not depends on the physical and mental health status of their citizens. The completed Human Genome project is a stepping stone for bioinformatics to make a huge impact on medical fields in terms of prevention, diagnosis and treatment (Oyelade et al. 2015). Bioinformatic applications on health has gained remarkable achievements such as mining data for genomics and proteomics, so that some genes could be identified as biological targets to diseases, and thus provided effective treatments. In addition, many national committees and groups have also encouraged the inclusion of genetic information in electronic health records (EHRs) (Sethi and Kimberly 2009). In addition, bioinformatics is also used in the exchange of healthcare information between patients and healthcare professionals through Internet technology providers and healthcare management facilities (Oyelade et al. 2015). Beside, with the current information, ehealth healthcare service has also developed strongly. Moreover, along with ehealth, there is telehealth which is a combination of technology and medical services, which helps patients to access medical services right at their home (Oyelade et al. 2015).

Biomedical researchers have received a lot of useful help from bioinformatics, which helps them understand the fundamental of biology and gene sequencing deeply (Karikari et al. 2015). There are many such healthcare applications, but especially bioinformatics plays a very important role in preventive medicine and epidemiology. Preventive medicine emerged in the late 1960s as health care was reformed, beginning with strategies for immunization, population screening, and preventative measures. Since that time, preventive medicine has been widely accepted as a means of improving the health of communities as well as reducing the health care costs borne by individuals (Clarke 2010). With the development of society and the emergence of more and more diseases and epidemics, epidemiology has always been a field that helps to provide insights into diseases and health evolution by observing and monitoring diseases in groups of populations from which to suggest possible situations, factors affecting the incidence rate and distribution of diseases (Frérot et al. 2018). For preventive medicine and epidemiology, to be strongly developed and to promote social development if using traditional methods, will face many difficulties. It can be seen that bioinformatics has played an important role in preventive medicine and epidemiology (Oyelade et al. 2015). Therefore, in this chapter has summarized the bioinformatics applications in preventive medicine and epidemiology to illustrate its capabilities and prospects.

17.2 Bioinformatics in Preventive Medicine

Preventive medicine is now more and more developed, especially since the COVID-19 pandemic, preventive medicine has been more and more focused on disease control and prevention. Accompanying the development of this industry is impossible not to mention the great support of the bioinformatics industry. Bioinformatics is applied to preventive medicine in screening and early detection of diseases, using machine learning to diagnose disease risk, developing health monitoring devices, etc. (Fig. 17.1). Current popular applications are bioinformatics applied to preventive medicine to support the early detection of diseases and provide prognosis and appropriate treatment.

The first is the support of bioinformatics in screening and early detection of diseases. Machine learning is an efficient method of storing, processing, and analyzing large chunks of data (Maity 2017). This greatly contributes to the early detection of diseases, thereby increasing the survival rate for patients (Kohli and Arora 2018). Guan Wang et al. used machine learning to screen and predict the risk of cardiovascular diseases after birth in women with pre-eclampsia in their study (Wang et al. 2021a). In addition, in the study of Zehra Karapinar Senturk, a machine learning algorithm was also used to select and classify features for early diagnosis of Parkinson's disease in a group of people at risk of the disease (Senturk 2020). Moreover, bioinformatics also helps screen newborns by creating a list of treatable diseases or conditions in infants (Wani et al. 2018). In a study of breast cancer screening in women, proteomic and bioinformatic tools were used to synthesize and analyze data to detect sensitivity and specificity of biomarkers present in women's serum for screening signs of breast cancer (Li et al. 2002). With other cancers, bioinformatics is also used for screening. As in pancreatic cancer, bioinformatics is applied along with a liquid biopsy to detect this disease early to give the most accurate prognosis (Ganasegeran and Abdulrahman 2020). In addition, in the study by Ye-Cheng Wang et al., databases and rate analysis were used to detect and screen biomarkers related to early stage liver cancer (Wang et al. 2021b). To detect early HIV-infected people with heart disease, Suraiya Rasheed et al. conducted a study

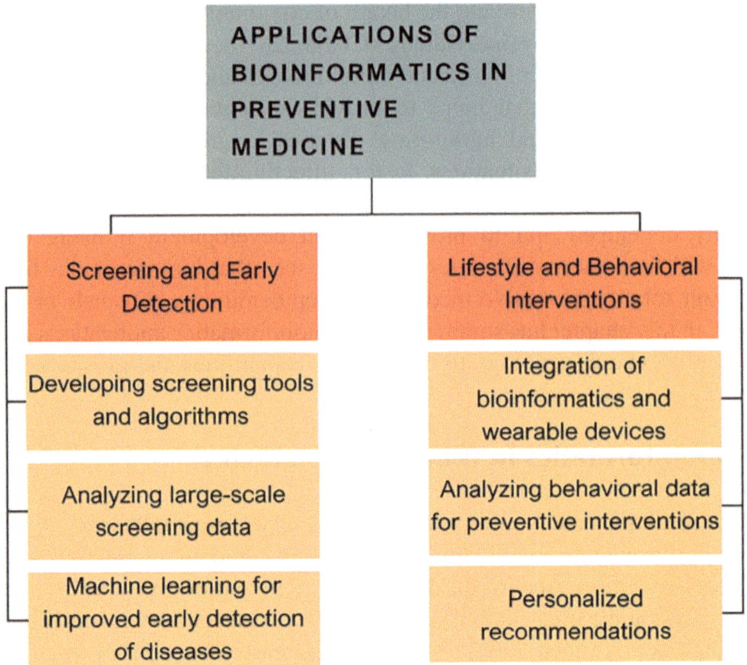

Fig. 17.1 Bioinformatics in preventive medicine. Bioinformatics applications are common in preventive medicine. Screening and early detection of diseases is applied bioinformatics to analyze and calculate data and use machine learning to screen for cancer, cardiovascular disease, etc. Development of monitoring devices Monitor the health and behavior of patients to promptly advise and intervene to change behavior and prevent the risk of disease

using bioinformatics tools for statistical analysis along with available databases to functional classification and expression characterization of proteins that may cause cardiac dysfunction in patients with chronic HIV infection (Rasheed et al. 2015).

Next, accompanied by great advances in modern technology today, the integration of bioinformatics together with wearable devices such as watches, glasses, rings, necklaces, bracelets, etc. to measure Health indicators are quite common. These devices are used different types of sensors to be able to measure basic human vital signs and record them along with security (Sabry et al. 2022). In the study of Chia-Tung Wu et al., they also combined wearables and air quality sensors to monitor and predict whether patients with chronic obstructive pulmonary disease experience exacerbations. of the disease within 7 days or not (Wu et al. 2021). In addition, wearable devices are also used to monitor and predict stroke rates in patients (Alex et al. 2022). Besides, the use of bioinformatics and digital technologies to promote or support behavior change is increasingly common and accepted in the diagnosis and treatment of patients (Michie et al. 2017). In the study by José A. Bauermeister et al., they used para data to describe the outcomes of HIV prevention and care interventions online (Bauermeister et al. 2017). In the study by Phillip J Hartin et al., they designed a mobile phone application "Grey Matters" to specifically provide a group of people with risk behaviors related to Alzheimer's disease to encourage and facilitate help. Behavior changing was found to reduce the risk of getting Alzheimer in the future (Hartin et al. 2016). In addition, in the study by Jacqueline Lorene Bender et al., they also systematically evaluated health behavior monitoring applications on mobile phones to monitor indicators and raise awareness to help improve health outcomes. Users change their behavior to prevent cancer (Bender et al. 2013). Or during a pandemic of infectious diseases—a threat to global health, bioinformatics and AI applications not only help the preventive medicine industry to prevent the threat of epidemics. But also create conditions to control people's health-seeking behavior and emotions during the epidemic period (Ganasegeran and Abdulrahman 2020).

It can be seen that bioinformatics has been playing a very important role in preventive medicine, especially for epidemics with complicated developments such as the most recent COVID-19 pandemic. The combination of bioinformatics and technology has created remoted healthcare applications such as eHealth, telehealth, etc., which are increasingly developed and widely applied in the community. Serving to monitor the health status of patients remotely, from which experts will answer and offer solutions to improve their health status. Thereby also reduces the cost of healthcare for the people and overcoming typical difficulties such as obstacles in moving to medical examination and treatment facilities as well as living too far from the primary medical facility (Koch 2006). During the COVID-19 pandemic, social distancing to prevent disease outbreaks has hindered people's medical examination and treatment. Telehealth has been fully utilized to minimize the risks and consequences of the epidemic. Telehealth has also contributed significantly to collect data from the number of new cases and symptoms of people with COVID-19 (Fig. 17.2). In addition, the pandemic has also caused great difficulties for people with chronic diseases such as diabetes or cancer, HIV, etc., which require periodic treatment at healthcare facilities. Since then, telehealth has also been of great help in capturing the status of these patients and providing remote care, helping them to receive treatment (Garfan et al. 2021).

Therefore, it can be seen that bioinformatics tools have many useful applications in the field of preventive medicine. These applications have been especially useful for healthcare workers and medical facilities offering appropriate precautions and treatment. Combined with the current and future digital age, bioinformatics has the potential to help people worldwide accelerate access to healthcare. The introduction of bioinformatics tools has benefited the healthcare industry in general and preventive medicine in particular, saving significant time in work and achieving high efficiency.

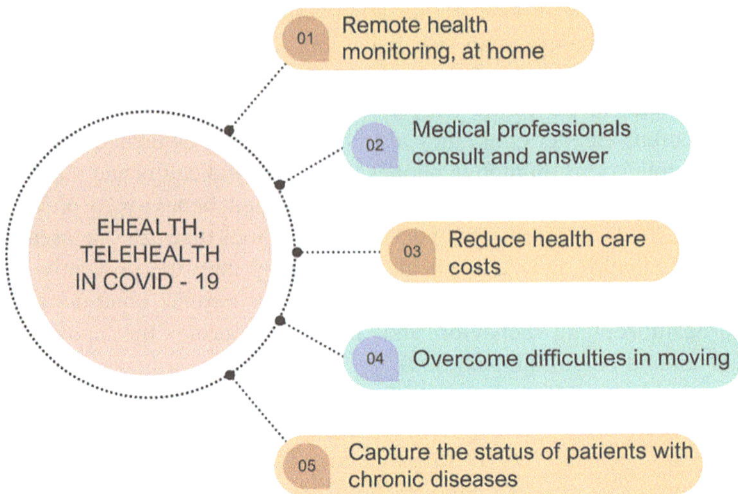

Fig. 17.2 Bioinformatics in COVID-9: Ehealth and Telehealth in COVID-19. In the context of a complicated pandemic, social distancing, the development of ehealth and telehealth models has been of great help to the pandemic. Using ehealth and telehealth to monitor people's health remotely or at home to advise and provide treatment solutions for them in a timely manner. Contribute to reducing healthcare costs and overcoming travel to medical facilities. Simultaneously, patients with long-lasting sicknesses that necessitate frequent subsequent monitoring are still carefully tracked

17.3 Bioinformatics in Epidemiology

In recent years, bioinformatics has been assessed as making many outstanding strides in the field of epidemiology. It has enabled epidemiologists to analyze big biological data efficiently and quickly. Bioinformatics also supports epidemiology in identifying genes involved in disease, detecting and predicting disease outbreaks, and seeking targeted therapies (Fig. 17.3). These applications have become essential in the context of the development of complex infectious diseases in the current era.

To date, the application of bioinformatics in epidemiology has gone through a long history. They stem from the development of gene sequencing technology. Understanding the location and function of genomes has radically changed the field of epidemiology. They allow researchers to identify and track pathogenic bacteria and infections (Köser et al. 2014; Gilchrist Carol et al. 2015). Genomic data is uploaded to existing databases to compare and determine the type of bacteria or viruses that are the cause of the outbreak. From there, epidemiologists can give appropriate preventive measures to the population. This application is clearly seen in some food-related diseases such as *Salmonella, E. coli* and *Listeria* (Lambert et al. 2015). Scientists can use bioinformatics tools to sequence pathogens, thereby accurately identifying pathogens, sources of contamination and preventing high-risk transmission routes (EFSA Panel on Biological Hazards (EFSA BIOHAZ Panel) et al. 2019). On the other hand, phylogenetic analysis has also been a widely applied

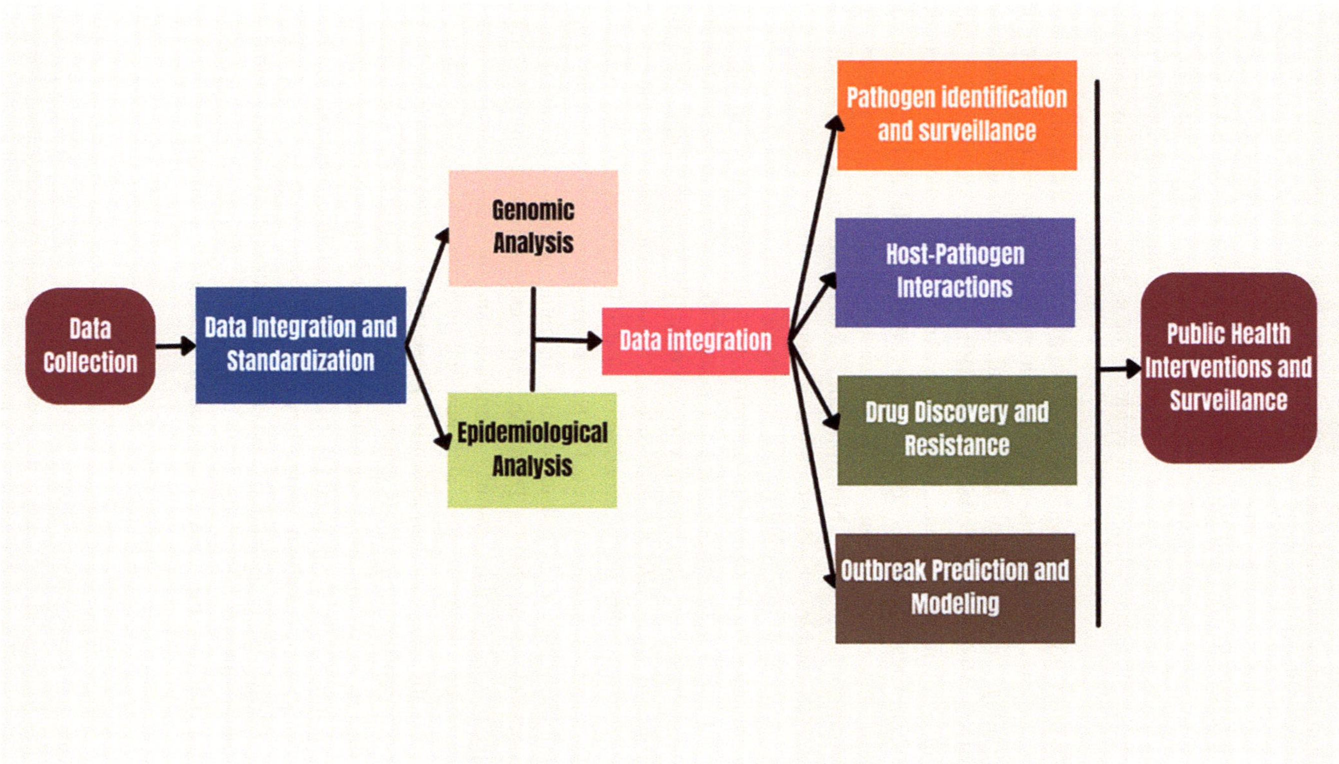

Fig. 17.3 Approaching of bioinformatics in epidemiology. Different types of data are collected from studies and investigations, such as genomics, clinical, and subclinical data. The data is then integrated and normalized for analysis. The genomic analysis involves using bioinformatics tools to compare pathogen gene sequences with existing databases, identifying genetic markers associated with virulence or transmissibility. Simultaneously, the epidemiological analysis examines demographic, geographical, and temporal information to identify patterns and trends in disease outbreaks. The integration of genomic and

bioinformatics method. This method helps to reconstruct the evolutionary relationship between different strains of pathogens. From there we can understand how pathogens change over time and find the source of the outbreak (Naser-Khdour et al. 2019). The COVID-19 pandemic is the clearest demonstration of it. Through sequencing the viral genomes from infected individuals worldwide, researchers have been able to identify and monitor different variants of the SARS-CoV-2, trace their transmission patterns, and evaluate their possible impact on infectiousness and vaccine effectiveness (Rothan and Byrareddy 2020; Junejo et al. 2020). This is one of the vital applications when a new epidemic takes place with many dangerous variants.

One of the other prominent applications of bioinformatics is the analysis of host-pathogen interactions through data analysis to identify protein-protein interactions and find signaling pathways (Jean Beltran et al. 2017). Bioinformatics databases such as STRING, VirHostNet, PHISTO, and HPIDB can provide extensive information about known interactions between pathogens and their host proteins (Valiente 2022). These databases compile experimentally validated and predicted protein-protein interactions, along with associated functional annotations and pathways. Additionally, some bioinformatics tools and approaches such as GWAS, and PRS enable the identification of genetic variations associated with increased or decreased susceptibility to specific microbes (Power et al. 2017). Likewise, techniques have been demonstrated to aid investigators in exploring whether variant genes in the nucleotide sequences of the genome impact vulnerability to specific pathologies. Uncommon variant genes, for instance within the genes commonly referred to as *BRCA1* and *BRCA2*, have been shown to elevate the susceptibility for malignant transformation of tissues located within the mammary glands and the uterus in females (Li et al. 2022). Furthermore, epidemiologists have also used bioinformatics as a tool to identify immune pathways involved in the infection process (Li et al. 2014). Therapies and vaccines will be invented when scientists have this information, thereby controlling and treating the disease better.

Nowadays, drug discovery and antimicrobial resistance are applications for which bioinformatics has excellent potential. These tools can find the most effective drug targets through the analysis of genetic data (You et al. 2022). The scientific inquiry conducted by Wang et al. identified three particular molecular markers, specifically an interleukin designated (IL-6), a matrix metalloproteinase numbered (MMP9), and a protein is known as pituitary tumor-transforming gene 1 (PTTG1), which they considered to potentially serve as indicators that could signal the presence of malignant growths arising from lung tissue as well as prospective points of interference that new treatments might target in order to combat this particular

Fig. 17.3 (continued) epidemiological data allows for a comprehensive understanding of various aspects, including pathogen identification and surveillance, host-pathogen interactions, drug discovery and resistance, and outbreak prediction and modeling. Finally, interventions are suggested based on the findings to prevent or minimize disease outbreaks

form of neoplastic disease (Wang et al. 2016). Furthermore, bioinformatics has also contributed to providing greater insight into antibiotic resistance. Bioinformatics tools can detect drug-resistant strains and their transmission through the integration of epidemiological and genetic data (Rodrigues et al. 2020; McInnes et al. 2020). The nucleic acid chains of a micro-organism scientifically referred to as *C. jejuni* along with another micro-organism known as *C. coli* underwent examination by Willi et al. The conclusions derived from this work identified numerous variations in the genetic composition of these micro-organisms that could possibly convey a capacity to withstand the impact of certain medicinal compounds intended to eliminate them (Quino et al. 2022). In summation, it has been exhibited that bioinformatics plays an important role for epidemiology in developing medications that aim at and support the management of drug-resistant infections.

With the advance of time, the sphere of application by bioinformatics has expanded to comprise the integration of data, computational modeling, and anticipatory investigation. This can potentially help applications of bioinformatics in the ahead-of-time identification and anticipation of sickness outbreaks, notably following the recent COVID-19 global pandemic. In recent years, bioinformatics played a highly integral part in examining the hereditary sequences of the SARS-CoV-2 virus to pursue its variations and keep track of their likely impact (Rothan and Byrareddy 2020; Junejo et al. 2020). Even more intriguingly, investigators have utilized bioinformatics to combine hereditary and epidemiological information into complex versions that are able to anticipate the dissemination of infectious sicknesses (Long et al. 2021). In 2021, I.F.F. dos Santos et al. used a susceptible–infected–removed (SIR) model to predict the short- and long-term development of the COVID-19 pandemic (dos Santos et al. 2021). The outcomes demonstrated that the adaptive SIR model displays powerful performance in replicating the dynamics of SARS-CoV-2 and projecting the trajectory of the outbreak in Brazil and other nations. Given the seriousness of this pandemic, an assortment of additional forecasting versions have emerged, consisting of the susceptible-exposed-infectious-recovered (SEIR) model (Heng and Althaus 2020) and the susceptible-infectious-recovered-deceased (SIRD) model (Chen 2022). These versions incorporate diverse factors, like population density, movement patterns, and intervention strategies, to assess various situations and assist informed decision-making for public health. To summarize, bioinformatics plays an indispensable role in swiftly determining and anticipating disease outbreaks.

As one of the speedily developing fields, the application of bioinformatics in epidemiology also faces numerous challenges that necessitate being tackled. One principal hurdle is combining and standardizing information from diverse resources, consisting of hereditary, clinical, and epidemiological data (National Academies of Sciences, Engineering, and Medicine 2016). Furthermore, ethical contemplations play an indispensable role in bioinformatics and epidemiological investigation, requiring privacy safeguards, informed assent, and responsible data-sharing practices. Another challenge lies in the computational, data analysis, and interpretation. Looking ahead, the potential of artificial intelligence (AI) and machine learning (ML) holds promise for enhancing data analysis in bioinformatics and epidemiology

(Ganasegeran and Abdulrahman 2020). Addressing these challenges will facilitate bioinformatics's advancement and application in strengthening disease surveillance and control.

17.4 Conclusion

The application of bioinformatics in both preventive medicine and epidemiology has achieved many visible achievements. If in the field of preventive medicine, bioinformatics is applied in the screening and early detection of disease risks through data analysis as well as community health behavioral interventions to devise methods then in epidemiology, bioinformatics is applied to finding genes directly related to causing pathogens and plays an important role in drug discovery and resistance by identifying resistance genes to develop therapeutics and improve drug resistance. In addition, bioinformatics is applied to the preventive medicine industry in the development of applications, and devices to monitor health and human behavior to receive remote healthcare consultation and care from doctors, and medical professionals to quickly have preventive solutions. In epidemiology, bioinformatics is also used to analyze phylogenetic detection to understand how pathogens and transmission patterns evolve, thereby monitoring populations at risk or source of disease outbreaks. Most recently, during the COVID-19 pandemic, bioinformatics, along with preventive medicine and epidemiology, have worked closely to help prevent epidemics and reduce morbidity and mortality. Bioinformatics and epidemiology through sequencing the genes of viruses from infected people to track variants to assess transmission patterns and their impact on the human body, thereby developing models of preventive vaccines. Bioinformatics helps preventive medicine develop remote health care and surveillance models in the context of complex pandemics such as eHealth and telehealth to detect cases early and provide appropriate treatment plans.

References

Alex SA et al (2022) Machine learning-based wearable devices for smart healthcare application with risk factor monitoring. In: Empowering sustainable industrial 4.0 systems with machine intelligence. IGI Global, pp 174–185

Bauermeister JA et al (2017) Addressing engagement in technology-based behavioural HIV interventions through paradata metrics. Curr Opin HIV AIDS 12(5):442–446

Bender JL et al (2013) A lot of action, but not in the right direction: systematic review and content analysis of smartphone applications for the prevention, detection, and management of cancer. J Med Internet Res 15(12):e2661

Bhardwaj R, Sharma M, Agrawal N (2021) Bioinformatics and its application areas. In: Computation in bioinformatics. Wiley, pp 121–137

Bioinformatics (2023). https://www.genome.gov/genetics-glossary/Bioinformatics

Chen M (2022) Analysis of SARS-CoV-2 high infection spread in India based on SIRD model and structural insights. In: 2022 2nd international conference on bioinformatics and intelligent computing. Association for Computing Machinery, Harbin, China, pp 381–384

Clarke JL (2010) Preventive medicine: a ready solution for a health care system in crisis. Popul Health Manag 13(S2):S-3

dos Santos IFF, Almeida GMA, de Moura FABF (2021) Adaptive SIR model for propagation of SARS-CoV-2 in Brazil. Physica A 569:125773

EFSA Panel on Biological Hazards (EFSA BIOHAZ Panel), Koutsoumanis K, Allende A et al (2019) Whole genome sequencing and metagenomics for outbreak investigation, source attribution and risk assessment of food-borne microorganisms. EFSA J 17(12):e05898

Frérot M et al (2018) What is epidemiology? Changing definitions of epidemiology 1978–2017. PLoS One 13(12):e0208442

Ganasegeran K, Abdulrahman SA (2020) Artificial intelligence applications in tracking health behaviors during disease epidemics. In: Hemanth DJ (ed) Human behaviour analysis using intelligent systems. Springer International Publishing, Cham, pp 141–155

Garfan S et al (2021) Telehealth utilization during the Covid-19 pandemic: a systematic review. Comput Biol Med 138:104878

Gauthier J, Vincent AT, Charette SJ, Derome N (2019) A brief history of bioinformatics. Brief Bioinform 20(6):1981–1996

Gilchrist Carol A et al (2015) Whole-genome sequencing in outbreak analysis. Clin Microbiol Rev 28(3):541–563

Hartin PJ et al (2016) The empowering role of mobile apps in behavior change interventions: the gray matters randomized controlled trial. JMIR Mhealth Uhealth 4(3):e4878

Heng K, Althaus CL (2020) The approximately universal shapes of epidemic curves in the susceptible-exposed-infectious-recovered (SEIR) model. Sci Rep 10(1):19365

Jean Beltran PM et al (2017) Proteomics and integrative omic approaches for understanding host–pathogen interactions and infectious diseases. Mol Syst Biol 13(3):922

Junejo Y et al (2020) Novel SARS-CoV-2/COVID-19: origin, pathogenesis, genes and genetic variations, immune responses and phylogenetic analysis. Gene Rep 20:100752

Karikari TK, Quansah E, Mohamed WMY (2015) Developing expertise in bioinformatics for biomedical research in Africa. Appl Transl Genom 6:31–34

Koch S (2006) Home telehealth—current state and future trends. Int J Med Inform 75(8):565–576

Kohli PS, Arora S (2018) Application of machine learning in disease prediction. IEEE, pp 1–4

Köser CU et al (2014) Rapid single-colony whole-genome sequencing of bacterial pathogens. J Antimicrob Chemother 69(5):1275–1281

Lambert D et al (2015) GeneSippr: a rapid whole-genome approach for the identification and characterization of foodborne pathogens such as priority Shiga toxigenic Escherichia coli. PLoS One 10(4):e0122928

Li J et al (2002) Proteomics and bioinformatics approaches for identification of serum biomarkers to detect breast cancer. Clin Chem 48(8):1296–1304

Li S et al (2014) Molecular signatures of antibody responses derived from a systems biology study of five human vaccines. Nat Immunol 15(2):195–204

Li S, Silvestri V, Leslie G et al (2022) Cancer risks associated with *BRCA1* and *BRCA2* pathogenic variants. J Clin Oncol 40(14):1529–1541

Long J, Khaliq AQM, Furati KM (2021) Identification and prediction of time-varying parameters of COVID-19 model: a data-driven deep learning approach. Int J Comput Math 98(8):1617–1632

Maity NG (2017) Sreerupa das, machine learning for improved diagnosis and prognosis in healthcare. IEEE, pp 1–9

McInnes RS et al (2020) Horizontal transfer of antibiotic resistance genes in the human gut microbiome. Curr Opin Microbiol 53:35–43

Michie SM et al (2017) Developing and evaluating digital interventions to promote behavior change in health and health care: recommendations resulting from an international workshop. J Med Internet Res 19(6):e7126

Naser-Khdour S et al (2019) The prevalence and impact of model violations in phylogenetic analysis. Genome Biol Evol 11(12):3341–3352

National Academies of Sciences, Engineering, and Medicine (2016) Big data and analytics for infectious disease research, operations, and policy. In: Alper J (ed) Proceedings of a workshop. The National Academies Press, Washington, DC, p 98

Oyelade J, Soyemi J, Isewon I, Obembe O (2015) Bioinformatics, healthcare informatics and analytics: an imperative for improved healthcare system. Int J Appl Inf Syst 8(5):1–6

Power RA, Parkhill J, de Oliveira T (2017) Microbial genome-wide association studies: lessons from human GWAS. Nat Rev Genet 18(1):41–50

Quino W et al (2022) Genomic analysis and antimicrobial resistance of *Campylobacter jejuni* and *Campylobacter coli* in Peru. Front Microbiol 12:802404

Rasheed S et al (2015) Possible biomarkers for the early detection of HIV-associated heart diseases: a proteomics and bioinformatics prediction. Comput Struct Biotechnol J 13:145–152

Rodrigues GL et al (2020) Frequency of antimicrobial resistance genes in *salmonella* from Brazil by *in silico* whole-genome sequencing analysis: an overview of the last four decades. Front Microbiol 11:1864

Rothan HA, Byrareddy SN (2020) The epidemiology and pathogenesis of coronavirus disease (COVID-19) outbreak. J Autoimmun 109:102433

Sabry F et al (2022) Machine learning for healthcare wearable devices: the big picture. J Healthc Eng 2022:4653923

Senturk ZK (2020) Early diagnosis of Parkinson's disease using machine learning algorithms. Med Hypotheses 138:109603

Sethi P, Kimberly T (2009) Translational bioinformatics and healthcare informatics: computational and ethical challenges. Perspect Health Inf Manag 6:1h

Valiente G (2022) The landscape of virus-host protein-protein interaction databases. Front Microbiol 13:827742

Wang LQ, Zhao LH, Qiao YZ (2016) Identification of potential therapeutic targets for lung cancer by bioinformatics analysis. Mol Med Rep 13(3):1975–1982

Wang G et al (2021a) A machine learning-based prediction model for cardiovascular risk in women with preeclampsia. Front Cardiovasc Med 8:736491

Wang Y-C et al (2021b) Bioinformatics screening of biomarkers related to liver cancer. BMC Bioinform 22(3):521

Wani MY et al (2018) Advances and applications of bioinformatics in various fields of life. Int J Fauna Biol Stud 5(2):3–10

Wu C-T et al (2021) Acute exacerbation of a chronic obstructive pulmonary disease prediction system using wearable device data, machine learning, and deep learning: development and cohort study. JMIR Mhealth Uhealth 9(5):e22591

You Y et al (2022) Artificial intelligence in cancer target identification and drug discovery. Signal Transduct Target Ther 7(1):156

Correction to: Advances in Bioinformatics

Vijai Singh and Ajay Kumar

Correction to:
Vijai Singh et al. (eds.), Advances in Bioinformatics,
https://doi.org/10.1007/978-981-99-8401-5

The original version of this book was inadvertently published with incorrect affiliations for authors Reshma Tendulkar, Mugdha Tendulkar, and Jayakumar Rajadas in Chapters 1, 15, and 12, and has been corrected as follows:

Reshma Tendulkar

Vivekanand Education Society's, College of Pharmacy, Mumbai, Maharashtra, India

Mugdha Tendulkar

K. J. Somaiya Medical College and Research Centre, Mumbai, Maharashtra, India

Jayakumar Rajadas

Advanced Drug Delivery and Regenerative Biomaterials Laboratory, Cardiovascular Institute and Pulmonary and Critical Care Medicine, Stanford University School of Medicine, Stanford University, Palo Alto, CA, USA

The affiliations of the chapter authors have been updated with this erratum.

The updated version of these chapters can be found at
https://doi.org/10.1007/978-981-99-8401-5_1
https://doi.org/10.1007/978-981-99-8401-5_12
https://doi.org/10.1007/978-981-99-8401-5_15

V. Singh, A. Kumar (eds.), *Advances in Bioinformatics*,
https://doi.org/10.1007/978-981-99-8401-5_18

MIX
Papier aus verantwortungsvollen Quellen
Paper from responsible sources
FSC® C105338

If you have any concerns about our products,
you can contact us on
ProductSafety@springernature.com

In case Publisher is established outside the EU,
the EU authorized representative is:
Springer Nature Customer Service Center GmbH
Europaplatz 3, 69115 Heidelberg, Germany

Printed by Libri Plureos GmbH
in Hamburg, Germany